AF389104

# Micro/Nanofluidic and Lab-on-a-Chip Devices for Biomedical Applications

# Micro/Nanofluidic and Lab-on-a-Chip Devices for Biomedical Applications

Editors

**Violeta Carvalho**
**Senhorinha Teixeira**
**João Ribeiro**

MDPI • Basel • Beijing • Wuhan • Barcelona • Belgrade • Manchester • Tokyo • Cluj • Tianjin

*Editors*

Violeta Carvalho
Department of Mechanical
Engineering
University of Minho
Guimarães
Portugal

Senhorinha Teixeira
Production and Systems
Department
University of Minho
Guimarães
Portugal

João Ribeiro
Department of Mechanical
Technology
Instituto Politécnico
de Bragança
Bragança
Portugal


*Editorial Office*
MDPI
St. Alban-Anlage 66
4052 Basel, Switzerland


This is a reprint of articles from the Special Issue published online in the open access journal *Micromachines* (ISSN 2072-666X) (available at: www.mdpi.com/journal/micromachines/special_issues/Micro_Devices_Biomedical_Applications).

For citation purposes, cite each article independently as indicated on the article page online and as indicated below:

LastName, A.A.; LastName, B.B.; LastName, C.C. Article Title. *Journal Name* **Year**, *Volume Number*, Page Range.

**ISBN 978-3-0365-6100-4 (Hbk)**
**ISBN 978-3-0365-6099-1 (PDF)**

# Contents

# About the Editors

**Violeta Carvalho**

Violeta Meneses Carvalho completed the Integrated Master in Biomedical Engineering in 2020 by the University of Minho and her area of expertise is Biomaterials, Rehabilitation, and Biomechanics. Currently, she is an invited assistant at the Production and Systems Department teaching statistics and numerical methods. At the same time, she is a PhD student in the Mechanical Engineering Doctoral program at the University of Minho focused on the development and optimization of innovative nanoparticles and a preclinical organ-on-a-chip platform for nanomedicine. In the scope of the PhD project, she is doing an internship at INL, International Iberian Nanotechnology Laboratory. She will also conduct experiments at Shin's Laboratory - Harvard Medical School through a Fullbright/FCT grant. Up to now, she has published 16 papers in indexed journals, four book chapters, one edited book, 10 conference papers, one conference poster, and six conference abstracts.

**Senhorinha Teixeira**

Senhorinha de Fátima Capela Fortunas Teixeira is Associate Professor in University of Minho, Researcher in University of Minho Centro ALGORITMI and Member of the Coordinating Committee of the Degree Program Integrated Master in Mechanical Engineering (MIEM) in University of Minho. From 2011 to 2015, she was Department Director, and from 2016 to 2021, Erasmus Academic Coordinator in Production and Systems Area in University of Minho. She completed the Bachelor in Systems Engineering and Informatics in 1984 by University of Minho and Doctor in Chemical Engineering in 1989 by University of Birmingham School of Chemical Engineering. She published almost 80 articles in journals and 20 book chapters and/or books. She organized 10 events, namely, she was a conference chair of ECOS 2018 - the 31st International Conference on Efficiency, Cost, Optimization, Simulation and Environmental Impact of Energy Systems, Guimarães, Portugal. She supervised 12 PhD theses and 70 MSc dissertations. She was Principal investigator in one project, Researcher in 13 projects and Supervisor in 22 projects. She works in the area of Engineering and Technology. In their professional activities, she interacted with more than 500 collaborators in co-authorship of scientific papers.

**João Ribeiro**

João Eduardo Pinto Castro Ribeiro completed his Ph.D. in Mechanical Engineering in 2006, his Master's (1998) and Licenciatura (1994) in the same speciality; he did all his academic career at the Faculty of Engineering of the University of Porto. He is an Adjunct Professor at the Polytechnic Institute of Bragança from 2000 to the present. He had been Course Director of Master in Industrial Engineering and President of the Scientific Committee of the same course for six years at the Polytechnic Institute of Bragança. He is currently the Coordinator of the Department of Mechanical Technology, a member of the course committee, and the scientific committee of the Master in Industrial Engineering at the Polytechnic Institute of Bragança. He has supervised more than 50 Master's students and nine research assistants in the field of Industrial and Mechanical Engineering. He is an integrated member of the Mountain Research Center (CIMO), and has published more than 40 articles in international journals. He wrote 12 book chapters and two books. He participates and/or participated as a Researcher in 10 projects and Technical development in four project(s), being PI of 2.

 *micromachines*

**MDPI**

*Editorial*

# Editorial for the Special Issue on Micro/Nanofluidic and Lab-on-a-Chip Devices for Biomedical Applications

**Violeta Meneses Carvalho** [1,2,3,*], **Senhorinha Teixeira** [2] and **João E. Ribeiro** [4,5,6]

1   MEtRICs, Mechanical Engineering Department, Campus de Azurém, University of Minho, 4800-058 Guimarães, Portugal
2   ALGORITMI Research Centre/LASI, Department of Production Systems, University of Minho, 4800-058 Guimarães, Portugal
3   Microelectromechanical Systems Research Unit (CMEMS-UMinho), School of Engineering, Campus de Azurém, University of Minho, 4800-058 Guimarães, Portugal
4   Campus de Santa Apolónia, Instituto Politécnico de Bragança, 5300-253 Bragança, Portugal
5   Centro de Investigação de Montanha (CIMO), Campus de Santa Apolónia, Instituto Politécnico de Bragança, 5300-253 Bragança, Portugal
6   Laboratório Associado Para a Sustentabilidade e Tecnologia em Regiões de Montanha (SusTEC), Campus de Santa Apolónia, Instituto Politécnico de Bragança, 5300-253 Bragança, Portugal
*   Correspondence: violeta.carvalho@dps.uminho.pt

**Citation:** Carvalho, V.M.; Teixeira, S.; Ribeiro, J.E. Editorial for the Special Issue on Micro/Nanofluidic and Lab-on-a-Chip Devices for Biomedical Applications. *Micromachines* **2022**, *13*, 1718. https://doi.org/10.3390/mi13101718

Received: 8 October 2022
Accepted: 9 October 2022
Published: 12 October 2022

**Publisher's Note:** MDPI stays neutral with regard to jurisdictional claims in published maps and institutional affiliations.



Micro/Nanofluidic and lab-on-a-chip devices have been increasingly used in biomedical research [1]. Because of their adaptability, feasibility, and cost-efficiency, these devices can revolutionize the future of preclinical technologies. Furthermore, they allow insights into the performance and toxic effects of responsive drug delivery nanocarriers to be obtained, which consequently allow the shortcomings of two/three-dimensional static cultures and animal testing to be overcome and help to reduce drug development costs and time [2–4]. With the constant advancements in biomedical technology, the development of enhanced microfluidic devices has accelerated, and numerous models have been reported.

Given the multidisciplinary of this Special Issue (SI), papers on different subjects were published making a total of 14 contributions, 10 original research papers, and 4 review papers. The review paper of Ko et al. [1] provides a comprehensive overview of the significant advancements in engineered organ-on-a-chip research in a general way while in the review presented by Kanabekova and colleagues [2], a thorough analysis of microphysiological platforms used for modeling liver diseases can be found. To get a summary of the numerical models of microfluidic organ-on-a-chip devices developed in recent years, the review presented by Carvalho et al. [5] can be read. On the other hand, Maia et al. [6] report a systematic review of the diagnosis methods developed for COVID-19, providing an overview of the advancements made since the start of the pandemic.

In the following, a brief summary of the research papers published in this SI will be presented, with organs-on-a-chip, microfluidic devices for detection, and device optimization having been identified as the main topics.

Some researchers focused on the development of advanced microfluidic devices which are devised to reconstruct the tissue architecture of organs biochemically and biophysically. For instance, Kuriu and co-workers [4] worked on the development of a microfluidic device to mimic the small intestine tract with villi to obtain insights into fluid flow by using particle image velocimetry. Through these experiments, it was possible to verify that microbeads tend to stick to the side surface of the villi, which can explain the relationship between fluid flow and the settlement of gut bacteria on the villi. Komen and colleagues [7], on the other hand, established an alternative to cancer xenografts due to ethical considerations and a lack of accuracy to predict physiological responses. The authors created a microfluidic device integrated with a U-shaped well for having a single spheroid and exposed it to a dynamic environment and to an in vivo-like concentration of oxaliplatin, a medication commonly used to treat colorectal cancer, and compared it to an in-vivo cancer xenograft. The effect of oxaliplatin on growth inhibition, proliferation, and apoptosis markers was evaluated. In terms of growth inhibition, the on-chip results were comparable to xenograft studies. Regarding the proliferation and apoptosis markers, a similar response was also observed which proved the potential of microfluidic devices to reduce the use of cancer xenografts for cancer research. Meanwhile, Callegari and colleagues [8] investigated the electrophysiological activity of neuronal populations to electrical stimulation. For this purpose, cortical and hippocampal neurons were established to reconstruct interconnected sub-populations. The results showed that cortical assemblies were more reactive than hippocampal ones. Through these results, the authors showed that the results depend on neuronal structure when electrical stimulation experiments are conducted. Despite the previous works presenting interesting results, the fabrication process of advanced microfluidic devices has constantly evolved, and the use of 3D (bio)printing has increased over the years. Lutsch et al. [9] presented an alternative to the regularly used PDMS casting method. The authors investigated different resins by conducting cytotoxicity, cytocompatibility, and HET CAM assay, but poly-(ethylene glycol)-diacrylate (PEGDA) stood out. This material provided excellent results and allowed for the acceleration and improvement of the current fabrication processes.

Other authors have focused on the use of microfluidic devices for detection purposes. Li et al. [10] designed an ease-of-use portable microfluidic device that combines microarray and microfluidics for point-of-care detection of several protein biomarkers in serum samples. The results were promising and showed similar outputs to those measured by the commercial method used at the clinic. Similarly, Sitkov [11] constructed a microfluidic biosensor system with peptide aptamers for protein biomarker detection of diseases in biological fluids. Through in silico techniques, the authors simulated the peptide aptamer for troponin T and its non-fluorescent digital twin and tested the device design. Since the simulation results were promising, a laboratory sample of the biosensor was designed and manufactured, but further experiments are needed. In a similar line of research, Zimmers and co-workers [12] conceived a novel diagnostic platform for the detection of target DNA. In brief, the target DNA is captured by magnetic beads and then loaded into a microfluidic reaction tape with the other reaction solutions. The device was able to detect 5 fM target DNA as well as *Schistosoma mansoni* DNA. Despite additional tests being needed, this technique requires less hands-on time and does not need an additional control reaction like current methods do. A different detection system was proposed by Saha and co-workers [13], but in this case for lactate quantification using a colorimetric assay. This can be used for evaluating the welfare of muscles and oxidative stress levels. The authors developed a wearable patch due to the relationship between sweat and blood lactate levels. For instance, during rest, the osmotic disc (hydrogel) can extract fluid from the skin via osmosis and deliver it to the paper, while during exercise, the paper can collect sweat even in the absence of the hydrogel patch. It was found that the molar concentration of lactate in sweat is correlated to sweat rate and that the measurements are more viable during high-intensity exercise. This constitutes an interesting technology that can be used for athlete monitoring.

The optimization of microfluidic devices has also been explored by some authors. Grigorev et al. [3] investigated how to achieve adequate flow rates for trapping single cells in a microfluidic chip. The authors applied a generative design methodology with an evolutionary algorithm and validated the device with experimental data. The experiments proved the efficiency of the device with 4 out of 4 RBCs trapped. On the other hand, Tsai and colleagues [14] focused on microfluidic devices for bacterial growth and how to guarantee microscope stability for long-term imaging of bacterial dynamics. For this purpose, an optimized integrated multi-level microfluidic chip was developed. The authors used a stabler microscopy immersion oil, and images were captured with a focally stable time-lapse for 72 h.

To conclude, we would like to acknowledge all the authors for their contribution to the success of this SI as well as the reviewers whose feedback helped to improve the quality of the published papers. We would also like to recognize Ms. Min Su from the Micromachines publishing office for her endless assistance and help in disseminating this SI.

**Conflicts of Interest:** The authors declare no conflict of interest.

# References


1. Ko, J.; Park, D.; Lee, S.; Gumuscu, B.; Jeon, N. Engineering Organ-on-a-Chip to Accelerate Translational Research. *Micromachines* **2022**, *13*, 1200. [CrossRef]
2. Kanabekova, P.; Kadyrova, A.; Kulsharova, G. Microfluidic Organ-on-a-Chip Devices for Liver Disease Modeling In Vitro. *Micromachines* **2022**, *13*, 428. [CrossRef] [PubMed]
3. Grigorev, G.V.; Nikitin, N.O.; Hvatov, A.; Kalyuzhnaya, A.V.; Lebedev, A.V.; Wang, X.; Qian, X.; Maksimov, G.V.; Lin, L. Single Red Blood Cell Hydrodynamic Traps via the Generative Design. *Micromachines* **2022**, *13*, 367. [CrossRef] [PubMed]
4. Kuriu, S.; Yamamoto, N.; Ishida, T. Microfluidic Device Using Mouse Small Intestinal Tissue for the Observation of Fluidic Behavior in the Lumen. *Micromachines* **2021**, *12*, 692. [CrossRef]
5. Carvalho, V.; Rodrigues, R.O.; Lima, R.A.; Teixeira, S. Computational Simulations in Advanced Microfluidic Devices: A Review. *Micromachines* **2021**, *12*, 1149. [CrossRef] [PubMed]
6. Maia, R.; Carvalho, V.; Faria, B.; Miranda, I.; Catarino, S.; Teixeira, S.; Lima, R.; Minas, G.; Ribeiro, J. Diagnosis Methods for COVID-19: A Systematic Review. *Micromachines* **2022**, *13*, 1349. [CrossRef]
7. Komen, J.; van Neerven, S.M.; Bossink, E.G.B.M.; de Groot, N.E.; Nijman, L.E.; van den Berg, A.; Vermeulen, L.; van der Meer, A.D. The Effect of Dynamic, In Vivo-like Oxaliplatin on HCT116 Spheroids in a Cancer-on-Chip Model Is Representative of the Response in Xenografts. *Micromachines* **2022**, *13*, 739. [CrossRef]
8. Callegari, F.; Brofiga, M.; Poggio, F.; Massobrio, P. Stimulus-Evoked Activity Modulation of In Vitro Engineered Cortical and Hippocampal Networks. *Micromachines* **2022**, *13*, 1212. [CrossRef]
9. Lutsch, E.; Struber, A.; Auer, G.; Fessmann, T.; Lepperdinger, G. A Poly-(Ethylene Glycol)-Diacrylate 3D-Printed Micro-Bioreactor for Direct Cell Biological Implant-Testing on the Developing Chicken Chorioallantois Membrane. *Micromachines* **2022**, *13*, 1230. [CrossRef] [PubMed]
10. Li, N.; Shen, M.; Xu, Y. A Portable Microfluidic System for Point-of-Care Detection of Multiple Protein Biomarkers. *Micromachines* **2021**, *12*, 347. [CrossRef]
11. Sitkov, N.; Zimina, T.; Kolobov, A.; Karasev, V.; Romanov, A.; Luchinin, V.; Kaplun, D. Toward Development of a Label-Free Detection Technique for Microfluidic Fluorometric Peptide-Based Biosensor Systems. *Micromachines* **2021**, *12*, 691. [CrossRef]
12. Zimmers, Z.A.; Boyd, A.D.; Stepp, H.E.; Adams, N.M.; Haselton, F.R. Development of an Automated, Non-Enzymatic Nucleic Acid Amplification Test. *Micromachines* **2021**, *12*, 1204. [CrossRef] [PubMed]
13. Saha, T.; Fang, J.; Mukherjee, S.; Knisely, C.T.; Dickey, M.D.; Velev, O.D. Osmotically Enabled Wearable Patch for Sweat Harvesting and Lactate Quantification. *Micromachines* **2021**, *12*, 1513. [CrossRef] [PubMed]
14. Tsai, H.F.; Carlson, D.W.; Koldaeva, A.; Pigolotti, S.; Shen, A.Q. Optimization and Fabrication of Multi-Level Microchannels for Long-Term Imaging of Bacterial Growth and Expansion. *Micromachines* **2022**, *13*, 576. [CrossRef]

**MDPI**

*Review*

# Engineering Organ-on-a-Chip to Accelerate Translational Research

Jihoon Ko [1,†], Dohyun Park [2,†], Somin Lee [3,†], Burcu Gumuscu [4] and Noo Li Jeon [1,3,5,*]

1 Department of Mechanical Engineering, Seoul National University, Seoul 08826, Korea; jihoonxx@snu.ac.kr
2 Bio-MAX Institute, Seoul National University, Seoul 08826, Korea; parkkhdh@gmail.com
3 Interdisciplinary Program in Bioengineering, Seoul National University, Seoul 08826, Korea; sml1003134@gmail.com
4 Biosensors and Devices Laboratory, Biomedical Engineering Department, Institute for Complex Molecular Systems, Eindhoven Artificial Intelligence Systems Institute, Eindhoven University of Technology, 5600 MB Eindhoven, The Netherlands; b.gumuscu@tue.nl
5 Institute of Advanced Machines and Design, Seoul National University, Seoul 08826, Korea
* Correspondence: njeon@snu.ac.kr; Tel.: +82-2-880-7111
† These authors contributed equally to this work.

**Abstract:** We guide the use of organ-on-chip technology in tissue engineering applications. Organ-on-chip technology is a form of microengineered cell culture platform that elaborates the in-vivo like organ or tissue microenvironments. The organ-on-chip platform consists of microfluidic channels, cell culture chambers, and stimulus sources that emulate the in-vivo microenvironment. These platforms are typically engraved into an oxygen-permeable transparent material. Fabrication of these materials requires the use of microfabrication strategies, including soft lithography, 3D printing, and injection molding. Here we provide an overview of what is an organ-on-chip platform, where it can be used, what it is composed of, how it can be fabricated, and how it can be operated. In connection with this topic, we also introduce an overview of the recent applications, where different organs are modeled on the microscale using this technology.

**Keywords:** organ-on-a-chip; microfabrication; microphysiological system; biophysical stimuli; biochemical stimuli; in vitro cell culture

**Citation:** Ko, J.; Park, D.; Lee, S.; Gumuscu, B.; Jeon, N.L. Engineering Organ-on-a-Chip to Accelerate Translational Research. *Micromachines* **2022**, *13*, 1200. https://doi.org/10.3390/mi13081200

Academic Editors: Violeta Carvalho, Senhorinha Teixeira and João Eduardo P. Castro Ribeiro

Received: 4 July 2022
Accepted: 20 July 2022
Published: 28 July 2022

**Publisher's Note:** MDPI stays neutral with regard to jurisdictional claims in published maps and institutional affiliations.



## 1. Introduction

Organ-on-a-Chip (OoC) can be described as a microfluidic device with small structures used for culturing cells. The small structures define the locations of cell growth, or create biochemical (e.g., growth factor) and biophysical (e.g., electrical, thermal, and mechanical) stimuli for generating physiologically relevant microenvironments. Therefore, OoC is an integrated system of microfluidic and bioengineering technology designed to reconstitute the tissue architecture of specific organs and recapitulate a key physiological function of a tissue. As a preclinical in-vitro model, testing the efficacy or toxicity of drugs and studying the pathophysiology of the tissue are two major applications of OoC. This novel experimental model with high human physiological relevance is built to compensate for the limitations of traditional 2D petri dish culture or animal models, narrowing the gap between preclinical and clinical results in drug developments and disease studies. The key functions of various organs, tissues, and pathologies have been modeled using OoC since the early 2010s [1]. Since then, the advancement in the field of OoC technology has accelerated and many novel models were reported as proof-of-concept studies in academics. Table 1 summarizes the advantages and disadvantages of the techniques being used to study cell microenvironments.

**Table 1.** Comparison between advantages and disadvantages of the available translational research model.

| Research Approach | Macro-Scale Techniques | Micro-Scale Techniques | Animal Models |
|---|---|---|---|
| Platform | Well plates | Microfluidic chips | Experiment animals |
| Types | 2D cell culture<br>Spheroid assay | Organ-on-a-chip | mouse, rabbit, monkey, murine, etc. |
| Advantages | A standardized platform, one size fits all principle<br>Well established process flow<br>Co-culture of different cell types<br>Cost-effective | Custom-designed per case<br>Precise control of cell microenvironment<br>Co-culture of different cell types<br>Fluid flow application<br>Dynamic cell-culture environment<br>Possibility to integrate with stimuli sources and measurement tools such as sensors<br>Reduced use of the reagents | Physiologically relevant results<br>Availability of in vivo conditions<br>Availability to observe the response of the organism as a complex entity |
| Disadvantages | Static (no flow) conditions<br>Physiologically less relevant model<br>Uniform distribution of reagents (i.e., no possibility to form gradients)<br>One culture condition can be tested at a time<br>Well plates are made of only one material, which is polystyrene | Commercially not available<br>Non-standardized process flow<br>Microfabrication requirement for prototyping<br>Not as budget friendly as the conventional cell culture platforms | The studies are subject to ethical concerns<br>Expensive and hard to maintain<br>Not possible to apply real-time monitoring in the absence of dedicated equipment<br>Complicated platform for mechanistic studies |

The use of OoC has just begun to expand toward clinics and pharmaceutical industries [2,3]. Along with the emerging interest in personalized and precision medicine as well as the advent of novel therapeutics such as cancer immunotherapy, the biomedical community started to positively consider the adoption of OoC as a promising tool for biomedical and pharmaceutical research [4]. Although technical and biological improvements are still needed to fulfill the criteria of its routine and universal use, global efforts from diverse professional fields started developing and applying this novel technology [3]. Based on this current tide in OoC community, recently published review papers focus on sharing opinions on what this society should pursue when developing the model successfully applied on preclinical study or drug discovery [5–7]. In this paper, we start by describing the fundamentals of the OoC, such as basic and advanced engineering techniques and biological components inside the OoC followed by the representative applications of OoC. This will guide the researchers from non-OoC community and provide better understanding on the potentials and advantages of OoC. Furthermore, the article will discuss about the current position and limitation of OoC in translational research and its mission and future perspective on successful bench-to-bedside cases.

## 2. Key Features of OoC

OoC is a microengineered device to recapitulate key functions of organs and tissues. In these platforms, the basic idea is not to generate organs themselves, but the functions of the organs are mimicked to study a particular aspect of the target organ. OoC platforms were invented for mitigating the limitations of both conventional cell culture plates and animal models (see Table 1 for an overview of the limitations). Although animal models can help us to understand various biological phenomena better, they are usually too complicated for mechanistic studies. Compared to in-vitro models, animal experiments are costly and time-consuming and it is challenging to observe phenomena happening in the deep tissue of the models. On the other hand, conventional cell culture plates allow- us to control the types and numbers of cells as well as apply controllable stimuli. Yet, integration of other relevant systems such as endocrine, neurological, and immunological considerations is still lacking in conventional cell culture plates. OoCs can provide well-defined, well-controllable, easy to observe, but still complex environments to display organ-specific functions. All OoC platforms have three fundamental characteristics: (i) the arrangement of cells in in-vivo-like layouts; (ii) the possibility to culture multiple cell types to reflect physiological relevance; and (iii) the presence of biochemical and biophysical stimulations to mimic the functions of tissues. To realize these characteristics, several technical and practical elements related to microfabrication and cell biology are applied, including the design, materials, the fabrication method of the platform, cell sources, the type of scaffold, and the type of the stimulus (Figure 1). In this paper, OoCs will be introduced in the context of these three characteristics and technical factors for producing these platforms.

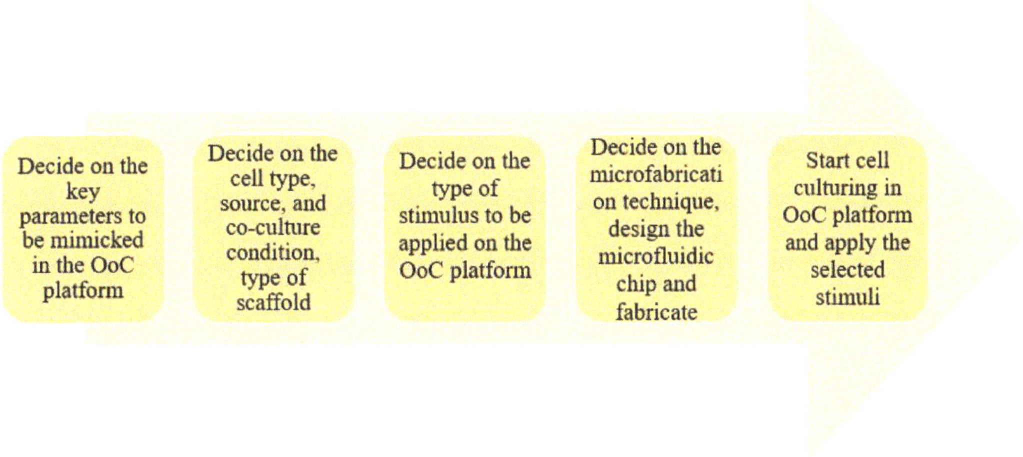

**Figure 1.** Typical prototyping steps followed to fabricate a microfluidic OoC platform.

### 3. Construction of an OoC Platform

Multiple cells interact with each other within micrometer-scaled environments in OoC systems. To build such systems, several microfabrication strategies can be used. Since the introduction of the first OoC device in the 2010s, the vast majority of OoC systems were fabricated using soft lithography [8,9]. Lithography is a combination of two words: "lithos" which means "stone", and "graphein" which means "to write" in Latin. Soft lithography is a process where a microchip layout is engraved in a soft material. In most cases, this soft material is polydimethylsiloxane (PDMS). Having several key properties including biocompatibility, optical transparency, gas permeability, and high-definition patternability make PDMS ideal for OoC and biological applications. Furthermore, the elasticity of PDMS enables OoC devices to utilize biophysical and mechanical stimuli to recapitulate the function of an organ. Fabrication of a PDMS-based OoC device starts with the design of the desired microchip layout in the software, such as AutoCAD, and transferring this layout to a photomask [10,11]. A photomask is an opaque surface printed on glass or film with transparent spots, or patterns, to allow light to pass through in a defined pattern. The photomask is used to fabricate a mold master. The mold master is usually made with photoresist upon a silicon substrate via a photomask and ultraviolet light exposure. A photoresist is a light-sensitive polymer, changing its molecular structure upon exposure to ultraviolet light and turning it into a soluble or insoluble material. A common photoresist type used in microfabrication is called SU-8. The solubility effect depends on the tone of the photoresist and in this way, the photomask with features can be replicated on a mold master with a positive or negative polarity. PDMS is a liquid polymer in the uncured form, upon mixing it with an activator solution and subsequently applying heat, it solidifies thanks to cross-linking reactions. Therefore, PDMS in the liquid form is poured on top of the mold master, and exposed to heat to structure the microchip layout into the PDMS in the solid or cross-linked form. The PDMS-based OoC device with a microchip layout is then punched to open all through holes that will be used for liquid injection ports (such as to inject cell culture media into the device). The final step is to bond the casted PDMS OoC device with a glass or other PDMS part via plasma bonding. Oxygen plasma exposed to the PDMS surface renders silanol groups (–OH) on the surface and immediate meeting of the surface with other oxidized PDMS or glass forms an irreversible Si–O–Si bond at the interface. This covalent bond prevents water from permeating into the glass-PDMS interface. Bonding of an engraved PDMS OoC device with a flat surface forms a microchannel where a hydrogel-cell mixture, cell suspension, or culture medium can be inserted. Such a microchannel is a basic design factor of the OoC devices. Figure 2a shows an example of the fabrication process of a PDMS-based device; the lung-on-a-chip device [12].

Although PDMS is an ideal material for OoC operations, it also has disadvantages [13,14]. For example, PDMS can absorb small and hydrophobic molecules. This hinders the use of PDMS-based OoCs from drug studies as the actual dose of drug introduced to cells will not be clear due to the absorbance effect. Apart from that PDMS is a soft and flexible material, which can be undesired for some of the applications requiring hard materials.

Another way of fabricating OoCs is based on injection molding (Figure 2b) [15,16]. This technique is mostly preferred in commercialized OoC systems because it allows for mass production. Injection-molded materials are mostly plastics, and therefore they are not flexible and do not absorb or permeate other molecules, but they are still optically transparent. Injection molding is a microfabrication process adapted from the industry. The molten material is injected into a previously machined mold, and the material gets solidified upon cooling down. The type of materials used in this process involve thermoplastics, elastomers, and polymers. Microstructured plastic parts are integrated into complete OoC devices by assembling with strews, solvent bonding, or thermal bonding [15]. Other bonding methods such as silicone adhesive bonding and plasma bonding with PDMS can also be used according to the materials selected to fabricate the main body. Some injection-molded OoC devices utilize the dimensions of conventional well-plates to be compatible with conventional liquid handlers or readout systems [17–22].

Another preferred fabrication method is 3D printing [23,24]. This technique offers the advantage of precise control over geometry on the microchannel geometry. Fabricating circular microchannels is challenging via soft lithography because the photoresist can be spread on a silicon substrate and crosslinked with a fixed height only. 3D printing techniques can control the spatial distribution via layer-by-layer assembly of the material [25]. It is also possible to print the cells in 3D on a substrate along with an extracellular matrix such as a hydrogel [26]. Printing ink can be made of natural (hydrogel, collagen, fibrin, alginate, etc.) or synthetic (polycaprolactone, silicone, gel-like Pluronic 127, etc.) material [27]. These inks can be 3D printed using several methods including (i) micro-extrusion printing, where the ink is directly deposited onto a substrate by using a micro-extrusion head, and (ii) inkjet printing, where the ink is distributed from an electrically heated or piezoelectric actuator nozzle and deposited onto a substrate in droplet form, and (iii) laser-assisted printing, where a laser beam writes on an ink coated on a substrate. The final resolution is the best when laser-assisted printing is used, while inkjet printing is a preferred technique to print the cells together with their supporting matrix. Complex architectures in human organs can be replicated more precisely, allowing for better recapitulation of the key tissue and organ-level functions. The 3D printing technique is also applicable for commercialized products and large-scale production of OoCs and, therefore, it also finds a place in the industry. Figure 2c demonstrates 3D printing methods to fabricate microfluidic devices with complex-shaped microchannels.

OoC devices can be fabricated using various techniques as summarized in this section. The fabrication method and material are selected according to the purpose of the study in the design step. PDMS is a suitable material to fabricate a few micrometer-scaled microchannels and 3D printing can be preferred to fabricate a microchannel network to mimic a complex 3D vascular structure.

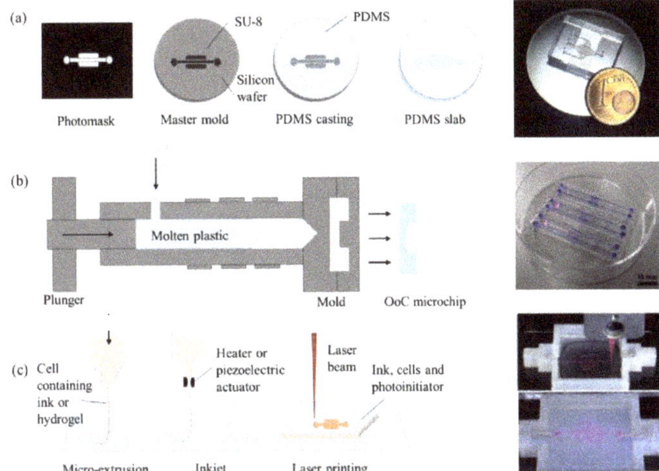


**Figure 2.** Fabrication methods of organ-on-a-chip models. (**a**) Schematic process of fabricating PDMS-based microfluidic devices. Microchip figure (**right**) reproduced from Ferraz et al., 2018 [11]. (**b**) Schematics and images of an injection molded microfluidic device. Microchip figure (**right**) reproduced from Virumbrales-Munoz et al., 2019 [16]. (**c**) Schematics and images of microchannels fabricated by a 3D printer. Complex-shaped channels can be fabricated with a 3D printer. Microchip figure (**right**) reproduced from Homan et al., 2016 [24].

## 4. Cell Microenvironments Mimicking In-Vivo

The first consideration in designing OoC devices is to arrange cells to exploit their key roles in organ function. Microchannels or pores are utilized to confine cells in a target area. Cells can be confined into droplets in droplet-based microfluidic devices or can be confined onto a surface made of hard (e.g., plastic) or soft (e.g., hydrogels) materials while this system is fed by parallel or coaxial flows. In the former case, the microfluidic device has at least one droplet-generator unit (commonly in the form of T, X, Ψ, or Y-shaped junctions) and a droplet splitting and merging unit for droplet handling (Figure 3a) [28–30]. Droplets are formed from a dispersed phase fluid into compartmentalized droplets surrounded by a continuous phase fluid. Droplets can be produced in various sizes and monodisperse (uniform size) forms and they are useful templates for drug delivery, nutrient delivery, and living cell encapsulation [31]. In the latter case, microchannels are engraved into a solid block of microchips, which may contain microgrooves with different heights, membrane-like barriers, micromechanical valves, and various micro-scale components [1,32,33]. Cells are seeded and grown in the vicinity of these structures depending on the OoC application. Parallel-flow-based microchips are useful models for external stimuli generation and spatial control of the cell culture layout. For example, neurons can be grown in microchips. Microchannels with 10 μm of width and 5 μm of height confined can prevent somas from intruding into the channels and only allow axons to grow along the channels, resulting in the perfect isolation of axons [33]. Similarly, a liver sinusoid model was built using a microfluidic endothelial-like barrier consisting of a parallel array of channels with a width of 2 μm and height of 1 μm [32]. The barrier was used for forming high resistance into the cell culture area and it also plays a key role in concentrating hepatocytes within the culture chamber, where the concentrated hepatocytes showed higher viability compared to the hepatocytes cultured in low density.

Capillary action is often used as a strategy to build the cellular microenvironment in OoC devices. Capillary action is governed by the interplay between the geometry or chemistry of a surface and the surface tension of a liquid. In microchips, capillary action ensures the spatial control of the liquid, i.e., liquids can be selectively patterned in desired spots. Micro bumps or micro pillaris are examples of elements to control the flow via surface geometry (Figure 3b,c) [34–37]. Microbumps and micropillars in the microchannels form narrow gaps that serve as gates, allowing or blocking the flow [34]. To allow for the flow into an adjacent microchannel, high bursting pressure should be applied. Using microscale gates, a hydrogel solution containing cells can fill a microchannel and another solution can fill the adjacent microchannel after cross-linking of the previously loaded hydrogel. As the hydrogels are physically connected, the cells in both hydrogels can chemically communicate and migrate into the other zones. Controlling the flow via geometrical change strategy was used in many OoC models for cell arrangement in several applications. Examples include vascular networks, angiogenesis models, blood-brain barrier models, and tumor extravasation models [38–41]. Along with geometrical changes, surface chemistry changes can be used for arranging cells in flow-based microchips (Figure 3d). Open microfluidic devices which have additional air-liquid interfaces other than loading ports use hydrophobically modified surfaces by air plasma to induce the wicking of liquids through a narrow gap [42,43]. Hydrogel solutions introduced to a microchip tend to fill areas with high wettability or hydrophilic areas and the difference in wettability enables the allocation of cells with in-vivo-like layouts within the devices. Use of different materials within a single device, air plasma, a hydrophobic or water-repellent coating is used for controlling wettability in microfluidic devices.

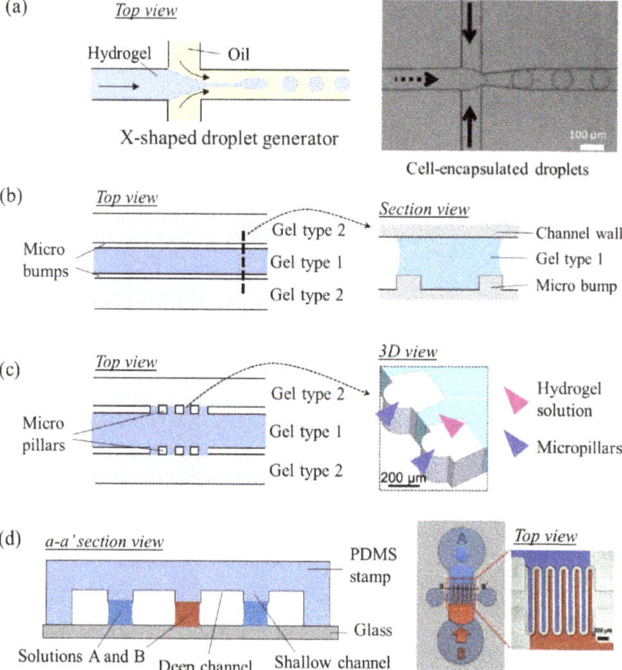



**Figure 3.** Methods used for positioning cells with in-vivo-like layouts. (**a**) Schematics of a droplet-based microfluidic device for culturing single-cells in droplets. The right panel figure is reproduced from Chan et al., 2013 [30]. (**b**) Schematics of a microchannel with micro bumps, allowing for the injection of different hydrogel types without physically dividing the channel. (**c**) Schematics of micropillars working as capillary valves. The right panel figure is reproduced from Hyung et al., 2021 [37]. (**d**) Schematics of capillarity-mediated liquid patterning. Shallow channels with hydrophilic regions attract the solution while hydrophobic regions repel the solution with water content. Shallow channels guide cells to be patterned in designed shapes. Figure reproduced from Lee et al. (2010) [43].

## 5. Selection of Cell Resource

The next critical element of an OoC device is the choice of cell resources. The same cell study performed in commercially available OoC devices may yield genetically and phenotypically different results if the cells from different resources are used [44]. Hence, cell resources should be selected according to the physiological relevance, sustainability, and purpose of the study. The most widely used cell resources are commercialized cell lines, primary cells, and induced pluripotent stem cells (iPSCs) from trusted supplier companies and cell banks.

Commercial cell lines often represent immortal cell lines that are easy to culture using established and reliable culture protocols. These cell lines grow in predictive ways, resulting in reproducible OoC models e.g., for optimization purposes. As commercial cell lines are popular in biological studies, there is vast information available in the literature about the morphology, gene profile, and molecular pathways belonging to the cell type of interest. Users, accordingly, take advantage to use such information to design experiments or compare the results observed in an OoC model. Some commercially available cells can be excellent sources of rare cell types that are difficult to differentiate from iPSCs. However, mutations originating from immortalization or multiple passages can decrease the physiological relevance of the OoC models when compared with the primary cells obtained from donors.

Primary cells directly obtained from donors are an excellent source when developing donor-specific OoC models. The National Center for Advancing Translational Sciences (NCATS; https://ncats.nih.gov/tissuechip, accessed on 1 August 2020) in the United States initiated a grant program to develop Clinical Trials on a Chip in 2020. The majority of granted proposals aimed at developing patient-specific OoC models, also called personalized medicine, using patient-derived cells. Due to the heterogeneity of patients, especially cancer patients, direct use of primary cells is an ideal approach to reconstitute diseased conditions within OoC devices, and it is believed that these cells can repeatedly generate responses against therapeutics. Even though primary cells have advantages in an identical gene profile to the donor, it is challenging to reuse or store the primary cells in the longer term (in the order of several weeks). When working with these cells in ex-vivo cultures, great care in isolation protocol, the composition of the cell culture medium, and the design of the microcellular environment are required to retain their original phenotypes.

Adult stem cells or iPSCs are increasingly used in OoC platforms because this cell type is infinitely sustainable and can be differentiated into various tissues, including diseased tissues via genetic engineering. For example, organoids representing different tissues derived from a single donor can be co-cultured in a multi-organ chip platform to show the systematic connection of organs. iPSC cells can be used to create fully developed tissues from scratch, although such studies require extensive time and resources. So far, the use of iPSCs has remained limited in developing OoC models to test therapeutics in a patient-specific manner because the protocols for differentiation and maturation are not standardized and have low reproducibility levels.

While commercialized cell lines are easy to access, they are not ideal for OoC studies requiring a high correlation with real tissues. Primary cells and iPSCs are becoming increasingly popular in OoC studies due to having a patient-specific, original gene profile and phenotype. On the other hand, these cells are difficult to access due to the non-frequent supply through biopsies or surgeries. Another drawback is that the differentiation of iPSCs is a time-consuming and laborious process.

## 6. Application of Stimuli

The ability to apply mechanical stimuli to cells is one of the most important features of OoC platforms. OoC platforms can generate precisely controlled mechanical stimuli that were challenging to be applied to conventional models. The capability of OoC platforms to apply a mechanical stimulus to the cells facilitated observing the response in a 3D environment and in real-time. This enabled the regulation of the growth of cells in OoC platforms and to recapitulate organ function in a more in-vivo-like fashion. This chapter describes how OoC platforms generate mechanical stimulus by exemplifying stretching, compression, flows, and shear stress as displayed in Figure 4.

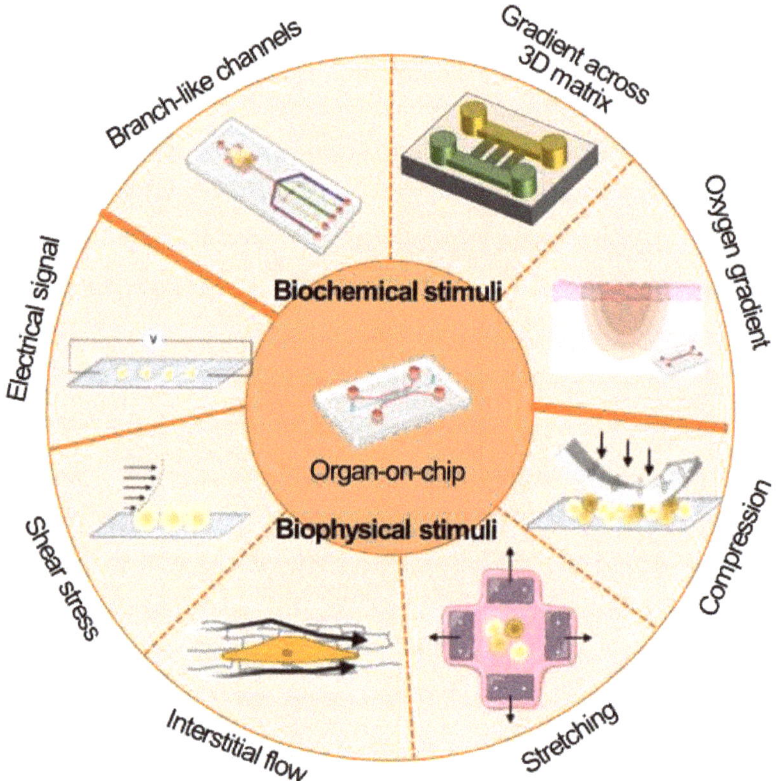


**Figure 4.** Biochemical and biophysical stimuli applied to organ-on-chip models.

Many OoC platforms utilized the elasticity of PDMS for generating mechanical stimuli of stretching and compression. A representative example of stretching is the lung-on-a-chip device [1]. The device consists of a middle channel vertically separated into two sections mediated by a porous membrane and two side channels on both sides of the middle channel. Negative pressure applied to the side channels deforms the thin walls of the middle channel, resulting in the stretch of the membrane in the middle channel. The inventors applied 10% of cyclic strain to the membrane where lung epithelial cells and microvascular endothelial cells are attached on each side respectively. The presence of cyclic strain significantly enhanced transmembrane uptake of nanoparticles into the endothelial layer as large as mouse model uptake. The article showed the importance of mechanical stress in modeling organs and demonstrated the usefulness of OoC devices to generate mechanical stress.

Also using the elasticity of PDMS, some OoC devices demonstrated responses of tissues against compression. Magdesian et al. demonstrated injury neuronal network by pressing the top of OoC device where neurons are cultured with a beaded AFM tip and observed reconnection of the network [45]. Ahn et al. demonstrated injury of microvessels induced by compression using an OoC platform [46]. An air channel is fabricated above a microvessel channel and the thin layer of PDMS between the two channels deforms to press the vessel zone when positive pressure is applied to the air channel. The compression is expected to be used for mimicking pressure-induced diseases such as glaucoma and asthma.

Various types of fluid flows exist in our body such as interstitial flow, intravascular flow, and transendothelial flow and they play essential roles in the differentiation, proliferation, migration, and gene expressions of cells [47]. OoC platforms are efficient tools for studying the effect of flows on cells. For example, Kim et al. reported interstitial flow regulates the angiogenic response and phenotype of endothelial cells. They found angiogenic sprouting was enhanced toward the reverse direction of the interstitial flow and the sprouts display abundant actin-rich filopodia at their distal edges [48]. Recently, Hajal et al. applied both luminal flow and trans-endothelial flow to the vascular network and demonstrated the increased potential of tumor extravasation [49]. These studies utilizing OoC platforms broadened our knowledge of cell biology under flow conditions.

There are many ways to generate flows within OoC platforms. Syringe pumps, pneumatic pumps, and peristaltic pumps are mainly used to generate flows by delivering culture medium. A kidney-on-a-chip platform utilized continuous medium flow over an epithelial monolayer generated by a syringe pump, resulting in 0.2 dyne/cm of shear stress applied to the cells [50]. The shear stress enhanced cell polarization and primary cilia formation compared to a static condition. In case precise control of flow is not required, OoC models often utilize gravity-driven flows. Pressure difference induced by the difference of heights of medium between medium reservoirs generates a flow to meet the balance of the heights [51]. Sometimes pipette tips or external accessories are inserted into the medium reservoir to extend the volume of the medium. This gravity-driven flow cannot be constant as the pressure difference is continuously reduced as the medium flows. Gravity-driven flow is also used to generate a pulsatile flow by placing an OoC platform on a rocker that flips tilting periodically [52].

Other than mechanical stimulation, microfluidic chips facilitate generating biochemical stimulation such as oxygen, [53,54]. nutrient, [55]. and growth factor gradient [56]. Brennan et al. comprehensively summarized microfluidic chips that adopted oxygen control techniques and oxygen sensors. They reviewed oxygen control methods including diffusion from a source fluid, separate gas perfusion, hydration layer, cellular consumption. Gradient formation of soluble factors is one of the significant advantages of microfluidic devices. Laminar flow at merging T or Y-shaped microchannels forms a diffusive profile. Using this phenomenon, various microfluidic chips for gradient formation were developed and they were applied for studying response of cells against biochemical cues such as cancer metastasis, immune response, axon guidance, and angiogenesis.

As mentioned so far, OoC platforms can apply mechanical/chemical stimulation to the tissues cultured in the platforms. External equipment can precisely control the stimuli and sometimes the stimuli can be applied to some specific part of the tissues. The controllable and selective stimuli are making OoC platforms attractive and powerful compared to other models.

Monitoring tools for drug toxicity in the cellular microenvironment.

Spatially and temporally resolved information about cell physiology and microenvironment as well as pharmacodynamic drug responses are monitored in the form of imaging, electrical signal measurement, and molecular measurements. Real-time monitoring in OoC devices is possible at the tissue level which is very difficult to observe in-vivo and difficult to reconstitute in simpler in-vitro models.

A tissue can be tracked by high-resolution imaging, which is capable of tracking single-cell level activities using microscopes. The tracking is made possible by staining the molecules of interest in such tissue constructs. Staining is performed using commercially available biomarkers which come in a range of different fluorescent colors. Mostly used biomarkers include primary and secondary antibodies, nucleus stains, cytoskeleton stains, and biomarkers.

Real-time information about the viability and metabolic activity of the tissue constructs and organoids is also observed using electrodes. The electrodes can be embedded in a microchannel thanks to microfabrication processes. Common usage areas of the electrodes include the measurement of transendothelial electrical resistance (TEER) while multi-electrode arrays (MEA) measure field potentials of cells (for example, for neural network characterization). A good example is the microfluidic blood-brain-barrier (μBBB) model that incorporates TEER electrodes on two channels separated by a porous membrane coated with endothelial cells and astrocytes on both sides, respectively [57]. Shear stress was induced by culture medium flow through the microchannel containing endothelial cells. The presence of the shear stress increased TEER levels in OoC co-culture compared to that of transwell co-culture. Transient drop and recovery of TEER were monitored in real-time in response to histamine exposure. The OoC device was used for testing barrier-enhancing or barrier-opening drugs to regulate drug delivery into the central nervous system.

The molecular analysis presents a multi-faceted way to collect information in OoC devices [58]. The supernatant of the cells or culture medium collected from the outlet is utilized for measurements in-chip and off-chip. For off-chip measurements, extraction of cells may become tricky for the devices fabricated with a permanent bonding method. In that case, an additional process of dissolving the extracellular matrix or detaching cells from the device surface is required. For profiling genes such as RNA sequencing, a sufficient number of cells is harvested from multiple OoC platforms experimented in the same condition due to the small number of cells contained in a single chip. Integrated sensors have been the workhorse for the in-chip measurements. Several types of sensors are presented to measure culture microenvironment (e.g., pH, oxygen level, nutrient content), mechanical stimulation (e.g., flow rate, compression, stretching), electrical stimulations (e.g., neural network signaling, cardiac signaling via pulse generation), chemical gradients (e.g., chemical factors, biomarkers, cell secretome ingredients).

## 7. Applications of OoCs

OoC technology in tissue engineering applications has been proved to demonstrate several advantages. OoC devices represent; physiologically-relevant systems mimicking key functions of tissues and organs; microfluidic devices that are compatible to apply biochemical and biophysical stimuli; spatiotemporally controllable 3D cell microenvironments. These powerful functions of OoC boost our knowledge in biological and pharmaceutical research. OoC devices are useful tools that can efficiently deliver reliably reproducible results for drug toxicity assessment and large-scale preclinical trials in the drug development pipeline.

## 8. Drug Development

The drug development process consists of five phases: (1) discovery and development, (2) preclinical research, (3) clinical development, (4) FDA review, and (5) FDA post-market safety monitoring. Developing a new drug takes 7 to 15 years (an average of 13.7 years). Overall, the probability of success for a drug from phase I to approval is 1 in 10,000, while the preclinical phase has the lowest success rate with 3% [59]. Due to the low success rates, the preclinical phase is called the "Death Valley" in the pharmaceutical industry. If a drug compound passes the preclinical test, the success rate can increase dramatically. A key requirement for passing the "Death Valley" is an intermediary research model that can accurately assess toxicity and predict human responses at the preclinical stage. Traditionally, toxicity and off-target effects are assessed with in-vitro 2D cell cultures or in-vivo animal experiments before clinical trials. In-vitro 2D cell cultures are seen as insufficient tools for drug testing due to the lack of complexity. Such tools typically adopt a simplified 2D model based on immortalized cell types, while heterogeneous interactions between cells cannot be recapitulated in 2D cell cultures. On the other hand, in-vivo animal models are too complex for preclinical research due to the differences in function at various organ/tissue levels, including immune system responses. Animal models may differ in tissue function when compared to humans, that is why the results from animal models may not be used for testing all drugs. OoC devices can provide sufficient complexity in cell cultures with the help of external stimuli, recapitulate the key functions of the tissues, and facilitate working with primary cells from patients and healthy individuals. In this way, OoC systems elicit appropriate responses to drug exposure. Table 2 summarizes the differences between animal models, 2D cell cultures, and OoC devices.

**Table 2.** Comparison of existing preclinical research models and OoC devices.

|  | Animal Model | 2D Cell Culture | OoC Device |
|---|---|---|---|
| Human tissue | No | Yes | Yes |
| Personalized medicine | No | Yes | Yes |
| Complexity | Yes | No | Limited |
| Control over microenvironment | No | Yes | Yes |
| Tissue-level function | Yes | Limited | Yes |
| Organ-level function | Yes | Limited | Limited |
| Real-time readouts | No | Limited | Yes |
| High-throughput, in parallel testing | No | Yes | Yes |
| Pharmaco-dynamics and-kinetics | Yes | No | Yes |

Cancer therapeutics requires a deeper understanding of not only cancer but the surrounding tumor microenvironment, which is not only made of cancer cells but also stromal cells, immune cells, blood vessels, and extracellular matrix [60]. Reconstructing a tumor microenvironment in OoC devices is regarded as a valid model for anticancer drug screening. After co-culture modalities of blood vessels and tumors, including peripheral stromal cells, have been established, increasing research attempts have been made to evaluate the emerging anticancer drugs, as well as strategies for chemotherapy, targeted therapy, and immunotherapy.

Tumor-on-chip technology has been developed extensively in the past decades. For example, an injection-molded 3D array was introduced for determining cell migration. The 3D array was filled with a collagen matrix, where natural killer cells migrated toward HeLa cells; therefore, spatiotemporal observation and analysis of the interaction between immune cells and cancer cells were made possible (Figure 5a) [19]. A tumor microenvironment consisting of a blood vessel and a lymph vessel can be fabricated using 3D printing. 3D-printed tubular structures serve as perfused microchannels, where complex mechanisms in molecular transport of anticancer drugs can be profiled between the vessels. A human microcirculatory system with stimulated neutrophils was also modeled in OoC to study systemic infection. This model facilitates the monitoring of dynamic interactions between intravascular tumor cells and neutrophils at high spatiotemporal resolution. The model was able to distinguish a chemokine-dependent neutrophil migration pattern that results in enhanced tumor cell extravasation (Figure 5b) [61]. In another example, organotypic endothelial cells were constructed in the form of lumen-like structures in a hydrogel environment. Using tumor endothelial cells, a patient-specific angiogenesis assay was developed to spatiotemporally evaluate the characteristics of cancer angiogenesis in each patient with kidney cancer (Figure 5c) [62].

Developing cell spheroids and organoids has also been explored in OoC devices. Table 3 provides an overview of on-chip cancer models that are used for drug treatment. A 3D vascularized ovarian cancer spheroid was exposed to Paclitaxel (Taxol®) and the uptake is examined through diffusivity measurements, functional flux analysis, and accumulation of fluorescently bound drugs using OoC. The presence of interstitial flow resulted in differences in responses corresponding to shrinkage and CD44 expression of vascularized tumors to Taxol [63]. A microtumor model supported by a microvascular network study explored efficacy assessments with FOLFOX (5-FU, Leucovorin, and Oxaliplatin), as a standard treatment chemotherapy drug. Tumor growth was significantly reduced in-vitro compared to the control "placebo" group [64]. Anticancer drugs were also assessed using OoC devices by focusing on the evaluation of tumors and blood vessels. In an OoC device, angiogenesis toward a glioblastoma tumor spheroid was monitored. In this system, inhibition of angiogenesis was observed when Bevacizumab (Avastin®), a representative anti-VEGF drug, was introduced to the system [18]. More details about microfluidic chips for anti-cancer drug screening are reviewed in the following articles [65,66]. None of these systems were used in preclinical research in drug discovery phases yet, although the emerging technology is increasingly being commercialized by industry at the moment (Figure 6).

**Table 3.** Examples of cancer-vascular models with anticancer drug screening.

| Cancer Therapy | OoC Device | Drug Treatment | Target Region | Reference |
|---|---|---|---|---|
| Chemotherapy | Ovarian cancer (A549, Skov3) spheroid and blood vessel coculture | Paclitaxel (5.0 µM) On day 7 of culture | Inhibited tumor cell proliferation by increasing intrinsic apoptosis | [63]. |
| | Breast cancer (MCF-7) spheroid and blood vessel coculture | Paclitaxel (0–500 ng/mL) On day 7 of culture | | [67]. |
| | Colorectal cancer (HCT116) cell and blood vessel coculture | FOLFOX 100 µM 5-FU; 10 µM leucovorin; 5 µM oxaliplatin On day 6–8 of culture | Inhibition of DNA synthesis in cancer cells by the formation of crosslinks in DNA | [66]. |
| Targeted therapy | Colorectal cancer (CRC-268) cell and blood vessel coculture | Bevacizumab (10 µg/mL) On day 7 of culture | Inhibiting the binding of VEGF to | [68] |
| | Glioblastoma (U87MG) cancer spheroid and blood vessel coculture | Bevacizumab (1.0 mg/mL) On day 3–7 of culture | Cell surface receptors | [18] |
| Immunotherapy | Glioblastoma (patient sample), blood vessel and tumor- associated macrophage (TAM) coculture | Nivolumab (1 µg/mL) BLZ945 (0.1 µg/mL) | PD-1 blockade (Nivolumab) CSF-1R inhibitor (BLZ945) | [69] |

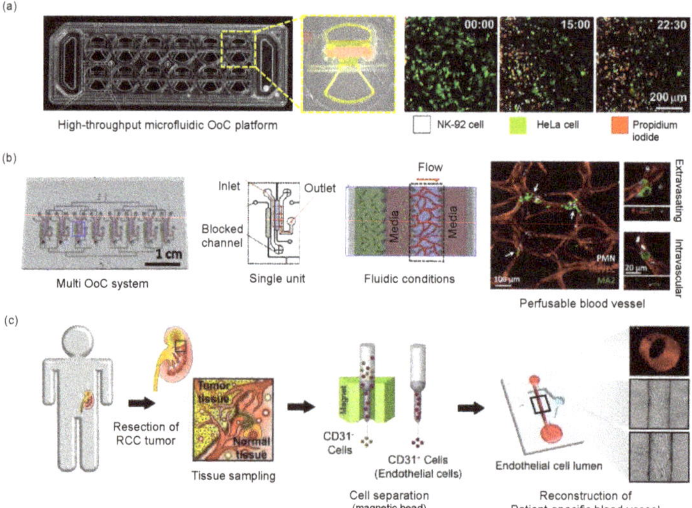


**Figure 5.** Modeling tumor microenvironment using microfluidic devices. (**a**) Large-scale producible microfluidic chip to utilize 3D cytotoxicity assay. Figure reproduced from Park et al., 2019 [21]. (**b**) Multiplexed microvascular network to quantify the dynamics of arrest and extravasation of tumor cells and inflammation-stimulated neutrophils in the microfluidics chip. Figure reproduced from Chen et al., 2018 [61]. (**c**) Isolation endothelial cells from cancer patient samples and reconstruction of patient-specific tumor vasculature models in the microfluidic device. Figure reproduced from Jimenez-Torres et al., 2019 [62].

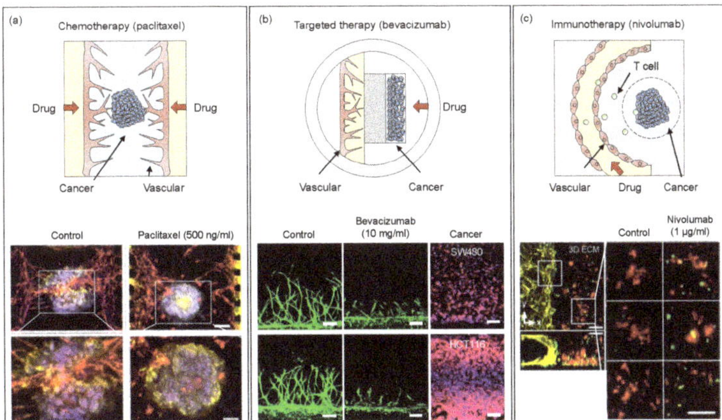

**Figure 6.** Examples of cancer-vascular models with anticancer drug screening. (**a**) On-chip tumor spheroid and blood vessel co-cultured microenvironment with biochemical and biophysical factors. Figure reproduced from Nashimoto et al., 2020 [67]. (**b**) A standardized microfluidic platform for high-throughput anti-angiogenic drug screening. Figure reproduced from Kim et al., 2021 [17]. (**c**) OoC device mimicking glioblastoma tumor niche including immune cells, brain microvessels, tumor-associated macrophages, and glioblastoma. Figure reproduced from Cui et al., 2020 [69].

## 9. OoC Devices Used for Drug Toxicity Assessment

Toxicity to human tissues and unknown safety issues lead to high failure rates in the drug candidate selection process due to unsystematic evaluation [70]. The absorption, distribution, metabolism, and excretion (ADME) of compounds in the body can be predicted using pharmacokinetic and pharmacodynamic modeling of the liver, kidney, blood vessel and heart, and other relevant tissues according to the target organ of the drug. OoCs can offer advantages over 2D cell culture platforms in the exploration of predictive models as discussed in the introduction section.

## 10. Liver Models

The liver is one of the first organs to experience drug-induced toxicity. For this reason, compound clearance on the liver must be assessed as early as in the product development phase. However, the recent ban on animal use in toxicity tests that were introduced in Europe resulted in the development of bioartificial liver systems [71]. Species-specific liver-chips using rat, dog, and human-derived hepatocytes interfaced with liver sinusoidal endothelial cells can be modeled with OoC [72]. Hepatic toxicity models mimicking hepato-cellular injury, steatosis, cholestasis, fibrosis, and species-specific toxicity are assessed when treated with the drug-candidate compounds. Such OoC devices predict liver toxicity and address the human relevance of liver toxicity found in animal studies [73]. For example, chronic hepatotoxicity testing can be performed using a perfusion-incubator-liver-chip fabricated via soft lithography [74]. This system can assess repeated dosing chronic hepatotoxicity with a perfusion incubator that uses $CO_2$ gas pressure to drive the perfusion medium. Liver-on-chip devices benefit from the use of 3D printing. Hollow microchannels fabricated using 3D printing and encapsulated in gelatin methacryloyl hydrogels were used as a substrate to co-culture HepaRG and HUVECs to build a model integrating the blood vessel and the hepatocyte layer [75]. This integrated model provides more in-vivo viability and permeability to perform reliable drug toxicity testing. High-throughput hepatotoxicity screening is also made possible in OoC platforms. A liver-on-a-chip incorporating the 96-well plate specification reconstituted a comprehensive liver microenvironment by co-culturing iPSC-derived hepatocytes, endothelial cells, and monoblasts together [76].

## 11. Kidney Models

Nephrotoxicity is considered a major reason for drug attrition in the pre-clinical phase. Approximately 20% of the drugs fail to pass nephrotoxicity tests and therefore cannot advance through clinical trials. To date, cell culture and animal models have been the workhorse of nephrotoxicity tests [77,78]. Inclusion of biophysical stimuli (in this case shear stress induced by the presence of a flow rate) in liver models was shown to enhance drug efflux and albumin uptake despite the application of flow in a unidirectional or bidirectional manner. The epithelium in kidney proximal tubules is continuously exposed to shear stress, which was successfully mimicked by the use of OoC systems [79]. A flow-inducible OoC device mimicking the dynamic culture of kidney organoids exhibited in-vitro glomerular development by inducing morphological maturation with vascularization [80]. It is also possible to model multiple organs with OoC technology. For example, an integrated liver-kidney-chip allowed the evaluation of drug-induced nephrotoxicity following liver metabolism in-vitro, where co-culture of hepatic and renal cells in a compartmentalized multi-layer was exposed to ifosfamide and verapamil [81]. The metabolites produced by liver metabolism were detected using mass spectrometry and were found to cause pronounced nephrotoxic effects on cell viability, lactate dehydrogenase leakage, and permeability of kidney cells (Figure 7) [82,83].

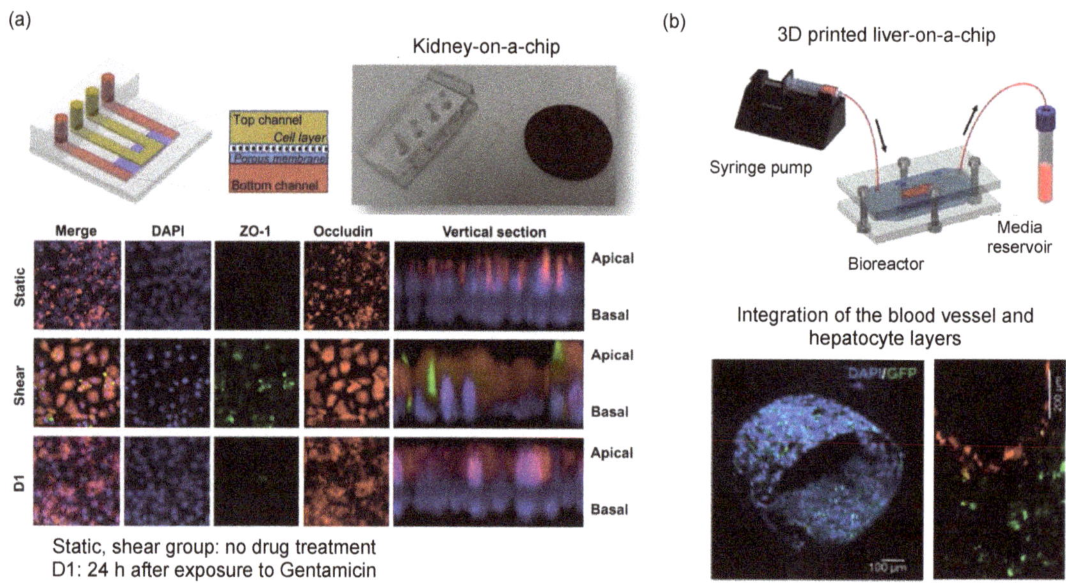


**Figure 7.** Liver and kidney-on-a-chip for hepatotoxicity and nephrotoxicity testing. (**a**) A study showing a pharmacokinetic profile of reducing nephrotoxicity of gentamicin in perfused kidney on-chip. Figure reproduced from Kim et al., 2016 [79]. (**b**) 3D printed liver-on-chip device for evaluating in-situ hepatocyte cell differentiation. Figure reproduced from Massa et al., 2017 [75].

## 12. Brain Models

The blood-brain barrier comprises endothelial cells, separating the blood from brain interstitial fluids. The endothelial cells also form a physical barrier by tight junction proteins that limit the permeation of ions and hydrophilic agents via paracellular pathways [84]. Brain-on-a-chip models mainly focus on creating (i) 2D cell configuration through structural constraints, (ii) porous membrane interface, and (iii) hydrogel-embedded 3D cell constructs separately to study distinct key tissue functions. 2D cell configurations benefit from the ability to fabricate microchannels with different heights, where neuron growth can be tracked. Microcompartments can isolate axons two-dimensionally by separating soma and axon in a neuron [33,85]. In this way, specific drug testing on axon and soma regions can be performed as well as anatomical studies focusing on the reconstruction of unidirectional axonal growth and myelination. In brain-on-chip devices, two or more cells can be cultured to investigate cell-cell communications via cytokine-mediated stimulants. The measurement techniques include off-chip cytokine level determination, protein-level validation, and imaging intracellular Ca2+ levels [37,86]. OoCs have also been built for constructing 3D neurovascular units using collagen and Matrigel, where the passage of small molecule drugs through the BBB was measured on the neurovascular unit [87]. Compound transport efficacy, molecular pathways, and toxicity of neuroactive drugs have been studied on such platforms. Metabolic fluxes and conversions through this neurovascular unit can analyze the role and response of the specific cell types found in the brain. Brain-on-chip devices can also be operated in a high-throughput manner by connecting the devices to high-content screening equipment [88,89].

## 13. Respiratory Models

Lung-on-a-chip was a monumental achievement in the development of OoC. In 2010, an OoC device that reconstructs the functional alveolar-capillary interface of the human lung was demonstrated [1]. This interface provides a comprehensive response at the organ level to bacteria and inflammatory cytokines thanks to the layered human alveolar epithelial and lung microvascular endothelial cells on the membrane. From a toxicological study point of view, the lung-mimicking model revealed that the cyclic mechanical strain featured the lung's toxic and inflammatory responses to silica nanoparticles. Subsequently, the research group developed a lung-on-a-chip targeting pulmonary edema as a disease model. This chip mimicked the alveolar-capillary interface of a human lung, reproducing drug toxicity–induced pulmonary edema observed in human cancer patients treated with interleukin-2 (IL-2) at similar doses in the same time frame [90]. Mechanical stimulation associated with physiological breathing motions plays a crucial role in the development of increased vascular leakage that leads to pulmonary edema. After formulating the molecular pathways, OoC devices have been used for the identification of potential new therapeutics, including angiopoietin-1 (Ang-1) and a new transient receptor potential vanilloid 4 (TRPV4) ion channel inhibitor (GSK2193874), which might prevent this life-threatening toxicity of IL-2.

## 14. Cardiovascular Models

The tissues in the cardiovascular system experience various types and levels of biophysical stimuli. For example, heart tissue experiences shear stress, tensile strain, and stretching when cardiac muscles pump blood into the vessels. Similarly in vessels, endothelial cells surrounding the inner wall of the vasculature are exposed to similar types of biophysical stimuli as well as hydrostatic pressure due to the pulsatile flow created by the heart. Among these stimuli, shear stress was found to be the most significant one, influencing cell morphology and proliferation characteristics (Figure 8). OoC models of the cardiovascular system have been studied with a focus on blood vessels and heart-on-chip platforms because deformation of the vascular barrier has a central role in many cardiovascular diseases. Once the vascular barrier architecture is created, many research questions could be studied on these platforms. For example, the adhesion of neutrophils inside lung microvessels was studied to mimic chronic obstructive pulmonary disease conditions. Shear stress at different pressure levels along with cyclic membrane stretching can also be applied on the microchannels to initiate rapid hematopoietic cell formations [91].

## 15. Intestine Models

The intestine is one of the organs with complex mechanics. Peristaltic motion is a highly synchronized contraction movement, which results in irregular tensile and compressive strains as well as shear stress. Such biophysical stimuli changes in different parts of the intestine because the viscosity of the medium keeps changing along the intestine as the digestion process continues. Intestine models have been a popular topic in OoC models because the intestine is the primary place to absorb drugs. A range of disease models including inflammatory bowel disease and colitis is interesting for drug toxicity testing [92]. Peristaltic motions in OoCs are generated in the form of cyclic stretching of PDMS, shear stress is generated via fluid flow in microchannels while co-culture of intestinal cells is performed using e.g., hydrogels or microchip architecture (Figure 8) [93]. In OoCs, epithelial barrier integrity, viability, mucus bilayer formation, and bacterial infection topics are studied as key parameters defining how good the mimicking conditions are [94–97]. In this context, OoCs do not focus on mimicking the entire intestine, but rather parts of it such as the small intestine, duodenum, and colon.

## 16. Musculoskeletal Models

The type of biophysical stimuli experienced by the musculoskeletal system changes depending on location. While connective tissues are exposed to extreme stretching, the response of the cells varies according to the stimuli [98]. Articular cartilage is an interesting tissue for OoCs as just a normal physical activity can create compression, shear stress, tensile stress, and osmolarity effects. Inappropriate application of forces results in osteoarthritis and tendinopathy that can also be modeled in OoC devices [99]. On the other hand, the musculoskeletal system is one of the most challenging and overlooked applications of OoC given the fact that the complex formation of tissues and biophysical stimuli. In a bone-marrow-on chip application, cell-seeded hydroxyapatite-coated zirconium oxide scaffolds maintained long-term culturing of multipotent hematopoietic stem and progenitor cells under fluid flow (Figure 8) [100].

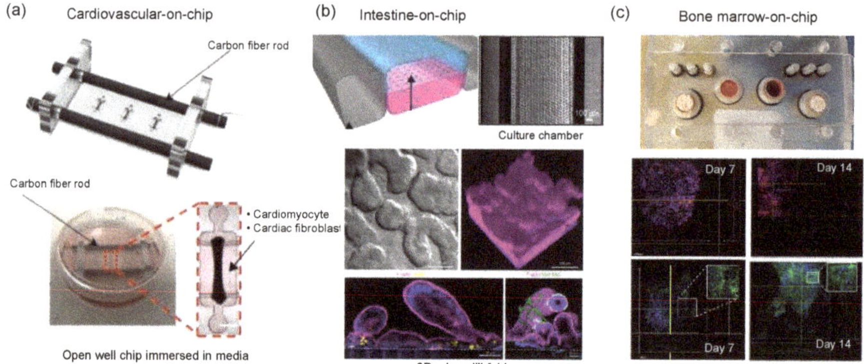


**Figure 8.** Cardiovascular, intestine and musculoskeletal OoC models. (**a**) Open well OoC device with carbon fiber rods and platinum wires. The voltage applied across the open well tissue culture-induced cardiac contractions. Figure reproduced from Mastikhina et al., 2020 [91]. (**b**) The primary human intestine chip contained human intestinal epithelium and intestinal microvascular endothelium co-cultured while the peristaltic motion was mimicked via an elastic membrane. Figure reproduced from Kasendra et al., 2018 [93]. (**c**) The bone-marrow on-chip model is cultured for up to four weeks under fluid flow. Figure reproduced from Sieber et al., 2018 [100].

## 17. Multi-Organ-on-Chip

Organ-to-organ interactions play a key role in identifying the toxicity of drugs. For example, a drug with high toxicity levels may be detoxified in the liver and converted to non-toxic compounds [101]. Conversely, a harmless drug may be metabolized in the liver into metabolites that are toxic to other tissues. OoC devices facilitate functional combinations such as modeling multiple organ ADMEs. OoCs are linked to each other using the elements required to scale the target organs, for example, dynamic flow fluid, cell composition, culture time point, and cellular compartment [102,103].

Recently, a tissue chip system connected by circulating vascular flow has been developed. In this system, the human heart, liver, bone, and skin tissue niches are cultured in their respective chambers and the endothelial barrier connecting them allows for interdependent functions. The interconnected tissues recapitulated the pharmacokinetic and pharmacodynamic profiles of doxorubicin and identified miRNA biomarkers [104]. A fully automated cell culture system was developed, interconnecting devices capable of continuous perfusion, medium exchange, fluid connection, sample collection, and in situ microscopy [105]. In another multi-organ microfluidic 'physiome-on-a-chip' platform as shown in Figure 9, [106]. Ten OoCs were linked to each other for 4 weeks to study pharmacokinetic analysis of diclofenac metabolism. Another multi-OoC was used for modeling anti-leukemia drug analysis on cancer-derived human bone marrow for investigating the toxicity effect of the liver using a multi-organ configuration. Similarly, multi-OoCs were used for studying multidrug-resistant hepatocytes and induced pluripotent stem cells–derived cardiomyocytes (Figure 9). Such pharmaceutical testing systems can establish a therapeutic window for comprehensive compound evaluation in human tissues. The accuracy of these systems is determined by evaluating repeated dose effects and off-target effects.

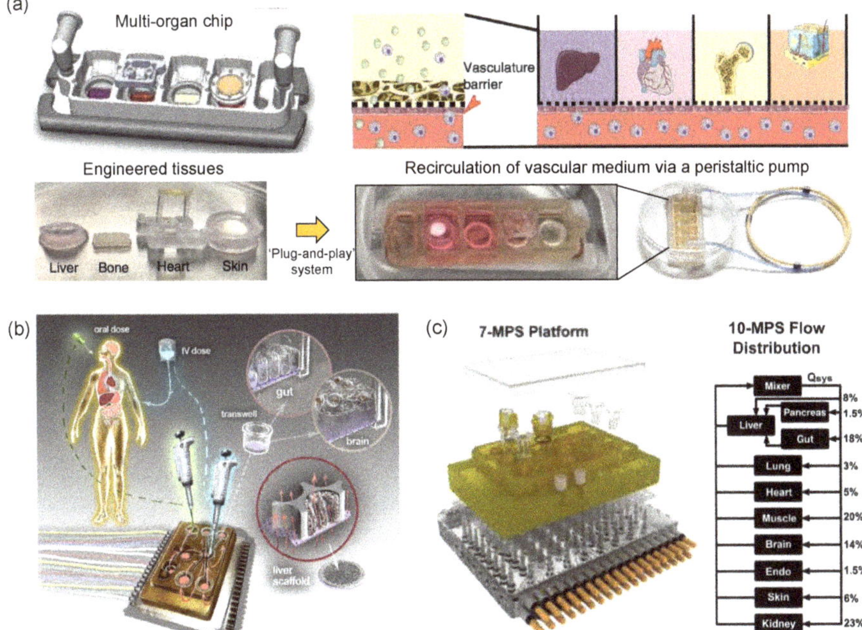


**Figure 9.** Modeling multi-organ systems on a chip. (**a**) The integrated multi-organ chip connects the liver, bone, heart, and skin along the vascular flow with the concept of a plug-and-play system to maintain tissue-specific niche. Figure reproduced from Ronaldson-Bouchard, K. et al., 2022 [104]. (**b**) Schematic diagram of a multi-organ on-chip system [106]. (**c**) Design and construction of multi-organ partitions within the microphysiological system platform. Multi-organ flow distribution design through the arrangement of major components such as channels, pumps, and reservoirs. Figure reproduced from Edington et al., 2018 [106].

## 18. Commercialization

Since OoC attracted the biomedical research community over a decade, the increasing interest has triggered the commercialization of the technology. Over 20 start-up companies including spin-offs from academic OC research groups actively supply products or provide services to OoC users [7,107,108]. The majority of the commercial systems incorporate microfluidics for the handling of cell microenvironment, therefore also introducing biophysical stimuli in the form of shear stress and interstitial flow. The minority of the commercial systems can produce mechanical strain such as breathing or peristalsis motion. The end-users of this commercialized platform range from individual researchers to pharmaceutical companies. Most of the commercialized products currently focus on drug toxicity testing and safety validation. Scaling-up in fabrication, validation of the reproducibility of OoC, and user-friendliness in the design of OoC devices are the major challenges to overcome before adopting the OoC technology in drug discovery phases.

## 19. Limitations

The technical level of OoC has reached the state of art, but the real-world application of the platform remains a challenge. Validation of reproducibility and accuracy of the OoC models, compared to in-vivo models or clinical trials, is required for full integration of the OoC technology into the drug discovery phases. Moreover, improved high-throughput operation and feasibility of mass production are key to reaching out to various end-users from pharma-industry associates to fundamental research scientists [109].

Innovative advancements have been achieved in the field of OoC in the past decade, while most of the models are developed as proof-of-concept. Bench-top-produced devices are a major bottleneck for commercial use in drug discovery research. PDMS-based OoC fabrication remains the most popular and traditional method to date, requiring dedicated fabrication equipment such as photo- and soft-lithography stations, clean bench stations, and an oxygen plasma treatment machine. During the device construction, liquid and hydrogel patterning inside the microfluidic channels require trained personnel due to the complexity in the design of the current OoC models [110]. Complications associated with microfabrication lead to poor interfacing, laborious production, and the burden of expenses for dedicated equipment.

Standardization is another problem being tackled in current OoC technology [111]. Reported OoC models vary extremely in the design and generation of physiologically relevant conditions. Although the OoC models are tested for reproducibility, it is difficult to control user-to-user reproducibility. Microfabrication and tissue culturing methods differ widely between the users, for example, the dimension of the microchannels, the initial number of cells in (co-)culture, and type of 3D matrix, etc. The standardization cannot be easily established in OoC devices as it was achieved for 2D cultures, because OoC devices have much more components and functionalities compared to 2D culture plates.

Analytical outcomes also lack full validation in OoC systems. Due to the performance variation of the monitoring tools, the range of the effective dosage of the drug applied in an OoC device may differ from in-vivo [72]. Scalability (i.e., analyzing the results from micron-scale environments for macro-scale systems) is also seen as a problem in OoC devices. The standardized references should be set up for analogical interpretation of drug test results from OoC. Compound concentrations are difficult to measure in PDMS-based platforms due to the adsorption of the small molecules into the PDMS material [12,112,113].

## 20. Future Considerations

The biggest expectation from OoC technology relies on its application in drug discovery phases. OoC devices successfully complement traditional models, while the simplicity and physiological relevance with organ-level complexity could be balanced depending on the purpose of a study. For instance, high-throughput, drug-toxicity-testing OoC models are expected to adopt a relatively simple structure similar to medium-throughput models designed to study advanced pathological functions in human disease [114].

OoC and organoids essentially have the same goal of recapitulating the functional and morphological properties of human organs in vitro. However, OoC technology relies on the development of artificial models to build structurally well-established systems in which cells and microenvironments can be precisely controlled. In contrast, organoids develop from autologous stem cells to conform a developmental program and reproduce key structural and functional properties of their in vivo architecture. Recently, as the concept of organoid-on-a-chip has emerged, the development of a platform that can help the maturation of organoids and further expand to large-scale experiments such as drug screening is becoming an important research field. It is now necessary to develop a platform that is more reproducible and more controllable through OoC technology rather than the classical organoid culture approach, extending it into the pipeline of drug discovery [115].

The critical point in adopting OoC in translational medicine is considering the phase of preclinical research progress. The early stage of the preclinical research, such as discovering drug candidates, would require an elaborate recapitulation of human physiology, which will enable the validation of the target drug mechanism. Mid-stage preclinical progress for narrowing down the specific target or dosage of the drug would require more strength in high-throughput property of the OoC to screen the efficacy efficiently. Multi-organ-on-chip could play on the last step of the preclinical studies to verify the toxicity and to provide better insight into the pharmacokinetics of the drug candidate [116].

Moreover, OoC models can facilitate personalized-medicine applications using cells obtained from real patients. In other words, OoC models can serve as a tool to predict the patient-specific response of a drug by interacting with patient-derived primary cells or stem cells in the cell microenvironment [6,117–119]. The personalized medicine approach is currently a popular application of OoC devices, and the validation and standardization studies are conducted accordingly.

Beyond the application of OoC in drug discovery, the incorporation of OoC models with gene editing strategies such as the CRISPR-Cas9 molecular system could help in the screening of cell phenotypes [2,72,111]. Moreover, omics-based analytics including proteomics and single-cell gene sequencing could be also implemented in OoC devices as monitoring tools [120–122].

## 21. Conclusions

OoC models serve as innovative tools to investigate the key functions of human tissues. These models are good alternatives to traditional preclinical models such as in-vivo animal models and simple 2D in-vitro models in the drug discovery process. Currently, the OoC technology is in active expansion to biomedical research, especially personalized medicine, while commercialization strategies are being applied by fulfilling the requirements of users [2,5,6,107,110,122].

Successful application of OoC devices requires a clear understanding of each component. Researchers interested in working with OoC models must be educated on several topics including materials, microfabrication methods, and structural design of the chip depending on the purpose of the study. For instance, plastic materials must be preferred when studying drug toxicity screening on simple tissue culture models [12,111–113]. If PDMS material is chosen for such a study, small molecule adsorption will limit the reliability of the model. Instead, PDMS will be handy when modeling complex, multi-layer tissue structures composed of a multi-cellular microenvironment. The selection of cell sources is another key point in developing OoC with highly physiological relevance. As the application of OoC expands to the biomedical and pharmaceutical field, usage of patient-derived cells and iPSCs is encouraged when recapitulation of patient-specific phenotype is needed [118,121,123–125]. The incorporation of biochemical and biophysical stimuli is also a critical factor to improve the functionality of the engineered tissues in OoCs [76]. During the construction of OoC devices, multiple monitoring techniques are integrated to collect real-time data which relates to the specific function or phenotype of the modeled tissue.

The birth of OoC technology stems from two major purposes. First, in-vitro models with patient-derived cells recapitulating tissue microenvironment or physiological function are introduced for studying the biology behind disease and pathology. Conventional animal models bear the risk of interspecies difference-based alterations in experimental results as well as the low-throughput nature of the models. Second, OoC can be used in the drug development process, both in drug toxicity and efficacy tests. The level of complexity in OoC models can be adjusted according to the research question and the phase of translational research. Particularly, recent trends in OoC research take attempts for commercializing high-throughput OoC models, which is appealing to pharmaceutical companies [5,7,126].

From developers to users, the OoC community grows rapidly. The interdisciplinarity of the community is the key factor for the successful translation of OoC to bench-to-bedside translation research.

**Author Contributions:** Conceptualization, J.K., D.P., S.L., B.G. and N.L.J.; investigation, J.K., D.P., S.L., B.G. and N.L.J.; resources, J.K., D.P., S.L., B.G. and N.L.J.; writing—original draft preparation, J.K., D.P., S.L., B.G. and N.L.J.; writing—review and editing, J.K., D.P., S.L., B.G. and N.L.J.; visualization, B.G. and J.K.; supervision, B.G. and N.L.J.; project administration, B.G. and N.L.J.; funding acquisition, B.G. and N.L.J. All authors have read and agreed to the published version of the manuscript.

**Funding:** This work was funded by grants from the National Research Foundation (NRF) of Korea [2021R1A3B1077481] awarded to N.L. Jeon and the Bio & Medical Technology Development Program of the NRF funded by K-BIO KIURI Center through the Ministry of Science and ICT (MSIT) [2020M3H1A1073304] awarded to D. Park. B.G. acknowledges the support of NWO Gravitation Fund on interactive polymer materials and the support of the institute of complex molecular systems at the Eindhoven University of Technology.

**Conflicts of Interest:** The authors declare no conflict of interest.

## References

1. Huh, D.; Matthews, B.D.; Mammoto, A.; Montoya-Zavala, M.; Hsin, H.Y.; Ingber, D.E. Reconstituting organ-level lung functions on a chip. *Science* **2010**, *328*, 1662–1668. [CrossRef] [PubMed]
2. Low, L.A.; Mummery, C.; Berridge, B.R.; Austin, C.P.; Tagle, D.A. Organs-on-chips: Into the next decade. *Nat. Rev. Drug Discov.* **2021**, *20*, 345–361. [CrossRef] [PubMed]
3. Marx, U.; Akabane, T.; Andersson, T.B.; Baker, E.; Beilmann, M.; Beken, S.; Brendler-Schwaab, S.; Cirit, M.; David, R.; Dehne, E.-M. Biology-inspired microphysiological systems to advance patient benefit and animal welfare in drug development. *Altex* **2020**, *37*, 365. [PubMed]
4. Sun, W.; Luo, Z.; Lee, J.; Kim, H.J.; Lee, K.; Tebon, P.; Feng, Y.; Dokmeci, M.R.; Sengupta, S.; Khademhosseini, A. Organ-on-a-chip for cancer and immune organs modeling. *Adv. Healthc. Mater.* **2019**, *8*, 1801363. [CrossRef] [PubMed]
5. Leung, C.M.; De Haan, P.; Ronaldson-Bouchard, K.; Kim, G.-A.; Ko, J.; Rho, H.S.; Chen, Z.; Habibovic, P.; Jeon, N.L.; Takayama, S. A guide to the organ-on-a-chip. *Nat. Rev. Methods Primers* **2022**, *2*, 33. [CrossRef]
6. Rothbauer, M.; Rosser, J.M.; Zirath, H.; Ertl, P. Tomorrow today: Organ-on-a-chip advances towards clinically relevant pharmaceutical and medical in vitro models. *Curr. Opin. Biotechnol.* **2019**, *55*, 81–86. [CrossRef]
7. Vulto, P.; Joore, J. Adoption of organ-on-chip platforms by the pharmaceutical industry. *Nat. Rev. Drug Discov.* **2021**, *20*, 961–962. [CrossRef]
8. Tajeddin, A.; Mustafaoglu, N. Design and fabrication of organ-on-chips: Promises and challenges. *Micromachines* **2021**, *12*, 1443. [CrossRef]
9. Xia, Y.; Whitesides, G.M. Soft lithography. *Angew. Chem. Int. Ed.* **1998**, *37*, 550–575. [CrossRef]
10. Gumuscu, B.; Bomer, J.G.; van den Berg, A.; Eijkel, J.C. Large scale patterning of hydrogel microarrays using capillary pinning. *Lab Chip* **2015**, *15*, 664–667. [CrossRef]
11. Ferraz, M.A.; Rho, H.S.; Hemerich, D.; Henning, H.H.; van Tol, H.T.; Hölker, M.; Besenfelder, U.; Mokry, M.; Vos, P.L.A.M.; Stout, T.A.E.; et al. An oviduct-on-a-chip provides an enhanced in vitro environment for zygote genome reprogramming. *Nat. Comm.* **2018**, *9*, 4934. [CrossRef] [PubMed]
12. Huh, D.; Kim, H.J.; Fraser, J.P.; Shea, D.E.; Khan, M.; Bahinski, A.; Hamilton, G.A.; Ingber, D.E. Microfabrication of human organs-on-chips. *Nat. Protoc.* **2013**, *8*, 2135–2157. [CrossRef] [PubMed]
13. Berthier, E.; Young, E.W.; Beebe, D. Engineers are from PDMS-land, Biologists are from Polystyrenia. *Lab Chip* **2012**, *12*, 1224–1237. [CrossRef] [PubMed]
14. Mukhopadhyay, R. When PDMS isn't the best. *Anal. Chem.* **2007**, *79*, 3248–3253. [CrossRef] [PubMed]
15. Becker, H.; Locascio, L.E. Polymer microfluidic devices. *Talanta* **2002**, *56*, 267–287. [CrossRef]
16. Virumbrales-Muñoz, M.; Ayuso, J.M.; Lacueva, A.; Randelovic, T.; Livingston, M.K.; Beebe, D.J.; Oliván, S.; Pereboom, D.; Doblare, M.; Fernández, L.; et al. Enabling cell recovery from 3D cell culture microfluidic devices for tumour microenvironment biomarker profiling. *Sci. Rep.* **2019**, *9*, 6199. [CrossRef]
17. Kim, S.; Ko, J.; Lee, S.R.; Park, D.; Park, S.; Jeon, N.L. Anchor-IMPACT: A standardized microfluidic platform for high-throughput antiangiogenic drug screening. *Biotechnol. Bioeng.* **2021**, *118*, 2524–2535. [CrossRef]

18. Ko, J.; Ahn, J.; Kim, S.; Lee, Y.; Lee, J.; Park, D.; Jeon, N.L. Tumor spheroid-on-a-chip: A standardized microfluidic culture platform for investigating tumor angiogenesis. *Lab Chip* **2019**, *19*, 2822–2833. [CrossRef]

19. Lee, S.; Kang, H.; Park, D.; Yu, J.; Koh, S.K.; Cho, D.; Kim, D.H.; Kang, K.S.; Jeon, N.L. Modeling 3D Human Tumor Lymphatic Vessel Network Using High-Throughput Platform. *Adv. Biol.* **2021**, *5*, 2000195. [CrossRef]

20. Lee, Y.; Choi, J.W.; Yu, J.; Park, D.; Ha, J.; Son, K.; Lee, S.; Chung, M.; Kim, H.-Y.; Jeon, N.L. Microfluidics within a well: An injection-molded plastic array 3D culture platform. *Lab Chip* **2018**, *18*, 2433–2440. [CrossRef]

21. Park, D.; Son, K.; Hwang, Y.; Ko, J.; Lee, Y.; Doh, J.; Jeon, N.L. High-throughput microfluidic 3D cytotoxicity assay for cancer immunotherapy (CACI-ImpacT platform). *Front. Immunol.* **2019**, *10*, 1133. [CrossRef] [PubMed]

22. Shin, N.; Kim, Y.; Ko, J.; Choi, S.W.; Hyung, S.; Lee, S.E.; Park, S.; Song, J.; Jeon, N.L.; Kang, K.S. Vascularization of iNSC spheroid in a 3D spheroid-on-a-chip platform enhances neural maturation. *Biotechnol. Bioeng.* **2022**, *119*, 566–574. [CrossRef] [PubMed]

23. Au, A.K.; Huynh, W.; Horowitz, L.F.; Folch, A. 3D-printed microfluidics. *Angew. Chem. Int. Ed.* **2016**, *55*, 3862–3881. [CrossRef] [PubMed]

24. Homan, K.A.; Kolesky, D.B.; Skylar-Scott, M.A.; Herrmann, J.; Obuobi, H.; Moisan, A.; Lewis, J.A. Bioprinting of 3D convoluted renal proximal tubules on perfusable chips. *Sci. Rep.* **2016**, *6*, 34845. [CrossRef] [PubMed]

25. Lee, B.; Kim, S.; Ko, J.; Lee, S.-R.; Kim, Y.; Park, S.; Kim, J.; Hyung, S.; Kim, H.-Y.; Jeon, N.L. 3D micromesh-based hybrid bioprinting: Multidimensional liquid patterning for 3D microtissue engineering. *NPG Asia Mater.* **2022**, *14*, 6. [CrossRef]

26. Derakhshanfar, S.; Mbeleck, R.; Xu, K.; Zhang, X.; Zhong, W.; Xing, M. 3D bioprinting for biomedical devices and tissue engineering: A review of recent trends and advances. *Bioact. Mater.* **2018**, *3*, 144–156. [CrossRef] [PubMed]

27. Gungor-Ozkerim, P.S.; Inci, I.; Zhang, Y.S.; Khademhosseini, A.; Dokmeci, M.R. Bioinks for 3D bioprinting: An overview. *Biomater. Sci.* **2018**, *6*, 915–946. [CrossRef]

28. Joensson, H.N.; Andersson Svahn, H. Droplet microfluidics—A tool for single-cell analysis. *Angew. Chem. Int. Ed.* **2012**, *51*, 12176–12192. [CrossRef]

29. Teh, S.-Y.; Lin, R.; Hung, L.-H.; Lee, A.P. Droplet microfluidics. *Lab Chip* **2008**, *8*, 198–220. [CrossRef]

30. Chan, H.F.; Zhang, Y.; Ho, Y.P.; Chiu, Y.L.; Jung, Y.; Leong, K.W. Rapid formation of multicellular spheroids in double-emulsion droplets with controllable microenvironment. *Sci. Rep.* **2013**, *3*, 3462. [CrossRef]

31. Vladisavljević, G.T.; Khalid, N.; Neves, M.A.; Kuroiwa, T.; Nakajima, M.; Uemura, K.; Ichikawa, S.; Kobayashi, I. Industrial lab-on-a-chip: Design, applications and scale-up for drug discovery and delivery. *Adv. Drug Deliv. Rev.* **2013**, *65*, 1626–1663. [CrossRef] [PubMed]

32. Lee, P.J.; Hung, P.J.; Lee, L.P. An artificial liver sinusoid with a microfluidic endothelial-like barrier for primary hepatocyte culture. *Biotechnol. Bioeng.* **2007**, *97*, 1340–1346. [CrossRef] [PubMed]

33. Taylor, A.M.; Blurton-Jones, M.; Rhee, S.W.; Cribbs, D.H.; Cotman, C.W.; Jeon, N.L. A microfluidic culture platform for CNS axonal injury, regeneration and transport. *Nat. Methods* **2005**, *2*, 599–605. [CrossRef] [PubMed]

34. Cho, H.; Kim, H.-Y.; Kang, J.Y.; Kim, T.S. How the capillary burst microvalve works. *J. Colloid Interface Sci.* **2007**, *306*, 379–385. [CrossRef] [PubMed]

35. Huang, C.P.; Lu, J.; Seon, H.; Lee, A.P.; Flanagan, L.A.; Kim, H.-Y.; Putnam, A.J.; Jeon, N.L. Engineering microscale cellular niches for three-dimensional multicellular co-cultures. *Lab Chip* **2009**, *9*, 1740–1748. [CrossRef]

36. Vulto, P.; Podszun, S.; Meyer, P.; Hermann, C.; Manz, A.; Urban, G.A. Phaseguides: A paradigm shift in microfluidic priming and emptying. *Lab Chip* **2011**, *11*, 1596–1602. [CrossRef]

37. Hyung, S.; Lee, S.-R.; Kim, J.; Kim, Y.; Kim, S.; Kim, H.N.; Jeon, N.L. A 3D disease and regeneration model of peripheral nervous system-on-a-chip. *Sci. Adv.* **2021**, *7*, eabd9749. [CrossRef]

38. Jeon, J.S.; Bersini, S.; Gilardi, M.; Dubini, G.; Charest, J.L.; Moretti, M.; Kamm, R.D. Human 3D vascularized organotypic microfluidic assays to study breast cancer cell extravasation. *Proc. Natl. Acad. Sci. USA* **2015**, *112*, 214–219. [CrossRef]

39. Kim, S.; Lee, H.; Chung, M.; Jeon, N.L. Engineering of functional, perfusable 3D microvascular networks on a chip. *Lab Chip* **2013**, *13*, 1489–1500. [CrossRef]

40. Lee, S.; Chung, M.; Lee, S.R.; Jeon, N.L. 3D brain angiogenesis model to reconstitute functional human blood-brain barrier in vitro. *Biotechnol. Bioeng.* **2020**, *117*, 748–762. [CrossRef]

41. Van Duinen, V.; Zhu, D.; Ramakers, C.; Van Zonneveld, A.; Vulto, P.; Hankemeier, T. Perfused 3D angiogenic sprouting in a high-throughput in vitro platform. *Angiogenesis* **2019**, *22*, 157–165. [CrossRef] [PubMed]

42. Berthier, J.; Brakke, K.A.; Berthier, E. *Open Microfluidics*; John Wiley & Sons: Hoboken, NJ, USA, 2016.

43. Lee, S.H.; Heinz, A.J.; Shin, S.; Jung, Y.G.; Choi, S.E.; Park, W.; Roe, J.-H.; Kwon, S. Capillary based patterning of cellular communities in laterally open channels. *Anal. Chem.* **2010**, *82*, 2900–2906. [CrossRef] [PubMed]

44. Zeilinger, K.; Freyer, N.; Damm, G.; Seehofer, D.; Knöspel, F. Cell sources for in vitro human liver cell culture models. *Exp. Biol. Med.* **2016**, *241*, 1684–1698. [CrossRef] [PubMed]

45. Magdesian, M.H.; Lopez-Ayon, G.M.; Mori, M.; Boudreau, D.; Goulet-Hanssens, A.; Sanz, R.; Miyahara, Y.; Barrett, C.J.; Fournier, A.E.; De Koninck, Y. Rapid mechanically controlled rewiring of neuronal circuits. *J. Neurosci.* **2016**, *36*, 979–987. [CrossRef] [PubMed]

46. Ahn, J.; Lee, H.; Kang, H.; Choi, H.; Son, K.; Yu, J.; Lee, J.; Lim, J.; Park, D.; Cho, M. Pneumatically actuated microfluidic platform for reconstituting 3d vascular tissue compression. *Appl. Sci.* **2020**, *10*, 2027. [CrossRef]

47. Freund, J.B.; Goetz, J.G.; Hill, K.L.; Vermot, J. Fluid flows and forces in development: Functions, features and biophysical principles. *Development* **2012**, *139*, 1229–1245. [CrossRef] [PubMed]
48. Kim, S.; Chung, M.; Ahn, J.; Lee, S.; Jeon, N.L. Interstitial flow regulates the angiogenic response and phenotype of endothelial cells in a 3D culture model. *Lab Chip* **2016**, *16*, 4189–4199. [CrossRef]
49. Hajal, C.; Ibrahim, L.; Serrano, J.C.; Offeddu, G.S.; Kamm, R.D. The effects of luminal and trans-endothelial fluid flows on the extravasation and tissue invasion of tumor cells in a 3D in vitro microvascular platform. *Biomaterials* **2021**, *265*, 120470. [CrossRef]
50. Jang, K.-J.; Mehr, A.P.; Hamilton, G.A.; McPartlin, L.A.; Chung, S.; Suh, K.-Y.; Ingber, D.E. Human kidney proximal tubule-on-a-chip for drug transport and nephrotoxicity assessment. *Integr. Biol.* **2013**, *5*, 1119–1129. [CrossRef]
51. Phan, D.T.; Wang, X.; Craver, B.M.; Sobrino, A.; Zhao, D.; Chen, J.C.; Lee, L.Y.; George, S.C.; Lee, A.P.; Hughes, C.C. A vascularized and perfused organ-on-a-chip platform for large-scale drug screening applications. *Lab Chip* **2017**, *17*, 511–520. [CrossRef]
52. Wang, Y.I.; Shuler, M.L. UniChip enables long-term recirculating unidirectional perfusion with gravity-driven flow for microphysiological systems. *Lab Chip* **2018**, *18*, 2563–2574. [CrossRef] [PubMed]
53. Brennan, M.D.; Rexius-Hall, M.L.; Elgass, L.J.; Eddington, D.T. Oxygen control with microfluidics. *Lab Chip* **2014**, *14*, 4305–4318. [CrossRef] [PubMed]
54. Sheidaei, Z.; Akbarzadeh, P.; Nguyen, N.-T.; Kashaninejad, N. A new insight into a thermoplastic microfluidic device aimed at improvement of oxygenation process and avoidance of shear stress during cell culture. *Biomed. Microdevices* **2022**, *24*, 15. [CrossRef] [PubMed]
55. Rosser, J.; Bachmann, B.; Jordan, C.; Ribitsch, I.; Haltmayer, E.; Gueltekin, S.; Junttila, S.; Galik, B.; Gyenesei, A.; Haddadi, B. Microfluidic nutrient gradient–based three-dimensional chondrocyte culture-on-a-chip as an in vitro equine arthritis model. *Mater. Today Bio* **2019**, *4*, 100023. [CrossRef] [PubMed]
56. Kim, S.; Kim, H.J.; Jeon, N.L. Biological applications of microfluidic gradient devices. *Integr. Biol.* **2010**, *2*, 584–603. [CrossRef] [PubMed]
57. Booth, R.; Kim, H. Characterization of a microfluidic in vitro model of the blood-brain barrier (μBBB). *Lab Chip* **2012**, *12*, 1784–1792. [CrossRef] [PubMed]
58. Kilic, T.; Navaee, F.; Stradolini, F.; Renaud, P.; Carrara, S. Organs-on-chip monitoring: Sensors and other strategies. *Microphysiological Syst.* **2018**, *2*, 1–32. [CrossRef]
59. Preziosi, P. Science, pharmacoeconomics and ethics in drug R&D: A sustainable future scenario? *Nat. Rev. Drug Discov.* **2004**, *3*, 521–526.
60. Gotwals, P.; Cameron, S.; Cipolletta, D.; Cremasco, V.; Crystal, A.; Hewes, B.; Mueller, B.; Quaratino, S.; Sabatos-Peyton, C.; Petruzzelli, L. Prospects for combining targeted and conventional cancer therapy with immunotherapy. *Nat. Rev. Cancer* **2017**, *17*, 286–301. [CrossRef]
61. Chen, M.B.; Hajal, C.; Benjamin, D.C.; Yu, C.; Azizgolshani, H.; Hynes, R.O.; Kamm, R.D. Inflamed neutrophils sequestered at entrapped tumor cells via chemotactic confinement promote tumor cell extravasation. *Proc. Natl. Acad. Sci. USA* **2018**, *115*, 7022–7027. [CrossRef]
62. Jiménez-Torres, J.A.; Virumbrales-Muñoz, M.; Sung, K.E.; Lee, M.H.; Abel, E.J.; Beebe, D.J. Patient-specific organotypic blood vessels as an in vitro model for anti-angiogenic drug response testing in renal cell carcinoma. *EBioMedicine* **2019**, *42*, 408–419. [CrossRef] [PubMed]
63. Haase, K.; Offeddu, G.S.; Gillrie, M.R.; Kamm, R.D. Endothelial regulation of drug transport in a 3D vascularized tumor model. *Adv. Funct. Mater.* **2020**, *30*, 2002444. [CrossRef] [PubMed]
64. Hachey, S.J.; Movsesyan, S.; Nguyen, Q.H.; Burton-Sojo, G.; Tankazyan, A.; Wu, J.; Hoang, T.; Zhao, D.; Wang, S.; Hatch, M.M. An in vitro vascularized micro-tumor model of human colorectal cancer recapitulates in vivo responses to standard-of-care therapy. *Lab Chip* **2021**, *21*, 1333–1351. [CrossRef] [PubMed]
65. Monjezi, M.; Rismanian, M.; Jamaati, H.; Kashaninejad, N. Anti-Cancer Drug Screening with Microfluidic Technology. *Appl. Sci.* **2021**, *11*, 9418. [CrossRef]
66. Shi, Y.; Cai, Y.; Cao, Y.; Hong, Z.; Chai, Y. Recent advances in microfluidic technology and applications for anti-cancer drug screening. *TrAC Trends Anal. Chem.* **2021**, *134*, 116118. [CrossRef]
67. Nashimoto, Y.; Okada, R.; Hanada, S.; Arima, Y.; Nishiyama, K.; Miura, T.; Yokokawa, R. Vascularized cancer on a chip: The effect of perfusion on growth and drug delivery of tumor spheroid. *Biomaterials* **2020**, *229*, 119547. [CrossRef]
68. Shirure, V.S.; Bi, Y.; Curtis, M.B.; Lezia, A.; Goedegebuure, M.M.; Goedegebuure, S.P.; Aft, R.; Fields, R.C.; George, S.C. Tumor-on-a-chip platform to investigate progression and drug sensitivity in cell lines and patient-derived organoids. *Lab Chip* **2018**, *18*, 3687–3702. [CrossRef]
69. Cui, X.; Ma, C.; Vasudevaraja, V.; Serrano, J.; Tong, J.; Peng, Y.; Delorenzo, M.; Shen, G.; Frenster, J.; Morales, R.-T.T. Dissecting the immunosuppressive tumor microenvironments in Glioblastoma-on-a-Chip for optimized PD-1 immunotherapy. *Elife* **2020**, *9*, e52253. [CrossRef]
70. Fischer, I.; Milton, C.; Wallace, H. Toxicity testing is evolving! *Toxicol. Res.* **2020**, *9*, 67–80. [CrossRef]
71. Hartung, T. Toxicology for the twenty-first century. *Nature* **2009**, *460*, 208–212. [CrossRef]
72. Jang, K.-J.; Otieno, M.A.; Ronxhi, J.; Lim, H.-K.; Ewart, L.; Kodella, K.R.; Petropolis, D.B.; Kulkarni, G.; Rubins, J.E.; Conegliano, D. Reproducing human and cross-species drug toxicities using a Liver-Chip. *Sci. Transl. Med.* **2019**, *11*, eaax5516. [CrossRef] [PubMed]

73. Gough, A.; Soto-Gutierrez, A.; Vernetti, L.; Ebrahimkhani, M.R.; Stern, A.M.; Taylor, D. Human biomimetic liver microphysiology systems in drug development and precision medicine. *Nat. Rev. Gastroenterol. Hepatol.* **2021**, *18*, 252–268. [CrossRef] [PubMed]

74. Yu, F.; Deng, R.; Hao Tong, W.; Huan, L.; Chan Way, N.; IslamBadhan, A.; Iliescu, C.; Yu, H. A perfusion incubator liver chip for 3D cell culture with application on chronic hepatotoxicity testing. *Sci. Rep.* **2017**, *7*, 14528. [CrossRef] [PubMed]

75. Massa, S.; Sakr, M.A.; Seo, J.; Bandaru, P.; Arneri, A.; Bersini, S.; Zare-Eelanjegh, E.; Jalilian, E.; Cha, B.-H.; Antona, S. Bioprinted 3D vascularized tissue model for drug toxicity analysis. *Biomicrofluidics* **2017**, *11*, 044109. [CrossRef] [PubMed]

76. Thompson, C.L.; Fu, S.; Heywood, H.K.; Knight, M.M.; Thorpe, S.D. Mechanical stimulation: A crucial element of organ-on-chip models. *Front. Bioeng. Biotechnol.* **2020**, *8*, 602646. [CrossRef] [PubMed]

77. Kim, S.Y.; Moon, A. Drug-induced nephrotoxicity and its biomarkers. *Biomol. Ther.* **2012**, *20*, 268. [CrossRef]

78. Wu, H.; Huang, J. Drug-induced nephrotoxicity: Pathogenic mechanisms, biomarkers and prevention strategies. *Curr. Drug Metab.* **2018**, *19*, 559–567. [CrossRef]

79. Kim, S.; LesherPerez, S.C.; Yamanishi, C.; Labuz, J.M.; Leung, B.; Takayama, S. Pharmacokinetic profile that reduces nephrotoxicity of gentamicin in a perfused kidney-on-a-chip. *Biofabrication* **2016**, *8*, 015021. [CrossRef]

80. Homan, K.A.; Gupta, N.; Kroll, K.T.; Kolesky, D.B.; Skylar-Scott, M.; Miyoshi, T.; Mau, D.; Valerius, M.T.; Ferrante, T.; Bonventre, J.V. Flow-enhanced vascularization and maturation of kidney organoids in vitro. *Nat. Methods* **2019**, *16*, 255–262. [CrossRef]

81. Li, Z.; Jiang, L.; Zhu, Y.; Su, W.; Xu, C.; Tao, T.; Shi, Y.; Qin, J. Assessment of hepatic metabolism-dependent nephrotoxicity on an organs-on-a-chip microdevice. *Toxicol. Vitr.* **2018**, *46*, 1–8. [CrossRef]

82. Mao, S.; Gao, D.; Liu, W.; Wei, H.; Lin, J.-M. Imitation of drug metabolism in human liver and cytotoxicity assay using a microfluidic device coupled to mass spectrometric detection. *Lab Chip* **2012**, *12*, 219–226. [CrossRef] [PubMed]

83. Zhang, J.; Wu, J.; Li, H.; Chen, Q.; Lin, J.-M. An in vitro liver model on microfluidic device for analysis of capecitabine metabolite using mass spectrometer as detector. *Biosens. Bioelectron.* **2015**, *68*, 322–328. [CrossRef] [PubMed]

84. Abbott, N.J.; Rönnbäck, L.; Hansson, E. Astrocyte–endothelial interactions at the blood–brain barrier. *Nat. Rev. Neurosci.* **2006**, *7*, 41–53. [CrossRef] [PubMed]

85. Park, J.W.; Vahidi, B.; Taylor, A.M.; Rhee, S.W.; Jeon, N.L. Microfluidic culture platform for neuroscience research. *Nat. Protoc.* **2006**, *1*, 2128–2136. [CrossRef]

86. Nierode, G.J.; Perea, B.C.; McFarland, S.K.; Pascoal, J.F.; Clark, D.S.; Schaffer, D.V.; Dordick, J.S. High-throughput toxicity and phenotypic screening of 3D human neural progenitor cell cultures on a microarray chip platform. *Stem Cell Rep.* **2016**, *7*, 970–982. [CrossRef] [PubMed]

87. Maoz, B.M.; Herland, A.; FitzGerald, E.A.; Grevesse, T.; Vidoudez, C.; Pacheco, A.R.; Sheehy, S.P.; Park, T.-E.; Dauth, S.; Mannix, R. A linked organ-on-chip model of the human neurovascular unit reveals the metabolic coupling of endothelial and neuronal cells. *Nat. Biotechnol.* **2018**, *36*, 865–874. [CrossRef]

88. Kane, K.I.; Moreno, E.L.; Hachi, S.; Walter, M.; Jarazo, J.; Oliveira, M.A.; Hankemeier, T.; Vulto, P.; Schwamborn, J.C.; Thoma, M. Automated microfluidic cell culture of stem cell derived dopaminergic neurons. *Sci. Rep.* **2019**, *9*, 1796. [CrossRef]

89. Wevers, N.R.; Kasi, D.G.; Gray, T.; Wilschut, K.J.; Smith, B.; van Vught, R.; Shimizu, F.; Sano, Y.; Kanda, T.; Marsh, G. A perfused human blood–brain barrier on-a-chip for high-throughput assessment of barrier function and antibody transport. *Fluids Barriers CNS* **2018**, *15*, 23. [CrossRef] [PubMed]

90. Huh, D.; Leslie, D.C.; Matthews, B.D.; Fraser, J.P.; Jurek, S.; Hamilton, G.A.; Thorneloe, K.S.; McAlexander, M.A.; Ingber, D.E. A human disease model of drug toxicity–induced pulmonary edema in a lung-on-a-chip microdevice. *Sci. Transl. Med.* **2012**, *4*, ra147–ra159. [CrossRef]

91. Mastikhina, O.; Moon, B.-U.; Williams, K.; Hatkar, R.; Gustafson, D.; Mourad, O.; Sun, X.; Koo, M.; Lam, A.Y.; Sun, Y. Human cardiac fibrosis-on-a-chip model recapitulates disease hallmarks and can serve as a platform for drug testing. *Biomaterials* **2020**, *233*, 119741. [CrossRef]

92. Yesil-Celiktas, O.; Hassan, S.; Miri, A.K.; Maharjan, S.; Al-kharboosh, R.; Quiñones-Hinojosa, A.; Zhang, Y.S. Mimicking human pathophysiology in organ-on-chip devices. *Adv. Biosyst.* **2018**, *2*, 1800109. [CrossRef]

93. Kasendra, M.; Tovaglieri, A.; Sontheimer-Phelps, A.; Jalili-Firoozinezhad, S.; Bein, A.; Chalkiadaki, A.; Scholl, W.; Zhang, C.; Rickner, H.; Richmond, C.A. Development of a primary human Small Intestine-on-a-Chip using biopsy-derived organoids. *Sci. Rep.* **2018**, *8*, 2871. [CrossRef] [PubMed]

94. Jalili-Firoozinezhad, S.; Gazzaniga, F.S.; Calamari, E.L.; Camacho, D.M.; Fadel, C.W.; Bein, A.; Swenor, B.; Nestor, B.; Cronce, M.J.; Tovaglieri, A. A complex human gut microbiome cultured in an anaerobic intestine-on-a-chip. *Nat. Biomed. Eng.* **2019**, *3*, 520–531. [CrossRef] [PubMed]

95. Maurer, M.; Gresnigt, M.S.; Last, A.; Wollny, T.; Berlinghof, F.; Pospich, R.; Cseresnyes, Z.; Medyukhina, A.; Graf, K.; Groeger, M. A three-dimensional immunocompetent intestine-on-chip model as in vitro platform for functional and microbial interaction studies. *Biomaterials* **2019**, *220*, 119396. [CrossRef] [PubMed]

96. Shin, W.; Kim, H.J. Intestinal barrier dysfunction orchestrates the onset of inflammatory host–microbiome cross-talk in a human gut inflammation-on-a-chip. *Proc. Natl. Acad. Sci. USA* **2018**, *115*, E10539–E10547. [CrossRef] [PubMed]

97. Shin, W.; Kim, H.J. 3D in vitro morphogenesis of human intestinal epithelium in a gut-on-a-chip or a hybrid chip with a cell culture insert. *Nat. Protoc.* **2022**, *17*, 910–939. [CrossRef] [PubMed]

98. Wang, J.; Khodabukus, A.; Rao, L.; Vandusen, K.; Abutaleb, N.; Bursac, N. Engineered skeletal muscles for disease modeling and drug discovery. *Biomaterials* **2019**, *221*, 119416. [CrossRef]

99. Gottardi, R. Load-induced osteoarthritis on a chip. *Nat. Biomed. Eng.* **2019**, *3*, 502–503. [CrossRef]
100. Sieber, S.; Wirth, L.; Cavak, N.; Koenigsmark, M.; Marx, U.; Lauster, R.; Rosowski, M. Bone marrow-on-a-chip: Long-term culture of human haematopoietic stem cells in a three-dimensional microfluidic environment. *J. Tissue Eng. Regen. Med.* **2018**, *12*, 479–489. [CrossRef]
101. Singh, D.; Cho, W.C.; Upadhyay, G. Drug-induced liver toxicity and prevention by herbal antioxidants: An overview. *Front. Physiol.* **2016**, *6*, 363. [CrossRef]
102. Rajan, S.A.P.; Aleman, J.; Wan, M.; Zarandi, N.P.; Nzou, G.; Murphy, S.; Bishop, C.E.; Sadri-Ardekani, H.; Shupe, T.; Atala, A. Probing prodrug metabolism and reciprocal toxicity with an integrated and humanized multi-tissue organ-on-a-chip platform. *Acta Biomater.* **2020**, *106*, 124–135. [CrossRef] [PubMed]
103. Theobald, J.; Ghanem, A.; Wallisch, P.; Banaeiyan, A.A.; Andrade-Navarro, M.A.; Taškova, K.; Haltmeier, M.; Kurtz, A.; Becker, H.; Reuter, S. Liver-kidney-on-chip to study toxicity of drug metabolites. *ACS Biomater. Sci. Eng.* **2018**, *4*, 78–89. [CrossRef] [PubMed]
104. Ronaldson-Bouchard, K.; Teles, D.; Yeager, K.; Tavakol, D.N.; Zhao, Y.; Chramiec, A.; Tagore, S.; Summers, M.; Stylianos, S.; Tamargo, M. A multi-organ chip with matured tissue niches linked by vascular flow. *Nat. Biomed. Eng.* **2022**, *6*, 351–371. [CrossRef] [PubMed]
105. Novak, R.; Ingram, M.; Marquez, S.; Das, D.; Delahanty, A.; Herland, A.; Maoz, B.M.; Jeanty, S.S.; Somayaji, M.R.; Burt, M. Robotic fluidic coupling and interrogation of multiple vascularized organ chips. *Nat. Biomed. Eng.* **2020**, *4*, 407–420. [CrossRef] [PubMed]
106. Edington, C.D.; Chen, W.L.K.; Geishecker, E.; Kassis, T.; Soenksen, L.R.; Bhushan, B.M.; Freake, D.; Kirschner, J.; Maass, C.; Tsamandouras, N. Interconnected microphysiological systems for quantitative biology and pharmacology studies. *Sci. Rep.* **2018**, *8*, 4530. [CrossRef]
107. Balijepalli, A.; Sivaramakrishan, V. Organs-on-chips: Research and commercial perspectives. *Drug Discov. Today* **2017**, *22*, 397–403. [CrossRef] [PubMed]
108. Zhang, B.; Radisic, M. Organ-on-a-chip devices advance to market. *Lab Chip* **2017**, *17*, 2395–2420. [CrossRef]
109. Junaid, A.; Mashaghi, A.; Hankemeier, T.; Vulto, P. An end-user perspective on Organ-on-a-Chip: Assays and usability aspects. *Curr. Opin. Biomed. Eng.* **2017**, *1*, 15–22. [CrossRef]
110. Danku, A.E.; Dulf, E.-H.; Braicu, C.; Jurj, A.; Berindan-Neagoe, I. Organ-On-A-Chip: A Survey of Technical Results and Problems. *Front. Bioeng. Biotechnol.* **2022**, *10*, 840674. [CrossRef]
111. Piergiovanni, M.; Leite, S.B.; Corvi, R.; Whelan, M. Standardisation needs for organ on chip devices. *Lab Chip* **2021**, *21*, 2857–2868. [CrossRef]
112. Campbell, S.B.; Wu, Q.; Yazbeck, J.; Liu, C.; Okhovatian, S.; Radisic, M. Beyond polydimethylsiloxane: Alternative materials for fabrication of organ-on-a-chip devices and microphysiological systems. *ACS Biomater. Sci. Eng.* **2020**, *7*, 2880–2899. [CrossRef] [PubMed]
113. Grant, J.; Özkan, A.; Oh, C.; Mahajan, G.; Prantil-Baun, R.; Ingber, D.E. Simulating drug concentrations in PDMS microfluidic organ chips. *Lab Chip* **2021**, *21*, 3509–3519. [CrossRef] [PubMed]
114. Lee, S.; Ko, J.; Park, D.; Lee, S.-R.; Chung, M.; Lee, Y.; Jeon, N.L. Microfluidic-based vascularized microphysiological systems. *Lab Chip* **2018**, *18*, 2686–2709. [CrossRef]
115. Park, S.E.; Georgescu, A.; Huh, D. Organoids-on-a-chip. *Science* **2019**, *364*, 960–965. [CrossRef] [PubMed]
116. Ahn, J.; Ko, J.; Lee, S.; Yu, J.; Kim, Y.; Jeon, N.L. Microfluidics in nanoparticle drug delivery; From synthesis to pre-clinical screening. *Adv. Drug Deliv. Rev.* **2018**, *128*, 29–53. [CrossRef]
117. Ajalik, R.E.; Alenchery, R.G.; Cognetti, J.S.; Zhang, V.Z.; McGrath, J.L.; Miller, B.L.; Awad, H.A. Human Organ-on-a-Chip Microphysiological Systems to Model Musculoskeletal Pathologies and Accelerate Therapeutic Discovery. *Front. Bioeng. Biotechnol.* **2022**, *10*, 846230. [CrossRef]
118. Van Den Berg, A.; Mummery, C.L.; Passier, R.; Van der Meer, A.D. Personalised organs-on-chips: Functional testing for precision medicine. *Lab Chip* **2019**, *19*, 198–205. [CrossRef]
119. Kim, H.; Sa, J.K.; Kim, J.; Cho, H.J.; Oh, H.J.; Choi, D.H.; Kang, S.H.; Jeong, D.E.; Nam, D.H.; Lee, H. Recapitulated Crosstalk between Cerebral Metastatic Lung Cancer Cells and Brain Perivascular Tumor Microenvironment in a Microfluidic Co-Culture Chip. *Adv. Sci.* **2022**, *2022*, 2201785. [CrossRef]
120. Ma, C.; Witkowski, M.T.; Harris, J.; Dolgalev, I.; Sreeram, S.; Qian, W.; Tong, J.; Chen, X.; Aifantis, I.; Chen, W. Leukemia-on-a-chip: Dissecting the chemoresistance mechanisms in B cell acute lymphoblastic leukemia bone marrow niche. *Sci. Adv.* **2020**, *6*, eaba5536. [CrossRef]
121. Xiao, Y.; Kim, D.; Dura, B.; Zhang, K.; Yan, R.; Li, H.; Han, E.; Ip, J.; Zou, P.; Liu, J. Ex vivo dynamics of human glioblastoma cells in a microvasculature-on-a-chip system correlates with tumor heterogeneity and subtypes. *Adv. Sci.* **2019**, *6*, 1801531. [CrossRef]
122. Ramadan, Q.; Zourob, M. Organ-on-a-chip engineering: Toward bridging the gap between lab and industry. *Biomicrofluidics* **2020**, *14*, 041501. [CrossRef] [PubMed]
123. Abulaiti, M.; Yalikun, Y.; Murata, K.; Sato, A.; Sami, M.M.; Sasaki, Y.; Fujiwara, Y.; Minatoya, K.; Shiba, Y.; Tanaka, Y. Establishment of a heart-on-a-chip microdevice based on human iPS cells for the evaluation of human heart tissue function. *Sci. Rep.* **2020**, *10*, 19201. [CrossRef] [PubMed]
124. Kim, S.; Lee, S.; Lim, J.; Choi, H.; Kang, H.; Jeon, N.L.; Son, Y. Human bone marrow-derived mesenchymal stem cells play a role as a vascular pericyte in the reconstruction of human BBB on the angiogenesis microfluidic chip. *Biomaterials* **2021**, *279*, 121210. [CrossRef] [PubMed]

125. Sances, S.; Ho, R.; Vatine, G.; West, D.; Laperle, A.; Meyer, A.; Godoy, M.; Kay, P.S.; Mandefro, B.; Hatata, S. Human iPSC-derived endothelial cells and microengineered organ-chip enhance neuronal development. *Stem Cell Rep.* **2018**, *10*, 1222–1236. [CrossRef] [PubMed]
126. Probst, C.; Schneider, S.; Loskill, P. High-throughput organ-on-a-chip systems: Current status and remaining challenges. *Curr. Opin. Biomed. Eng.* **2018**, *6*, 33–41. [CrossRef]

*Review*

# Microfluidic Organ-on-a-Chip Devices for Liver Disease Modeling In Vitro

Perizat Kanabekova [1], Adina Kadyrova [2] and Gulsim Kulsharova [1,*]

[1] School of Engineering and Digital Sciences, Nazarbayev University, Nur-Sultan 010000, Kazakhstan; perizat.kanabekova@nu.edu.kz

[2] Department of Biological Sciences, School of Sciences and Humanities, Nazarbayev University, Nur-Sultan 010000, Kazakhstan; adina.kadyrova@nu.edu.kz

* Correspondence: gulsim.kulsharova@nu.edu.kz

**Abstract:** Mortality from liver disease conditions continues to be very high. As liver diseases manifest and progress silently, prompt measures after diagnosis are essential in the treatment of these conditions. Microfluidic organs-on-chip platforms have significant potential for the study of the pathophysiology of liver diseases in vitro. Different liver-on-a-chip microphysiological platforms have been reported to study cell-signaling pathways such as those activating stellate cells within liver diseases. Moreover, the drug efficacy for liver conditions might be evaluated on a cellular metabolic level. Here, we present a comprehensive review of microphysiological platforms used for modelling liver diseases. First, we briefly introduce the concept and importance of organs-on-a-chip in studying liver diseases in vitro, reflecting on existing reviews of healthy liver-on-a-chip platforms. Second, the techniques of cell cultures used in the microfluidic devices, including 2D, 3D, and spheroid cells, are explained. Next, the types of liver diseases (NAFLD, ALD, hepatitis infections, and drug injury) on-chip are explained for a further comprehensive overview of the design and methods of developing liver diseases in vitro. Finally, some challenges in design and existing solutions to them are reviewed

**Keywords:** organ-on-chip; liver-on-chip; liver disease; microfluidic devices

**Citation:** Kanabekova, P.; Kadyrova, A.; Kulsharova, G. Microfluidic Organ-on-a-Chip Devices for Liver Disease Modeling In Vitro. *Micromachines* **2022**, *13*, 428. https://doi.org/10.3390/mi13030428

Academic Editors: Violeta Carvalho, Senhorinha Teixeira and João Eduardo P. Castro Ribeiro

Received: 8 February 2022
Accepted: 8 March 2022
Published: 10 March 2022

**Publisher's Note:** MDPI stays neutral with regard to jurisdictional claims in published maps and institutional affiliations.

## 1. Introduction

Today, the issue of liver disease is very common. Only in 2017, more than 1.5 billion people had some kind of liver condition, and more than half of them were affected by non-alcoholic fatty liver disease (NAFLD) [1]. Every year, more than 2 million people die due to liver disease [2]. The prevalence of death due to cirrhosis has increased from less than 2% to almost 2.5% starting in 1990 [3]. Statistical data show that the number of deaths due to liver disease in Europe and Central Asia increased primarily due to increases in ALD (alcoholic liver disease). In developing countries, the primary cause of liver disease is hepatitis infections, which can be prevented and treated, but in limited availability [3].

It is difficult to overstate the importance of the liver in a human body. The liver is a key organ in maintaining the metabolism and digestion, immunity, and detoxification of the organism. The liver cells produce substances that are significant for its function. For example, the bile salts, which are produced by liver cells (hepatocytes), emulsify the fats, therefore, facilitating its digestion [4]. Moreover, the liver is responsible for the synthesis of the majority of plasma proteins such as globulins, which are part of the immune system, and coagulation factors, which are involved in blood clotting. Next, the liver is responsible for drug metabolism. Hepatocytes modify the compounds with the CYP450 group of enzymes and then bind to carrier molecules so that they can be excreted in blood or bile for further removal through kidneys and intestine [5]. Although the liver has the essential capacity to regenerate, those functions can be affected by chronic exposure to a range of chemicals such as saturated and trans-fatty acids, alcohol, or chronic inflammation due to viral disease. Such negative exposure to tissue leads to liver diseases.

A common feature of different liver diseases is that the continuous impact of function significantly increases the risks of the replacement of hepatocytes by fibrous tissue.

In turn, fibrosis and continuous inflammation are essential risk factors for cirrhosis and hepatocellular carcinoma (HCC) as shown in Figure 1 [6]. As the diseases are asymptomatic in the early stages and manifest the symptoms mainly in later stages, timely diagnosis and measures are essential in the course of diseases.

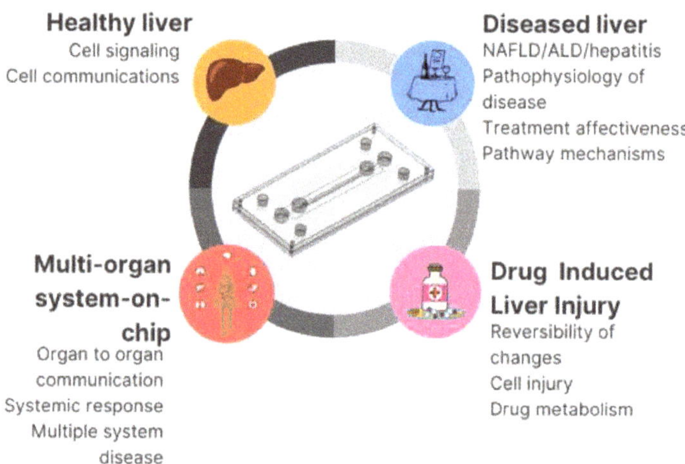


**Figure 1.** The use of microfluidic devices in different aspects of healthy and liver disease modeling.

Different in vitro and in vivo models to study the pathophysiology of liver diseases exist. There are several in vitro models-cell culture models, including 2D monoculture and co-culture, 3D cultures, spheroid/organoid cells, human precision-cut liver slices, immortalized cell lines, or sandwich-cultured cell lines [7,8]. Each type of model has its own benefits and limitations. For example, monocultures lack interaction with non-parenchymal cells, limited in control of uniformity in the generation of spheroids and organoids [8]. Moreover, most of those models lack oxygen-induced differences in zonation of hepatic sinusoids [7]. Animal models such as mouse studies allow mimicking liver diseases through diet; for example, a high-fat diet induces NAFLD-like condition or through induction of genetic models [8]. Liver fibrosis induction in mice can also be induced by using hepatotoxins, and hepatitis infection might be initiated by infecting transgenic mice [9]. However, genetic models have different etiology from human NAFLD, making them unfavorable for studying the onset and currency of the condition. Alternatively, rat or minipig models, or primates can be used for studies. The main limitation in terms of non-rodent models is that their metabolic pathways were found to be different, while primate models are much more expensive and raise ethical issues [8].

Emerging technologies and engineering techniques such as microfluidics can facilitate the demand for new in vitro approaches for studying liver diseases. Among such tools, microfluidic organ-on a-chip platforms have seen tremendous growth due to their potential to reduce drug development costs and time. The development of microfluidic devices allows the growth and maintenance of cell lines with the evaluation of function. Biocompatible materials and flow conditions mimic the microphysiological environment, contributing to conducting experiments with different animal cells. The size of microchannels not only reduces the amount of reagent use but also enhances the quick cellular response due to the ability to emulate the microphysiological environment by controlling spatiotemporal parameters such as cell–cell/cell–matrix interaction, shear stresses, excretion of metabolites [10]. Another important aspect of the microsizes of channels is cell-to-liquid ratio, since the volume of the circulating liquid is generally much higher than the volume scale of the tissue is [11]. The composition of the circulating medium can be modified during the flow experiments to achieve more physiological liquid-to-cell ratios [12]. Additionally, drug metabolites can reach higher concentrations due to small volumes of microchannels. Fluidic experiment eluents or medium containing metabolites can be coupled to other devices or microchannels implemented elsewhere on the same platform. Overall, combining microfluidics with tissue engineering can allow achieving even complex architecture of an organ [13,14].

Due to the principal advantages in mimicking microphysiological systems and repeating the microstructure of the human tissue, organs-on-chips have significant potential to replace animal testing models, which are known to exhibit different responses in questions of drug toxicity studies [15]. Moreover, currently drug development relies significantly on animal testing to evaluate the potential drug compound, while organ-on-chip platforms might provide the information based on human cell tests [16]. It opens further potential for personalized medicine, where the cells of patients could be used to access the drug sensitivity in conditions like cancer [17]. These devices are also significant in pathogenesis studies and cell signaling/communication experiments [10].

Organ-on-a-chip technology led to the design of a wide range of tissues on microfluidic devices, including kidneys, heart, and liver [18]. Despite the complexity of biomimicking, the designs of developed devices even allow recapitulating the physiological structures as brain–blood barrier and air–blood barrier in brain and lungs in a wide range of models. For example, the liver-on-chip models range from early models based on monoculture of hepatic cells to more complicated 3D cultures, spheroids, and organoids. The control of fluid flow can potentially allow the development of metabolic zonation [15]. In addition, the possibility to integrate several types of cells with control of biomechanical properties aids in rebuilding the physiological environment of cells [15].

Recently, the exposure of hepatocytes to specific compounds led to the development of disease models on chips. Those are essential for studying the pathological processes at the cellular level in detail. Turning to liver-on-chip devices, the models for conditions like Non-Alcoholic Fatty Liver Disease (NAFLD), Alcoholic Liver Disease (ALD), hepatitis infections, and Drug-Induced-Liver-Injury (DILI) have been developed.

Several reviews have been reported which included some aspects of liver-on-a-chip platforms for liver disease modelling but gave incomplete and brief descriptions of designed diseases on-chips. For example, Hassan et al. reviewed the importance of different cells in the construction of liver-on-chip-platforms and metabolic zonation within the liver [19]. Another review describing the role of non-parenchymal cells in progression indicated the NAFLD as a multi-organ disease in which the functionality of other organs is compromised by the liver state [20]. Similarly, significance of non-parenchymal cells in NAFLD pathophysiology and simulation of different liver compartments was reviewed by Deng, Wei, et al. [21]. The application of liver-on-chip technology to drug discovery was reviewed by Clapp et al., who described the potential use of technology to build models of NAFLD and HBV infections [22]. Concerning the design of microfluidic devices, Moradi with his team presented a detailed review of existing liver-on-chip models, categorizing them by gradient, membrane, or zonation use [23].

Liver diseases can be induced by high drug compounds as well. Microfluidic platforms designed to study the DILI were reviewed by Lin and Khetani, including models involving other organs such as the kidney, nervous system, and skin to evaluate the toxicity of compounds on other organs [24]. The fibrosis modelling using microphysiological platforms was reported by categorizing studies by cell culture techniques such as 2D, 3D models, bioprinted cultures, and describing models by the type of tissue [25]. Van Grunsven et al. presented in vitro fibrosis models, including methods to generate 3D spheroids and microfluidics developed for drug toxicity studies, which can potentially be used for investigating fibrogenesis within the liver [26]. The concept of fibrogenesis in different tissues including lungs, liver, and heart are reported in a review by Hayward et al., as a way for cancer pathogenesis as it is an essential step in the initiation of oncological processes [27].

Here, we present a comprehensive review of microfluidic organs-on-chip platforms used for studying the pathophysiology of different liver diseases from both medical and engineering points of view. The review is aimed at highlighting different perspectives on disease models, methods, and the outcomes of the studies. It also includes other important aspects of organ-on-chip devices such as cell culture models, challenges related to the design of microphysiologic platforms, and the most recent solutions.

## 2. Organ-on-a-Chip Considerations

### 2.1. Cell Culture Models

The scientific world is used to working with regular 2D cell culture for in vitro cell-based experiments. Despite the simplicity of use, those cells cannot mimic the in vivo microenvironment, as a result, the functions and processes within the tissue cannot be fully achieved [28]. To overcome the limitations, different cell culture techniques were developed to better recapitulate the features of the tissue such as the interaction between the cells and metabolic gradients. Table 1 represents the advantages, disadvantages, and current applications of existing cell techniques.

**Table 1.** Cell culture methods.

| In Vitro Cell Culturing Models | Advantages | Limitations | Applications |
|---|---|---|---|
| 2-D (monolayer cultures) | • Easy to grow Low Cost Accessibility [23] | • Loss of tissue specific functions; no cell signaling limited cell–cell/cell–matrix interactions disturbed cellular, mechanical, chemical morphology Uncontrolled access to oxygen and other compounds Viable for limited amount of time [23] | 2D culture |
| 3-D (organoids, multicellular spheroids) | • Cell–cell communication Physiological relevance Response to stimuli Cell polarity and phenotype Ability to differentiate, proliferate Tissue viability Drug metabolism [29] | • Cell maturation problems Static: Irrelevant to cell migration and transport Loss of cell shape Hard to collect cellular components for biochemical and genetical analysis No shear stress Necrotic core formation Uncontrolled oxygen and media gradient [30] | Organ Development Tissue Morphogenesis Drug Toxicity tests Creation of cancer organoids-customizing individual cancer treatments Study effect of pathogens [31] Organoids and spheroids, adapted from Velasco et al. [32] under the terms of the creative commons attribution license |

**Table 1.** *Cont.*

| In Vitro Cell Culturing Models | Advantages | Limitations | Applications |
|---|---|---|---|
| Organ-on-a-chip (microfluidic devices) | • Flow mimicking cell conditions<br>Chemical gradients<br>Easily manipulated cell conditions<br>Relevant amount of shear stress<br>Tissue specific relevant response<br>Recruitment of immune cells<br>Relevant mechanical microenvironment [30] | • High cost<br>Complexity in manufacturing chips<br>More difficult to culture cells inside, required perfusion system<br>The materials of the chip non-specifically absorb drugs [21] | Analysis<br>Recreation interfaces of cellular tissues<br>Study for disease mechanisms<br>Drug toxicity tests/development<br>Could be used in personalized medicine<br><br>Microfluidic chip |

### 2.1.1. 2D Cell Culture

The cell monolayers are extensively used in studies of drug pharmacokinetics and are essential in the preclinical stage of drug safety/toxicity investigation [33]. Here, the exposure of cells to media and oxygen is homogeneous throughout the surface, and therefore, hypoxic conditions cannot be mimicked. This might be essential in studies of cancer pathophysiology and tissues with metabolic zonation and may affect cell proliferation [34]. On the other hand, 2D cell culture experiments are better in terms of reproducibility, and the cells are easier to handle [25].

### 2.1.2. 3D Cell Culture

The three-dimensional techniques can be scaffold-based and scaffold-free ones depending on their use of external support [33,35]. Scaffold, designed from natural materials, plays the role of extracellular matrix and aids in cell adhesion, interaction, and exchange of required gases and nutrients. Hydrogels, solid-state scaffolds, might be implemented for the recapitulation of 3D format. Non-scaffold-based cultures include spheroid cultures and organoid cultures [33]. Any type of 3D culture technique allows controlling the concentration gradient of compounds that affect the behavior of the cell such as communication signals, migration to other places, and expression of specific genes [34].

### 2.1.3. Spheroids

The self-aggregation of the cell using techniques such as hanging drop plates, spheroid microplates, magnetic levitation, microfluidic devices, and bioprinting generates the spheroid structured cells, the shape of which prevents the adhesion to flat surfaces [36]. Those cells as free-floating aggregates, shown in Table 1, were found to better recreate the features of in vivo tissue and excellent simulation of conditions such as low-oxygen state, dormancy, and activation of inhibition of cell death pathway [34]. They can be generated from most cell lines, primary or tumor cells and do not require growth factors or ECM for maintenance [37]. Although spheroids can exist in plates where they were generated, some studies reported that incorporation into scaffolds increases the efficacy in designing tumor models [35].

### 2.1.4. Organoids

Organoids are closer to their original tissue by functionality and by histological characteristics. In the majority of cases, they are generated from different stem cells and need ECM and growth factors for maintenance [37]. Organoids can be prepared by the same methods as spheroids, but those might lack resembling mechanical forces and physiological cell signals.

To address these issues, microfluidic devices such as organs-on-chips can recreate the microphysiological environment by providing dynamic flow conditions [38]. Organs-on-chips aim to mimic closely the anatomy, physiology, and functionality of a human organ and can be a better model for studying pathophysiology of organ diseases [39].

Table 1 represents the cell culture methods by describing the advantages and disadvantages of using them. It also includes the potential applications of the technique and provides sample images of cell culture grown.

## 2.2. Cell Sources

Physiologically, the liver tissue consists of hepatic cells and non-parenchymal hepatic cells such as stellate cells, Kupffer cells, and endothelial cells. Their role is essential in drug metabolism and disease pathophysiology [40].

There are several cell sources that might be used for in vitro human liver cell culture models. Those are the cell lines, primary hepatocytes (human and animal), and hepatocytes derived from stem cells (induced pluripotent, adult stem cells, and human embryonic stem cells).

Primary human hepatocytes can be ideal for the final drug metabolism testing. While setting up the experiment, they are less likely to be used due to limited availability and inability to proliferate in vitro [40]. In comparison, hepatic cell lines can be propagated and effectively used in experiments, although there is lower expression of enzymes that can be found in primary hepatic cells [41]. The usage of stem cells often raises ethical questions due to sourcing. These cells are in low availability and lack common protocol for use, which makes them more difficult to handle, and they have limited capacity to proliferate.

## 2.3. Dynamic Flow Considerations

Organ-on-a-chip technology is not only based on various cell types or tissue models, but it also heavily incorporates engineering aspects such as spatially guided placement of cells, presence of flow, mechanical stimulation (shear stress), environmental control ($O_2$, $CO_2$, pH, and nutrients), and integration of sensors [42].

Among them, particularly for liver-on-a-chip platforms, fluid control is of the utmost importance. The fluid control in liver-on-a-chip platforms enables the development of metabolic zonation emulating in vivo conditions of the liver. In in vivo conditions, liver cells are exposed to a gradient of oxygen and hormones [15], and inside the microchannels, this can be emulated by exposure of cells to different axial oxygen gradients [43,44]. The flow rate of the perfusion media varies depending on the research application from 2 nL/min to 5 mL/min [45]. However, the flow rate of a liver-on-a-chip system needs to be chosen very carefully, since shear force caused by flow strongly impacts cells' intrinsic physiology and hepatocyte's function overall [46]. Shear rates higher than (5 dyn/cm$^2$) 0.5 Pa have been shown to reduce the hepatocyte function [47]. The physiologically relevant range for shear stress inside a single microfluidic device was reported to vary from ~0 to 0.03 dyn/cm$^2$ (3 mPa) based on the study done with intestinal epithelial cells [48].

For microfluidic cell culture, oxygen is essential and is mainly transferred using medium flow. Therefore, flow rate contributes not only to the shear stress experienced by perfusion culture, but also is a key element in dissolved oxygen transport to hepatocytes. Previously, flow experiments have been conducted to measure the impact of different medium flow rates ranging from 1 μL/min to 10 μL/min on oxygen consumption of hepatocytes [49]. The study showed that the minimum saturation of cells occurred at 3 μL/min flow rates [49]. Ehrlich et al. reported that an oxygen consumption rate of hepatocytes should be at least 5 nmol/min/$10^6$ cells in designing physiological design parameters of liver-on-a-chip platforms in a physiological system suggesting that the device should be able to achieve flow rates higher than 22 μL/min/$10^6$ cells [15]. Overall, there should be a balance between a high enough flow rate to ensure sufficient oxygen transport to hepatocytes, but it needs to be balanced with shear stress values.

Dynamic flow culture in organ-on-a-chip platforms also depends on the design of microchannels and cell chambers. Although there are some variations in the designs of fluidic systems for liver-on-a-chip platforms, most of the reported research follow one of the main three designs [15] (1) a simple flat-plate design where cells are cultured on the bottom part of a microchannel [39,47]. (2) A packed-bed design in which hepatocytes aggregates are placed within the 3D chamber [50] and (3) hollow-fiber design in which cells aggregate around fibers delivery oxygen and nutrients [51]. There are several reviews specifically focused on fabrication and design perspectives of organ-on-a-chip devices in the literature [15,52].

## 3. Liver-on-a-Chip Platforms Modeling NAFLD

NAFLD is a chronic condition, which develops due to abnormal lipid accumulation or steatosis, as a result of a fatty diet, lifestyle, and hormonal changes [53]. With progress, the inflammatory component joins the pathological steatotic process, causing steatohepatitis, which consequently leads to fibrogenesis. Histologically, the transition from healthy cells to different degrees of steatosis with subsequent ballooning, fibrosis, and cirrhosis takes place [53]. This condition is asymptomatic in the early stages and might be diagnosed accidentally. The signs and symptoms develop when liver decompensation occurs. As NAFLD is associated with other metabolic conditions such as insulin resistance, obesity, and others, it might manifest and be diagnosed earlier. In addition, the chronic use of drugs such as methotrexate, glucocorticoids, chemotherapeutic drugs, and tetracycline increases the risk of drug-induced hepatic steatosis [53].

One of the benefits of microfluidic devices is not only the possibility of long-term cell maintenance of 2D and 3D cultures but also that some of them are designed to prepare these cell cultures. For example, the formation of 3D organoids from HepRG progenitor cells was possible by using SteatoChip by Teng et al. (Figure 2A), which represented better recapitulation of liver functions in comparison to conventional 2D cultures [54]. The NASH on-chip, designed by Freag et al., was based on 3D coculture of cells in hydrogel prepared from the collagen, which were studied for biochemical hallmarks of the disease, shown in Figure 2B (2021) [55]. In turn, this validated the significance of 3D structures in the recapitulation of spatial organization to enhance cell communication with controlled conditions. Next, the study by Wang et al. designed the Liver Organoid on Chip, which allowed the formation of 3D aggregates of the liver organoids as well and maintained the long-term perfusion environment as shown in Figure 2C (2020) [56].

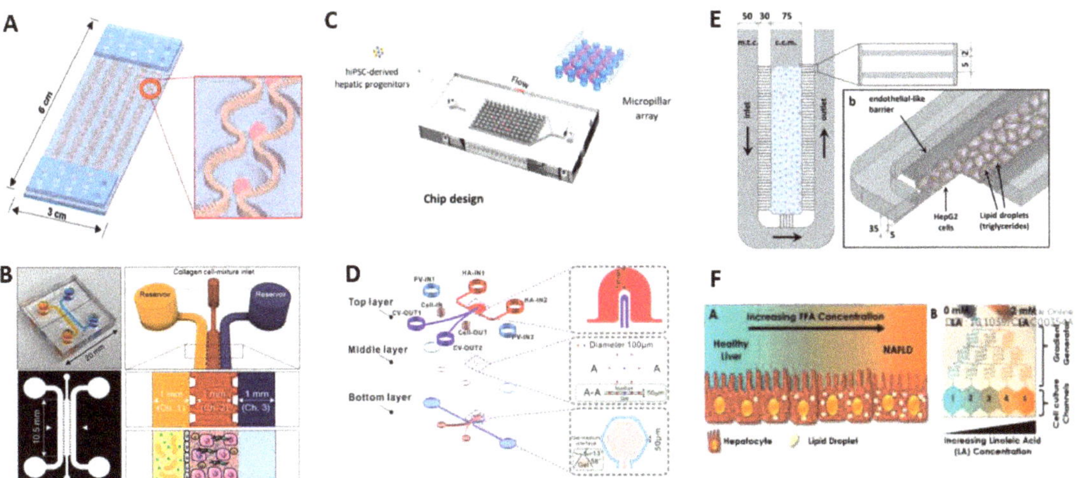

**Figure 2.** Modeling NAFLD using microfluidic devices. (**A**) Steatochip recapitulating liver microstruc-

ture, for growth and differentiation of HepRG cells [54] (reproduced with permission from Teng et al., 2021). (**B**) NASH-on-chip, prepared with hydrogel channel with hepatic, Kupffer and stellate cells [55] (reproduced with permission from Freag et al., 2020). (**C**) NAFLD-on-chip inducing differentiation of hiPSC to organoids followed by steatosis progression [56] (reproduced with permission from Wang et al., 2020 Copyright: ACS Biomater. Sci. Eng. 2020, 6, 5734−5743). (**D**) Schematic illustration of liver lobule chip [57] (reproduced with permission from Du et al., 2021). (**E**) NAFLD on-on-a-chip in 3D and schematic view with HepG2 cell culture [58] (reproduced with permission from Gori et al., 2016; this is an open access article distributed under the terms of the Creative Commons Attribution License). (**F**) NAFLD model on chip generating the gradient of liloleic acid [59] (reproduced with permission from Bulutoglu et al., 2019).

In addition to preparing cell cultures and modelling the NAFLD-on-chip devices, microfluidic devices allow the evaluation of drug efficacy on a cellular level. For example, in a study by Gori, Giannitelli hypolipidemic drugs were applied on NAFLD-on-chip hepatic cells, and lipid-lowering activity was confirmed observing lower levels of steatosis [58]. Potentially, the disease-on-chip devices allow not only conducting drug toxicity studies but also drug discovery and drug effectiveness studies as a part of drug approval steps.

The microfluidic devices might be used not only to mimick the microphysiological systems but also to study basic cellular processes in the progression of a condition. For example, Slaughter et al. designed the two-chambered device accommodating liver cells and adipocytes cells (2021) [60]. In this study, the correlation between NAFLD and obesity and diabetes was established by tracking levels of pro-inflammatory cytokines, adipokines, and insulin resistance markers. Although one of the major challenges in the design of liver-on-chip and therefore disease-on-chip platforms is the induction of bile canaliculi, some studies design devices mimic this structure. The NAFLD-on-chip presented in Figure 2D, designed by dual 'blood' supply to hepatocytes and recapitulating liver lobule by hexagonally shaped chamber bile canaliculi microstructure development, was successful due to beneficial distribution of hepatocytes and microvessels [57]. Another way to form the 3D cell spheroids was used in a study where AggreWell plates were used to generate cell culture for Steatosis Disease-on-Chip with an optimized ratio of different cells [61]. Moreover, reversal of the pathological process in case of removal of damaging factors was observed, and the effect of anti-steatotic drugs was evaluated. Similarly, Suurmond et al. induced steatosis in spheroid cells generated using AggreWell plates and explored the feedback signaling between them in the pathophysiology of NASH [62]. The three-culture spheroid microphysiological platform was used to induce the NAFLD with progression to NASH by measuring the deposition of the collagen and markers of fibrotic tissue [63]. The increase of inflammatory cytokines and activation of stellate cells were correlated with transcriptomic analysis and gene expression. Moreover, the authors investigated the reversibility of the condition by evaluating the positive effect of treatment drugs in an ongoing study [63].

NAFLD-on-chip models were designed by exposing hepatocyte cells to different fatty acids such as linoleic acid, palmitic acid, or oleic acid for a fixed period. Increased steatosis was observed in cells that were exposed to higher levels of acid. For example, the device which is shown in Figure 2E replicated liver sinusoid by using a system of a grid of microchannels instead of endothelial vascular cells. In this study, treatment of cells with fatty acid in chip conditions resulted in increased cell viability in comparison to static 2D culture, therefore highlighting the benefits of microfluidic devices in animal cell study experiments [58]. Interestingly, one of the studies revealed that the metabolic zonation allowed correlating oxygen concentration to steatosis. Therefore, in the microfluidic device (Figure 2F), the cells with the same concentration of fatty acids, but lower levels of oxygen, had a higher level of steatosis [59].

One of the studies recreated liver acinus with its vascularized form and in coupling with pancreas and adipose tissue from induced pluripotent stem cells [64]. Moreover, the fluorescent protein biosensors were introduced into the device to evaluate the release of reactive oxygen species and insulin resistance. This large-scale study presented metrics for liver-specific biomarkers, which identify the progression from NAFLD to steatohepatitis with the experimental timeline [64].

Overall, microfluidic liver-on-a-chip platforms mimicking NAFLD could be applied to not only to study pathogenesis of disease and drug efficacy but also to emulate other similar steatotic conditions such as metabolic dysfunction-associated fatty liver disease (MAFLD). However, a potential challenge in modeling this condition may be the difficulty in recapitulating systemic diseases such as obesity, diabetes, and hyperlipidemia, which are the diagnostic criteria for MAFLD.

## 4. Liver-on-a-Chip Platforms Modeling ALD

Ethanol is metabolized by alcohol dehydrogenase enzymes present in hepatocytes, resulting in the formation of formaldehyde, which is further metabolized to acetate. Excessive and chronic alcohol consumption leads to the tissue injury associated with ethanol metabolism. One of the earliest reactions of hepatocytes to this type of stress is a fat accumulation of hepatocytes that causes steatosis. It develops in 90% of individuals who consume 4–5 standard doses per day for a long-term perspective (standard dose = 14 g of pure alcohol) and more quickly among binge drinkers [65]. The steatosis progresses to steatohepatitis if lifestyle modifications did not take place with further development of fibrosis and cirrhosis.

Despite the similarities, in NAFLD, the more prominent fatty degeneration of cells can be observed, while in ALD, the inflammatory component is more significant. Moreover, perivascular changes are more frequent in ALD [66]. Although alcoholic drinks have different concentrations of ethanol, it is the effect of ethanol on cells which causes health consequences on liver tissue.

According to many scholars, one of the early indicators of liver function decrease is an alteration of bile canaliculi. In a study by Nawroth et al., the ECM scaffolding was conducted to improve the integrity of bile canaliculi structure, which further allowed to track alteration in canaliculi network in response to alcohol exposure in microfluidic devices as shown in Figure 3A (2021) [67]. As a result, in response to alcohol, the expansion and decreased branching of bile canaliculi was observed, which goes along with the currency of ALD in humans with signs of cholestasis or decreased bile flow.

In addition to an increased inflammatory response in cells to alcohol, the study with the microfluidic device shown in Figure 2B revealed that the higher accumulation of collagen can be observed in liver-on-chip samples with co-culture in comparison to mono-cultures [68]. This proves the involvement of stellate and Kupffer cells and induction of fibrosis of the tissue in the pathophysiology of liver diseases.

The more complicated microfluidic device, which can be reassembled, was designed by Zhou et al. and included a TGF-β aptasensor to track its secretion from stellate cells and hepatocytes (2015) [69]. The experiment revealed that there is the primary release of cytokine by hepatocyte, which further activates the stellate cells. The release of TGF-β by stellate cells is more significant and mediates the inflammatory component of the ALD. Here the reconfigurable device, represented in Figure 3C, allowed controlling variables in ALD-on-chip such as presence and amount of stellate cells in the co-culture used [69]. Another device designed 3D liver-sinusoid-on-a-chip as shown in Figure 3D to investigate the pathophysiological process of ALD [70]. Here, different cells formed layers recapitulating liver structure and were separated by porous PC membrane to mimic canaliculi and acinus structurally. The damage caused by alcohol was estimated by evaluating DNA damage by reactive oxygen species and established hepatic sinusoid based on results of biomarkers [70].

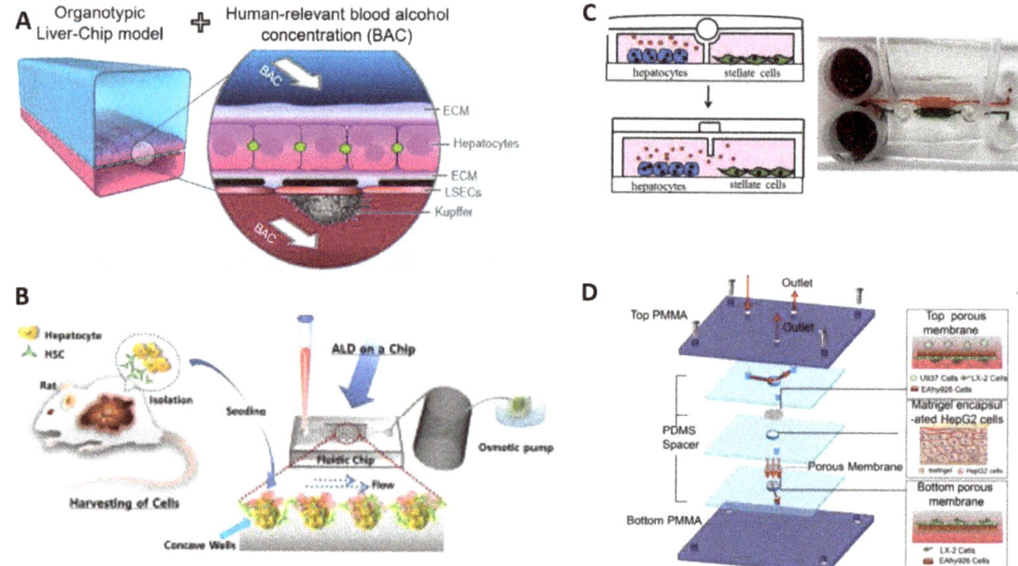

**Figure 3.** Modeling Alcoholic Liver Disease using microfluidic devices. (**A**) ALD Liver-Chip model with approach of exposing to alcohol [67] (reproduced with permission from Nawroth et al., 2020 Copyright: Cell Reports 36, 109393, July 20, 2021). (**B**) ALD model prepared by culturing primary hepatocytes with stellate cells in PDMS based Liver-on-Chip [68] (reproduced with permission from Lee et al., 2016). (**C**) Liver injury-on-a-chip representing non-mixing channels for co-culture growth [69], adapted from Zhou et al., 2015. (**D**) Liver-chip-based ALD design based on PMMA device with PDMS spaces and PC membrane to maintain cell lines [70] (reproduced with permissions from Deng et al., 2019).

## 5. Viral Hepatitis Pathophysiology Using Liver-on-a-Chip Devices

There are different viruses, which upon entry into the human body can cause hepatitis infection. Hepatitis A virus (HAV) and Hepatitis E virus (HEV) may enter the organism through consuming contaminated water or food [71]. Hepatitis B, C, and D are transmitted through blood (sharing needles, blood transfusion) and sexual contact. Moreover, hepatitis B infection can be transmitted vertically (from mother to child). While HAV and HEV are not likely to cause chronic infection and treatment is symptomatic, other viral infections are treated by slowing down viral replication only.

The highest probability for HBV to become chronic is during early childhood, although it may happen during adulthood too. In comparison, chronic HCV cases are more common among adults [71]. The injury to hepatocyte cells happens due to immune response to viral particles and infected cells. Chronic inflammation leads to continued hepatocyte death with subsequent liver degeneration. Next, activation of the stellate cells plays a role in fibrogenesis of the tissue, which is followed by cirrhosis and potentially has an increased risk of developing hepatocellular carcinoma [72].

Early studies implemented simple microchannel devices to grow cell culture and achieved transfection efficacy sufficient for HBV infection to be higher than in standard tissue culture [73]. Later devices implemented more complex designs with greater number of cells cell types used in studies [73]. For example, the viral replication of HBV in human liver sinusoid on a chip was investigated by detecting viral proteins in the co-culture of primary human hepatocytes and non-parenchymal cells in a study by Kang et al. as shown in Figure 4A (2017) [74]. Here, the cells in the structure maintained viability for almost four weeks, which might potentially be prolonged to study the persistence of viral disease and progression to chronic one. As 3D cultures are known for better recapitulation of the microphysiological environment, one study presented an essential protocol for the preparation of chip and infecting cells with HBV [75]. In this study, the synthesis of cccDNA was achieved indicating the possibility of generating chronic forms of the infection. In another study, the authors investigated the immune response of the cells to the HBV virus and outlined that there was reduced expression of interferons, which are essential for innate immunity [76]. Moreover, the role of Kupffer cells in HBV infection seems to depend on exogenous stimuli of lipopolysaccharides in generating an early response, indicating that the disease is not only affecting the liver tissue but also other organs, primarily the lymphatic system.

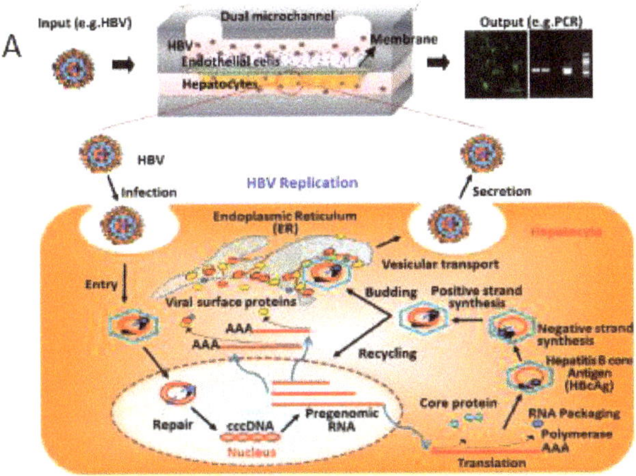


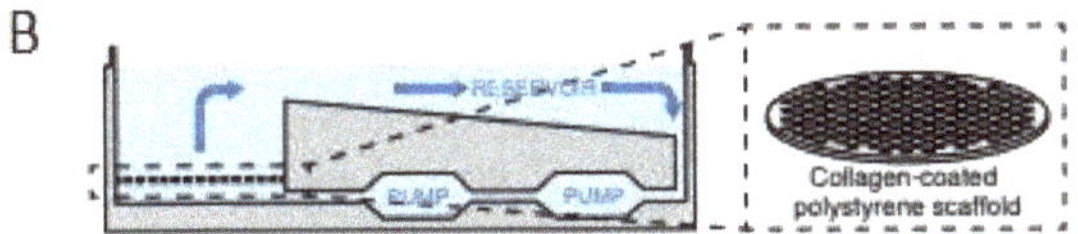


**Figure 4.** Modeling hepatitis infections using microfluidic devices. (**A**) Human Liver Sinusoid on a Chip design to infect the hepatocytes within the device with HBV virus [74], adapted with permission from Kang et al., 2017. Copyright 2017 by Kang et al.; licensee MDPI, Basel, Switzerland. (**B**) 3D microfluidic liver culture, grown in bioreactor, where circulation is created using pumps [76], adapted with permission from Ortega-Prieto et al., 2018, Copyright © Ortega-Prieto et al., 2018.

## 6. Liver Drug-Induced Injury (DILI) Using Liver-on-Chip Devices

The majority of compounds ingested and absorbed in intestine such as alcohol and drugs are carried by the blood to the liver, making it susceptible to primary injuries by those compounds [77]. Hepatotoxicity was described for a range of compounds, including those that were initially considered to be harmless such as dietary supplements, vitamins, and products such as beans. The biotransformation of compounds can be explained by the alteration of the properties of the compound to activate the drug or to metabolize it for further excretion. However, this can generate byproducts such as reactive oxygen species, which might damage the liver tissue.

Liver fibrosis is a response of the tissue to continuous injury. The external stimuli transform the stellate cells into myofibroblasts, which in turn activate the cascade of changes in the structure of the tissue. Some of those processes are degradation of the external matrix, recruiting other fibroblast-like cells from organisms by chemotaxis, fibrogenesis, and others [78]. The risk factors for this condition are NAFLD, ALD, and hepatic viral infections as described above as well as a range of genetic diseases, multifactor diseases as diabetes mellitus, liver transplantation, autoimmune hepatitis, primary biliary cholangitis, and others [79].

Several liver microfluidic platforms were designed to recreate the DILI in in vitro conditions. For example, one study presented a 3D co-culture device to investigate the liver injury due to acetaminophen with supplementary drugs such as rifampin, omeprazole and ciprofloxacin by measuring the release of lactate dehydrogenase by cells as well as urea, albumin synthesis, CYP, and UGT enzyme activities [80]. The time- and dose-dependent effect of acetaminophen on 3D liver organoids from induced pluripotent stem cells was also studied by Wang et al., where cell injury was evaluated by cell viability assay and functionality of CYP family of enzymes [81]. A more complicated 3D liver sinusoid-on-chip was designed using porous membranes mimicking the environment around sinuses to evaluate the toxicity of acetaminophen in comparison to 2D culture [82]. The results of the study confirmed previous studies in that 3D cultures are more effective in the recapitulation of the liver sinus microenvironment.

To investigate the mechanism and process of DILI, some studies conducted experiments where the ongoing evaluation of liver cells took place. In a study by Bavli et al., the mitochondrial dysfunction in a microphysiological platform was investigated (2016) [83]. The authors used oxygen, lactate, and glucose sensors and recorded a rapid decrease in the oxygen uptake by cells with a subsequent decrease in glucose uptake and lactate production in response to rotenone, with dose-dependent changes for other drugs. Another study investigated drug-to-drug interactions in 3D biomimetic microtissue to evaluate CYP enzyme family induction and inhibition and cell functionality in the chip.

As the hepatotoxic compounds may affect other organs in similar or different ways, some microphysiological platforms involve additional systems. For example, the 3D liver/lung-on-chip platform maintained for four weeks was exposed to aflatoxin B1 [84]. During the experiment, the cilia beating frequency of the bronchial culture and its viability were evaluated along with measuring the activity of liver cell metabolism and injury. Similarly, the effect of inhaled aflatoxin B1 on lung spheroid cells in the presence of liver spheroids was found to be decreased due to the metabolizing effect of the CYP450 enzyme [85]. Another stem cell-derived 3D multi-organoid chip involved the heart, lung, nerves, colon, testis, and blood vessels by creating chips for every tissue with common circulating media [86]. The effect of 10 different drugs, including compounds like vitamin C and ibuprofen, was evaluated on spheroids by measuring specific cell biomarkers at different time points of the experiment. Moreover, the toxicity of drugs such as chemotherapy prodrugs was estimated in the presence and absence of liver organoids. This goes along with existing literature on ability of the liver cells to metabolize the drug, causing the downstream toxicity effect on cardiac tissues [86].

To study the pathophysiology of the fibrosis, Jang et al. used methotrexate to treat 3D layer-membrane co-culture in liver-on-chip [87]. Here, activated stellate cells resulted in an increase in the release of interferon-$\gamma$, which has a strong association with inflammatory processes and fibrosis. In designing the fibrotic model, the improved cell delivery of non-parenchymal cells in devices with multilayer construction was observed when cells were encapsulated in gelatin bioink during cell printing [88].

## 7. Challenges in Designing Liver Disease-on-Chip Models

Microfluidic devices allow for a wide range of applications and liver-on-chip platforms reported by many scholars reveal functional units of the organ. This allows the recreation of liver parts for detailed study of the metabolism of healthy cells as well as the effect of drugs on cells. Next, it provides the base for the design of disease models such as NAFLD, ALD, or viral hepatitis infections.

It is clear that the creation of disease models is a developing perspective of organ-on-chip devices, as the field still has to address numerous challenges in creating a healthy liver model. Some of these challenges are related to the design of microphysiological structures such as bile canaliculi, and others are related to maintaining and handling cell cultures. However, some of those challenges were solved by simple additional devices or updates in existing ones. For example, the recreation of bile canaliculi and the creation of a stable and long-term 3D culture with immune Kupffer cells or stellate cells in designing healthy liver-on-chip is quite difficult. Mainly, this is due to the fact that bile cells have specific polarity with expression of transport proteins on lumen size, while Kuppffer and stellate cells cannot be randomly introduced within the device, requiring equal distribution and being located in between the hepatocyte cells. In the generation of 3D culture cells or spheroid cells, one of the essential criteria is equal sizing and variability of cells [15]. As a result, some commercially available plates designed to generate spheroids or required diameter with minimal variability were designed. Moreover, 3D liver tissues can be bioprinted, for example, by NovoGen Bioprinter, which allows the generation of tissue that can be maintained for 28 days. Further, this tissue was effective in the recapitulation of drug-induced liver injury experimentally [89]. Using the same NovoGen Bioprinter, the 3D bioprinted tissue was created with primary human hepatocytes and non-parenchymal cells in the study by Norona with her team [90]. Here, the recapitulation of the features of the native liver was studied by evaluating E-cadherin expression by cells, and the induction of fibrosis was done using methotrexate and thioacetamide.

Apart from using different cell culture techniques, there is a wide range of cell types that can be used in microfluidic devices to mimic liver functions. Each type of cell culture has its own limitations. For example, HepaRG cells are known for lower metabolic activity, which can be enhanced by adding 2% dimethyl sulfoxide. However, in a co-culture, this can result in the proliferation of non-parenchymal hepatic cells [91]. The way to facilitate cellular proliferation was suggested by Zuo et al., who generated HepatoCells from primary human hepatocytes, which allowed higher reproducibility by knocking out specific genes [92]. Besides the proliferative capacity, the scholars experience difficulties in maintaining the spheroids homogenous for a prolonged time as with time they tend to lose their volume and decrease in size [93]. In this study, the authors revealed that flow conditions within the microfluidic device have a positive impact on spheroids, resulting in stable volume and stable excretion of albumin for more than a week in comparison to static culture conditions. However, metabolic activity and function of cells including the production of albumin are reported to vary over time [94,95]. For example, Bhise et al. observed a drop in albumin at day 21 in the 30-day long-term hepatocyte culture in a bioprinted liver-on-a-chip device [94]. Overall, while many reports show that liver-on-chip provides a dynamic flow environment for hepatocytes resulting in higher albumin synthesis and urea excretion (detoxification) compared with static cultures [96], values of albumin secretion are normally significantly lower than in vivo conditions. For example, albumin production for primary cells has been reported to be 20–30 µg/(day·$10^6$ cells) [15,90] which is typically much higher than albumin levels reported by cell line based liver-on-a-chip platforms (0.6–80 pg/cell/day for HepG2 cells/spheroids) [94]. Since most of the liver-on-a-chip platforms employ cell lines for reproducibility purposes, obtaining real, in vivo-like metabolic activity and biomarker levels remains to be a challenge for the field overall.

One of the other challenges in designing liver-on-a-chip platforms is drug absorption by the materials used to prepare microfluidic devices. Polydimethylsiloxane (PDMS) is one of the most commonly used materials used in microfluidic devices due to its low-cost, optical transparency, elasticity, and gas permeability [39]. However, its main disadvantage is its ability to nonspecifically bind drug compounds, affecting the concentrations of the drugs to which hepatic cells were exposed, which has been well established in the literature [97]. This challenge of organ-on-a-chip devices is being addressed by use of alternative materials and substitutes for PDMS [98].

Moreover, some scholars report the issues related to unequal oxygen concentration within the organ-on-chip different from metabolic zonation within the in-vivo liver. Within the liver, the inflow to the sinusoid comes from arterial blood (rich in oxygen) and portal vein (rich in nutrients) and goes towards the central vein. This facilitates the metabolism of drugs/chemicals to be biotransformed and maintains the oxygen delivery even to less oxygenated zones, which is essential for the expression of some genes [91]. Earlier, it was suggested to decrease the seeding density of cells in vitro to allow the media to accommodate greater amounts of oxygen [99]. Interestingly, although perfused state suggested oxygen delivery, there were limitations in flow rate and capacity. Hence, the introduction of molecules with high oxygen-carrying capacity, mimicking hemoglobin in humans, was suggested [15]. Next, it implemented the introduction of a range of sensors inside of the chip, comprehending the device overall. For example, Farooqi et al. introduced trans-epithelial electrical resistance (TEER) sensors to follow up on the differentiation of cells within microfluidic devices and the integrity of the epithelial layer [100]. Different types of sensors were effectively introduced by de Hoyos-Vega et al., where enzymatic assays to analyze media from hepatocytes, beads-based immunoassays to evaluate secretion of specific growth factors, and electrochemical biosensor detecting the biomarkers released by hepatocytes were tested in microfluidic liver-on-chip [101].

Another challenge related to organ-on-chip platforms is the inability to replicate communication with other organs due to the fact that physiologically, the organs are connected with a wide range of systems. For example, liver organ itself has a dual nutrient supply: the hepatic artery carries oxygenated blood from heart through the general circulation and hepatic portal vein carries the deoxygenated blood with nutrients absorbed from small intestine. Moreover, the hormonal regulation of gastrointestinal tract is far more complicated, making it difficult to recapitulate the communication with other systems. However, there are already some multisystem organ devices such as Gut-Liver Chip that were designed to mimic the correlation between the gastrointestinal tract and liver [102]. The correlation between liver and gut is essential as the majority of ingested compounds are absorbed by the intestine and carried to the liver by the vascular system. Esch et al. recapitulated the liver-intestine model to track the absorption of carboxylated polystyrene nanoparticles by the gut, which resulted in liver injury through increased levels of transaminases [103]. Similarly, a four-organ-chip mimicking intestine, kidney, skin, and liver was designed by Maschmayer et al., where the liver was represented by 3D spheroids [104]. This chip maintained the microphysiological system for 4 weeks and was characterized to be optimal for administration, distribution, metabolism, and excretion (ADME) of drugs for further studies. The different four-organ systems including the heart, muscle, nervous system, and liver were designed to evaluate the response of systems to doxorubicin, atorvastatin, valproic acid, acetaminophen, and N-acetyl-m-aminophenol [105]. The toxicity of drugs on different systems was evaluated by measuring specific tissue biomarkers such as contractile frequency of the cardiomyocytes, electrophysiological recording of neurons, skeletal muscle viability, and metabolic byproducts of the liver after two-day exposure to the components.

In addition to challenges related to design of the device, there are questions that appear within the work such as the most optimal source of cells, using scaffolds or connecting platforms with each other. Next, scientists work on preparation of media that would be universal for multiple types of cells and mimic the blood [18]. For now, in experiments, scientists mix tissue specific media, which might contain byproducts that are toxic for other organs.

Although there are some challenges related to the design and fabrication of the liver-on-chip and disease design, the microfluidic devices have a great potential in disease modelling, drug development, and safety testing. Moreover, it has a promising future in personalized medicine, where the patient-derived tissue would be placed in microphysiological platforms to evaluate drug sensitivity/toxicity.

One of the beneficial potentials and future perspectives of microfluidic devices was reported by Ewart et al., who analyzed Liver-Chip for its ability to predict DILI [106]. This study reported 80% sensitivity of compound toxicities, which has a significant potential in drug development, especially preclinical stage. Moreover, the economic value model in the pharmaceutical industry for increased precision of preclinical studies estimated approximately USD 3 billion in benefits for Liver-Chip studies and for the use of other organ-chips with the same sensitivity up to USD 24 billion in benefits [106].

## 8. Conclusions

Overall, microfluidic devices have a great potential to replace animal experiments for drug testing and disease modelling. Organ-on-a-chip technology has been drawing attention as a new generation alternative to existing in vitro models for drug candidate screening in drug development [107]. The devices can provide cells with a microenvironment, which resembles the in vivo situation more closely [108]. Currently, organ-on-chip devices offer opportunities such as the co-culture of multiple cells in dynamic conditions with a recreation of organ-to-organ communication and integration of biosensors. However, despite the big progress made in organ-on-a-chip platforms used for emulating healthy and liver diseases on-chip, further studies need to be conducted to address existing challenges. One challenge is in variation and inconsistency between different manufacturing batches and different laboratories that produce liver-on-a-chip models. Use of various cell sources and the resulting discrepancy in metabolic function of cells and biomarker values from in vivo conditions continue to make difficult to reproduce not only liver diseases but also a healthy liver environment "on-chip". PDMS, which has been used for the majority of reported liver models, is also a limitation, but it is actively being replaced by emerging and alternative material substitutes. Complexity and cost of fabrication of organ-on-a-chip is another aspect of interest for the development of robust and reproducible platforms in the future. Despite the limitations, modeling healthy and liver diseases using microfluidic platforms can be accelerated in the near future due to advances in microfabrication and engineering tools as well as due to raising interest within the research community to organ-on-a-chip technology. Modeling liver diseases using liver-on-a-chip platforms can open doors to coupling the platforms with other organ-emulating systems. Multi-organ systems would allow to conduct cross-organ studies and give a more holistic view on liver diseases overall.

**Author Contributions:** Conceptualization, P.K. and G.K., writing—original draft preparation, P.K., G.K. and A.K.; writing—review and editing, G.K. All authors have read and agreed to the published version of the manuscript.

**Funding:** This research was funded by the Nazarbayev University Faculty-development research grant (080420FD1910), the Ministry of Education and Science of the Republic of Kazakhstan Grant for young researchers, (AP09058308), and social policy grant awarded to G.K.

**Data Availability Statement:** Not applicable.

**Acknowledgments:** We acknowledge Elvira Darbayeva for finding numerical values of shear stress and oxygen transport for hepatocytes as part of her literature review for an MSc thesis.

**Conflicts of Interest:** The authors declare no conflict of interest. The funders had no role in the design of the study; in the collection, analyses, or interpretation of data; in the writing of the manuscript; or in the decision to publish the results.

## References

1. Moon, A.M.; Singal, A.G.; Tapper, E.B. Contemporary Epidemiology of Chronic Liver Disease and Cirrhosis. *Clin. Gastroenterol. Hepatol.* **2019**, *18*, 2650–2666. [CrossRef] [PubMed]
2. Asrani, S.K.; Devarbhavi, H.; Eaton, J.; Kamath, P.S. Burden of liver diseases in the world. *J. Hepatol.* **2019**, *70*, 151–171. [CrossRef] [PubMed]
3. Sepanlou, S.G.; Safiri, S.; Bisignano, C.; Ikuta, K.S.; Merat, S.; Saberifiroozi, M.; Poustchi, H.; Tsoi, D.; Colombara, D.V.; Abdoli, A.; et al. The global, regional, and national burden of cirrhosis by cause in 195 countries and territories, 1990–2017: A systematic analysis for the Global Burden of Disease Study 2017. *Lancet Gastroenterol. Hepatol.* **2020**, *5*, 245–266. [CrossRef]
4. Chiang, J.Y.L.; Ferrell, J.M. Bile Acid Metabolism in Liver Pathobiology. *Gene Expr.* **2018**, *18*, 71–87. [CrossRef] [PubMed]
5. Corsini, A.; Bortolini, M. Drug-induced liver injury: The role of drug metabolism and transport. *J. Clin. Pharm.* **2013**, *53*, 463–474. [CrossRef] [PubMed]
6. Schuppan, D.; Afdhal, N.H. Liver cirrhosis. *Lancet* **2008**, *371*, 838–851. [CrossRef]

7.  Collins, S.D.; Yuen, G.; Tu, T.; Budzinska, M.A.; Spring, K.J.; Bryant, K.; Shackel, N.A. In vitro models of the liver: Disease modeling, drug discovery and clinical applications. In *Hepatocellular Carcinoma*; Tirnitz-Parker, J.E.E., Ed.; Western Sidney University: Brisbane, QLD, Australia, 2019.
8.  Soret, P.-A.; Magusto, J.; Housset, C.; Gautheron, J. In Vitro and In Vivo Models of Non-Alcoholic Fatty Liver Disease: A Critical Appraisal. *J. Clin. Med.* **2020**, *10*, 36. [CrossRef]
9.  Liu, Y.; Meyer, C.; Xu, C.; Weng, H.; Hellerbrand, C.; Dijke, P.T.; Dooley, S. Animal models of chronic liver diseases. *Am. J. Physiol. Gastrointest. Liver Physiol.* **2013**, *304*, G449–G468. [CrossRef]
10. Wu, Q.; Liu, J.; Wang, X.; Feng, L.; Wu, J.; Zhu, X.; Wen, W.; Gong, X. Organ-on-a-chip: Recent breakthroughs and future prospects. *Biomed. Eng. Online* **2020**, *19*, 9. [CrossRef]
11. Zhang, B.; Radisic, M. Organ-on-a-chip devices advance to market. *Lab Chip* **2017**, *17*, 2395–2420. [CrossRef]
12. Van Midwoud, P.M.; Verpoorte, E.; Groothuis, G.M.M. Microfluidic devices for in vitro studies on liver drug metabolism and toxicity. *Integr. Biol.* **2011**, *3*, 509–521. [CrossRef] [PubMed]
13. Baudoin, R.; Griscom, L.; Monge, M.; Legallais, C.; Leclerc, E. Development of a renal microchip for in vitro distal tubule models. *Biotechnol. Prog.* **2007**, *23*, 1245–1253. [CrossRef] [PubMed]
14. Dash, A.; Inman, W.; Hoffmaster, K.; Sevidal, S.; Kelly, J.; Obach, R.S.; Griffith, L.; Tannenbaum, S.R. Liver tissue engineering in the evaluation of drug safety. *Expert Opin. Drug Metab. Toxicol.* **2009**, *5*, 1159–1174. [CrossRef] [PubMed]
15. Ehrlich, A.; Duche, D.; Ouedraogo, G.; Nahmias, Y. Challenges and Opportunities in the Design of Liver-on-Chip Microdevices. *Annu. Rev. Biomed. Eng.* **2019**, *21*, 219–239. [CrossRef] [PubMed]
16. Quan, Y.; Sun, M.; Tan, Z.; Eijkel, J.C.T.; Berg, A.V.D.; van der Meer, A.; Xie, Y. Organ-on-a-chip: The next generation platform for risk assessment of radiobiology. *RSC Adv.* **2020**, *10*, 39521–39530. [CrossRef]
17. Caballero, D.; Reis, R.L.; Kundu, S.C. Engineering Patient-on-a-Chip Models for Personalized Cancer Medicine. *Adv. Exp. Med. Biol.* **2020**, *1230*, 43–64.
18. Low, L.A.; Mummery, C.; Berridge, B.R.; Austin, C.P.; Tagle, D.A. Organs-on-chips: Into the next decade. *Nat. Rev. Drug Discov.* **2021**, *20*, 345–361. [CrossRef]
19. Hassan, S.; Sebastian, S.; Maharjan, S.; Lesha, A.; Carpenter, A.M.; Liu, X.; Xie, X.; Livermore, C.; Zhang, Y.S.; Zarrinpar, A. Liver-on-a-Chip Models of Fatty Liver Disease. *Hepatology* **2020**, *71*, 733–740. [CrossRef]
20. De Chiara, F.; Ferret-Miñana, A.; Ramón-Azcón, J. The Synergy between Organ-on-a-Chip and Artificial Intelligence for the Study of NAFLD: From Basic Science to Clinical Research. *Biomedicines* **2021**, *9*, 248. [CrossRef]
21. Deng, J.; Wei, W.; Chen, Z.; Lin, B.; Zhao, W.; Luo, Y.; Zhang, X. Engineered Liver-on-a-Chip Platform to Mimic Liver Functions and Its Biomedical Applications: A Review. *Micromachines* **2019**, *10*, 676. [CrossRef]
22. Clapp, N.; Amour, A.; Rowan, W.C.; Candarlioglu, P.L. Organ-on-chip applications in drug discovery: An end user perspective. *Biochem. Soc. Trans.* **2021**, *49*, 1881–1890. [CrossRef] [PubMed]
23. Moradi, E.; Jalili-Firoozinezhad, S.; Solati-Hashjin, M. Microfluidic organ-on-a-chip models of human liver tissue. *Acta Biomater.* **2020**, *116*, 67–83. [CrossRef] [PubMed]
24. Lin, C.; Khetani, S.R. Advances in Engineered Liver Models for Investigating Drug-Induced Liver Injury. *Biomed. Res. Int.* **2016**, *2016*, 1829148. [CrossRef] [PubMed]
25. Sacchi, M.; Bansal, R.; Rouwkema, J. Bioengineered 3D Models to Recapitulate Tissue Fibrosis. *Trends Biotechnol.* **2020**, *38*, 623–636. [CrossRef] [PubMed]
26. Van Grunsven, L.A. 3D in vitro models of liver fibrosis. *Adv. Drug Deliv. Rev.* **2017**, *121*, 133–146. [CrossRef] [PubMed]
27. Hayward, K.L.; Kouthouridis, S.; Zhang, B. Organ-on-a-Chip Systems for Modeling Pathological Tissue Morphogenesis Associated with Fibrosis and Cancer. *ACS Biomater. Sci. Eng.* **2021**, *7*, 2900–2925. [CrossRef]
28. Huh, D.; Hamilton, G.A.; Ingber, D.E. From 3D cell culture to organs-on-chips. *Trends Cell. Biol.* **2011**, *21*, 745–754. [CrossRef]
29. Antoni, D.; Burckel, H.; Josset, E.; Noel, G. Three-dimensional cell culture: A breakthrough in vivo. *Int. J. Mol. Sci.* **2015**, *16*, 5517–5527. [CrossRef]
30. Xinaris, C.; Brizi, V.; Remuzzi, G. Organoid Models and Applications in Biomedical Research. *Nephron* **2015**, *130*, 191–199. [CrossRef]
31. Bhatia, S.N.; Ingber, D.E. Microfluidic organs-on-chips. *Nat. Biotechnol.* **2014**, *32*, 760–772. [CrossRef]
32. Velasco, V.; Shariati, S.A.; Esfandyarpour, R. Microtechnology-based methods for organoid models. *Microsyst. Nanoeng.* **2020**, *6*, 76. [CrossRef] [PubMed]
33. Joseph, J.S.; Malindisa, S.T.; Ntwasa, M. Two-dimensional (2D) and three-dimensional (3D) cell culturing in drug discovery. In *Cell Culture*; IntechOpen: London, UK, 2019.
34. Langhans, S.A. Three-Dimensional in Vitro Cell Culture Models in Drug Discovery and Drug Repositioning. *Front. Pharm.* **2018**, *9*, 6. [CrossRef] [PubMed]
35. Ho, W.J.; Pham, E.A.; Kim, J.W.; Ng, C.W.; Kim, J.H.; Kamei, D.T.; Wu, B.M. Incorporation of multicellular spheroids into 3-D polymeric scaffolds provides an improved tumor model for screening anticancer drugs. *Cancer Sci.* **2010**, *101*, 2637–2643. [CrossRef] [PubMed]
36. Białkowska, K.; Komorowski, P.; Bryszewska, M.; Miłowska, K. Spheroids as a Type of Three-Dimensional Cell Cultures-Examples of Methods of Preparation and the Most Important Application. *Int. J. Mol. Sci.* **2020**, *21*, 6225. [CrossRef]

37. Gunti, S.; Hoke, A.T.K.; Vu, K.; London, N.R., Jr. Organoid and Spheroid Tumor Models: Techniques and Applications. *Cancers* **2021**, *13*, 874. [CrossRef]
38. Kim, S.K.; Kim, Y.H.; Park, S.; Cho, S.-W. Organoid engineering with microfluidics and biomaterials for liver, lung disease, and cancer modeling. *Acta Biomater.* **2021**, *132*, 37–51. [CrossRef]
39. Kulsharova, G.; Kurmangaliyeva, A.; Darbayeva, E.; Rojas-Solórzano, L.; Toxeitova, G. Development of a Hybrid Polymer-Based Microfluidic Platform for Culturing Hepatocytes towards Liver-on-a-Chip Applications. *Polymers* **2021**, *13*, 3215. [CrossRef]
40. Beckwitt, C.H.; Clark, A.M.; Wheeler, S.; Taylor, D.L.; Stolz, D.B.; Griffith, L.; Wells, A. Liver 'organ on a chip'. *Exp. Cell Res.* **2018**, *363*, 15–25. [CrossRef]
41. Zeilinger, K.; Freyer, N.; Damm, G.; Seehofer, D.; Knöspel, F. Cell sources for in vitro human liver cell culture models. *Exp. Biol. Med.* **2016**, *241*, 1684–1698. [CrossRef]
42. Zhang, B.; Korolj, A.; Lai, B.F.L.; Radisic, M. Advances in organ-on-a-chip engineering. *Nat. Rev. Mater.* **2018**, *3*, 257–278. [CrossRef]
43. Korin, N.; Bransky, A.; Dinnar, U.; Levenberg, S. A parametric study of human fibroblasts culture in a microchannel bioreactor. *Lab Chip* **2007**, *7*, 611–617. [CrossRef] [PubMed]
44. Mehta, G.; Mehta, K.; Sud, D.; Song, J.; Bersano-Begey, T.; Futai, N.; Heo, Y.S.; Mycek, M.-A.; Linderman, J.J.; Takayama, S. Quantitative measurement and control of oxygen levels in microfluidic poly(dimethylsiloxane) bioreactors during cell culture. *Biomed. Microdevices* **2007**, *9*, 123–134. [CrossRef] [PubMed]
45. Kaarj, K.; Yoon, J.Y. Methods of Delivering Mechanical Stimuli to Organ-on-a-Chip. *Micromachines* **2019**, *10*, 700. [CrossRef] [PubMed]
46. Goldstein, Y.; Spitz, S.; Turjeman, K.; Selinger, F.; Barenholz, Y.; Ertl, P.; Benny, O.; Bavli, D. Breaking the Third Wall: Implementing 3D-Printing Technics to Expand the Complexity and Abilities of Multi-Organ-on-a-Chip Devices. *Micromachines* **2021**, *12*, 627. [CrossRef]
47. Tilles, A.W.; Baskaran, H.; Roy, P.; Yarmush, M.L.; Toner, M. Effects of oxygenation and flow on the viability and function of rat hepatocytes cocultured in a microchannel flat-plate bioreactor. *Biotechnol. Bioeng.* **2001**, *73*, 379–389. [CrossRef]
48. Delon, L.; Guo, Z.; Oszmiana, A.; Chien, C.-C.; Gibson, R.; Prestidge, C.; Thierry, B. A systematic investigation of the effect of the fluid shear stress on Caco-2cells towards the optimization of epithelial organ-on-chip models. *Biomaterials* **2019**, *225*, 119521. [CrossRef]
49. Rennert, K.; Steinborn, S.; Gröger, M.; Ungerböck, B.; Jank, A.-M.; Ehgartner, J.; Nietzsche, S.; Dinger, J.; Kiehntopf, M.; Funke, H.; et al. A microfluidically perfused three dimensional human liver model. *Biomaterials* **2015**, *71*, 119–131. [CrossRef]
50. Griffith, L.G.; Swartz, M.A. Capturing complex 3D tissue physiology in vitro. *Nat. Rev. Mol. Cell Biol.* **2006**, *7*, 211–224. [CrossRef]
51. Zeilinger, K.; Schreiter, T.; Darnell, M.; Söderdahl, T.; Lübberstedt, M.; Dillner, B.; Knobeloch, D.; Nüssler, A.K.; Gerlach, J.C.; Andersson, T.B. Scaling down of a clinical three-dimensional perfusion multicompartment hollow fiber liver bioreactor developed for extracorporeal liver support to an analytical scale device useful for hepatic pharmacological in vitro studies. *Tissue Eng. Part C Methods* **2011**, *17*, 549–556. [CrossRef]
52. Tajeddin, A.; Mustafaoglu, N. Design and Fabrication of Organ-on-Chips: Promises and Challenges. *Micromachines* **2021**, *12*, 1443. [CrossRef]
53. Benedict, M.; Zhang, X. Non-alcoholic fatty liver disease: An expanded review. *World J. Hepatol.* **2017**, *9*, 715–732. [CrossRef] [PubMed]
54. Teng, Y.; Zhao, Z.; Tasnim, F.; Huang, X.; Yu, H. A scalable and sensitive steatosis chip with long-term perfusion of in situ differentiated HepaRG organoids. *Biomaterials* **2021**, *275*, 120904. [CrossRef] [PubMed]
55. Freag, M.S.; Namgung, B.; Fernandez, M.E.R.; Gherardi, E.; Sengupta, S.; Jang, H.L. Human Nonalcoholic Steatohepatitis on a Chip. *Hepatol. Commun.* **2021**, *5*, 217–233. [CrossRef] [PubMed]
56. Wang, Y.; Wang, H.; Deng, P.; Tao, T.; Liu, H.; Wu, S.; Chen, W.; Qin, J. Modeling Human Nonalcoholic Fatty Liver Disease (NAFLD) with an Organoids-on-a-Chip System. *ACS Biomater. Sci. Eng.* **2020**, *6*, 5734–5743. [CrossRef]
57. Du, K.; Li, S.; Li, C.; Li, P.; Miao, C.; Luo, T.; Qiu, B.; Ding, W. Modeling nonalcoholic fatty liver disease on a liver lobule chip with dual blood supply. *Acta Biomater.* **2021**, *134*, 228–239. [CrossRef]
58. Gori, M.; Simonelli, M.C.; Giannitelli, S.M.; Businaro, L.; Trombetta, M.; Rainer, A. Investigating Nonalcoholic Fatty Liver Disease in a Liver-on-a-Chip Microfluidic Device. *PLoS ONE* **2016**, *11*, e0159729. [CrossRef]
59. Bulutoglu, B.; Rey-Bedón, C.; Kang, Y.B.; Mert, S.; Yarmush, M.L.; Usta, O.B. A microfluidic patterned model of non-alcoholic fatty liver disease: Applications to disease progression and zonation. *Lab Chip* **2019**, *19*, 3022–3031. [CrossRef]
60. Slaughter, V.L.; Rumsey, J.W.; Boone, R.; Malik, D.; Cai, Y.; Sriram, N.N.; Long, C.J.; McAleer, C.W.; Lambert, S.; Shuler, M.L.; et al. Validation of an adipose-liver human-on-a-chip model of NAFLD for preclinical therapeutic efficacy evaluation. *Sci. Rep.* **2021**, *11*, 13159. [CrossRef]
61. Lasli, S.; Kim, H.J.; Lee, K.; Suurmond, C.A.E.; Goudie, M.; Bandaru, P.; Sun, W.; Zhang, S.; Zhang, N.; Ahadian, S.; et al. A Human Liver-on-a-Chip Platform for Modeling Nonalcoholic Fatty Liver Disease. *Adv. Biosyst.* **2019**, *3*, e1900104. [CrossRef]
62. Suurmond, C.A.E.; Lasli, S.; van den Dolder, F.W.; Ung, A.; Kim, H.J.; Bandaru, P.; Lee, K.; Cho, H.J.; Ahadian, S.; Ashammakhi, N.; et al. In Vitro Human Liver Model of Nonalcoholic Steatohepatitis by Coculturing Hepatocytes, Endothelial Cells, and Kupffer Cells. *Adv. Healthc. Mater.* **2019**, *8*, e1901379. [CrossRef]

63. Kostrzewski, T.; Snow, S.; Battle, A.L.; Peel, S.; Ahmad, Z.; Basak, J.; Surakala, M.; Bornot, A.; Lindgren, J.; Ryaboshapkina, M.; et al. Modelling human liver fibrosis in the context of non-alcoholic steatohepatitis using a microphysiological system. *Commun. Biol.* **2021**, *4*, 1080. [CrossRef]

64. Saydmohammed, M.; Jha, A.; Mahajan, V.; Gavlock, D.; Shun, T.Y.; DeBiasio, R.; Lefever, D.; Li, X.; Reese, C.; E Kershaw, E.; et al. Quantifying the progression of non-alcoholic fatty liver disease in human biomimetic liver microphysiology systems with fluorescent protein biosensors. *Exp. Biol. Med.* **2021**, *246*, 2420–2441. [CrossRef]

65. Osna, N.A.; Donohue, T.M.; Kharbanda, K.K. Alcoholic liver disease: Pathogenesis and current management. *Alcohol Res. Curr. Rev.* **2017**, *38*, 147–161.

66. Toshikuni, N.; Tsutsumi, M.; Arisawa, T. Clinical differences between alcoholic liver disease and nonalcoholic fatty liver disease. *World J. Gastroenterol.* **2014**, *20*, 8393–8406. [CrossRef]

67. Nawroth, J.C.; Petropolis, D.B.; Manatakis, D.V.; Maulana, T.I.; Burchett, G.; Schlünder, K.; Witt, A.; Shukla, A.; Kodella, K.; Ronxhi, J.; et al. Modeling alcohol-associated liver disease in a human Liver-Chip. *Cell Rep.* **2021**, *36*, 109393. [CrossRef]

68. Lee, J.; Choi, B.; No, D.Y.; Lee, G.; Lee, S.-R.; Oh, H.; Lee, S.-H. A 3D alcoholic liver disease model on a chip. *Integr. Biol.* **2016**, *8*, 302–308. [CrossRef]

69. Zhou, Q.; Patel, D.; Kwa, T.; Haque, A.; Matharu, Z.; Stybayeva, G.; Gao, Y.; Diehl, A.M.; Revzin, A. Liver injury-on-a-chip: Microfluidic co-cultures with integrated biosensors for monitoring liver cell signaling during injury. *Lab Chip* **2015**, *15*, 4467–4478. [CrossRef]

70. Deng, J.; Chen, Z.; Zhang, X.; Luo, Y.; Wu, Z.; Lu, Y.; Liu, T.; Zhao, W.; Lin, B. A liver-chip-based alcoholic liver disease model featuring multi-non-parenchymal cells. *Biomed. Microdevices* **2019**, *21*, 57. [CrossRef]

71. Castaneda, D.; Gonzalez, A.J.; Alomari, M.; Tandon, K.; Zervos, X.B. From hepatitis A to E: A critical review of viral hepatitis. *World J. Gastroenterol.* **2021**, *27*, 1691. [CrossRef]

72. Barry, A.; Baldeosingh, R.; Lamm, R.; Patel, K.; Zhang, K.; Dominguez, D.A.; Kirton, K.J.; Shah, A.P.; Dang, H. Hepatic Stellate Cells and Hepatocarcinogenesis. *Front. Cell Dev. Biol.* **2020**, *8*, 709. [CrossRef]

73. Sodunke, T.R.; Bouchard, M.J.; Noh, H. Microfluidic platform for hepatitis B viral replication study. *Biomed. Microdevices* **2008**, *10*, 393–402. [CrossRef]

74. Kang, Y.B.; Rawat, S.; Duchemin, N.; Bouchard, M.; Noh, M. Human Liver Sinusoid on a Chip for Hepatitis B Virus Replication Study. *Micromachines* **2017**, *8*, 27. [CrossRef]

75. Ortega-Prieto, A.M.; Skelton, J.K.; Cherry, C.; Briones-Orta, M.A.; Hateley, C.A.; Dorner, M. "Liver-on-a-Chip" Cultures of Primary Hepatocytes and Kupffer Cells for Hepatitis B Virus Infection. *J. Vis. Exp.* **2019**, *144*, e58333. [CrossRef]

76. Ortega-Prieto, A.M.; Skelton, J.K.; Wai, S.N.; Large, E.; Lussignol, M.; Vizcay-Barrena, G.; Hughes, D.; Fleck, R.; Thursz, M.; Catanese, M.T.; et al. 3D microfluidic liver cultures as a physiological preclinical tool for hepatitis B virus infection. *Nat. Commun.* **2018**, *9*, 682. [CrossRef]

77. Gu, X.; Manautou, J.E. Molecular mechanisms underlying chemical liver injury. *Expert. Rev. Mol. Med.* **2012**, *14*, e4. [CrossRef]

78. Aydin, M.M.; Akcali, K.C. Liver fibrosis. *Turk. J. Gastroenterol.* **2018**, *29*, 14–21. [CrossRef]

79. Bataller, R.; Brenner, D.A. Liver fibrosis. *J. Clin. Investig.* **2005**, *115*, 209–218. [CrossRef]

80. Deng, J.; Zhang, X.; Chen, Z.; Luo, Y.; Lu, Y.; Liu, T.; Wu, Z.; Jin, Y.; Zhao, W.; Lin, B. A cell lines derived microfluidic liver model for investigation of hepatotoxicity induced by drug-drug interaction. *Biomicrofluidics* **2019**, *13*, 024101. [CrossRef]

81. Wang, Y.; Wang, H.; Deng, P.; Chen, W.; Guo, Y.; Tao, T.; Qin, J. In situ differentiation and generation of functional liver organoids from human iPSCs in a 3D perfusable chip system. *Lab Chip* **2018**, *18*, 3606–3616. [CrossRef]

82. Tian, T.; Chen, C.; Sun, H.; Hui, J.; Ge, Y.; Liu, T.; Mao, H. A 3D bio-printed spheroids based perfusion in vitro liver on chip for drug toxicity assays. *Chin. Chem. Lett.* **2021**. [CrossRef]

83. Bavli, D.; Prill, S.; Ezra, E.; Levy, G.; Cohen, M.; Vinken, M.; Vanfleteren, J.; Jaeger, M.; Nahmias, Y. Real-time monitoring of metabolic function in liver-on-chip microdevices tracks the dynamics of mitochondrial dysfunction. *Proc. Natl. Acad. Sci. USA* **2016**, *113*, E2231–E2240. [CrossRef] [PubMed]

84. Bovard, D.; Sandoz, A.; Luettich, K.; Frentzel, S.; Iskandar, A.; Marescotti, D.; Trivedi, K.; Guedj, E.; Dutertre, Q.; Peitsch, M.C.; et al. A lung/liver-on-a-chip platform for acute and chronic toxicity studies. *Lab Chip* **2018**, *18*, 3814–3829. [CrossRef] [PubMed]

85. Schimek, K.; Frentzel, S.; Luettich, K.; Bovard, D.; Rütschle, I.; Boden, L.; Rambo, F.; Erfurth, H.; Dehne, E.-M.; Winter, A.; et al. Human multi-organ chip co-culture of bronchial lung culture and liver spheroids for substance exposure studies. *Sci. Rep.* **2020**, *10*, 7865. [CrossRef] [PubMed]

86. Skardal, A.; Aleman, J.; Forsythe, S.; Rajan, S.; Murphy, S.; Devarasetty, M.; Zarandi, N.P.; Nzou, G.; Wicks, R.; Sadri-Ardekani, H.; et al. Drug compound screening in single and integrated multi-organoid body-on-a-chip systems. *Biofabrication* **2020**, *12*, 025017. [CrossRef]

87. Jang, K.-J.; Otieno, M.A.; Ronxhi, J.; Lim, H.-K.; Ewart, L.; Kodella, K.R.; Petropolis, D.B.; Kulkarni, G.; Rubins, J.E.; Conegliano, D.; et al. Reproducing human and cross-species drug toxicities using a Liver-Chip. *Sci. Transl. Med.* **2019**, *11*, eaax5516. [CrossRef]

88. Lee, H.; Kim, J.; Choi, Y.; Cho, D.-W. Application of Gelatin Bioinks and Cell-Printing Technology to Enhance Cell Delivery Capability for 3D Liver Fibrosis-on-a-Chip Development. *ACS Biomater. Sci. Eng.* **2020**, *6*, 2469–2477. [CrossRef]

89. Nguyen, D.G.; Funk, J.; Robbins, J.B.; Crogan-Grundy, C.; Presnell, S.C.; Singer, T.; Roth, A.B. Bioprinted 3D Primary Liver Tissues Allow Assessment of Organ-Level Response to Clinical Drug Induced Toxicity In Vitro. *PLoS ONE* **2016**, *11*, e0158674. [CrossRef]

90. Norona, L.M.; Nguyen, D.G.; Gerber, D.A.; Presnell, S.C.; LeCluyse, E.L. Modeling Compound-Induced Fibrogenesis In Vitro Using Three-Dimensional Bioprinted Human Liver Tissues. *Toxicol. Sci. Off. J. Soc. Toxicol.* **2016**, *154*, 354–367. [CrossRef]
91. Ishida, S. Requirements for designing organ-on-a-chip platforms to model the pathogenesis of liver disease. In *Organ-on-a-Chip*; Elsevier: Amsterdam, The Netherlands, 2020; pp. 181–213.
92. Zuo, R.; Li, F.; Parikh, S.; Cao, L.; Cooper, K.; Hong, Y. Evaluation of a Novel Renewable Hepatic Cell Model for Prediction of Clinical CYP3A4 Induction Using a Correlation-Based Relative Induction Score Approach. *Drug Metab. Dispos.* **2017**, *45*, 198–207. [CrossRef]
93. Lee, S.A.; Kang, E.; Ju, J.; Kim, D.S.; Lee, S.H. Spheroid-based three-dimensional liver-on-a-chip to investigate hepatocyte-hepatic stellate cell interactions and flow effects. *Lab Chip* **2013**, *13*, 3529–3537. [CrossRef]
94. Bhise, N.S.; Manoharan, V.; Massa, S.; Tamayol, A.; Ghaderi, M.; Miscuglio, M.; Lang, Q.; Zhang, Y.S.; Shin, S.R.; Calzone, G.; et al. A liver-on-a-chip platform with bioprinted hepatic spheroids. *Biofabrication* **2016**, *8*, 014101. [CrossRef] [PubMed]
95. Wagner, I.; Materne, E.-M.; Brincker, S.; Süßbier, U.; Frädrich, C.; Busek, M.; Sonntag, F.; Sakharov, D.A.; Trushkin, E.V.; Tonevitsky, A.G.; et al. A dynamic multi-organ-chip for long-term cultivation and substance testing proven by 3D human liver and skin tissue co-culture. *Lab Chip* **2013**, *13*, 3538–3547. [CrossRef] [PubMed]
96. Prodanov, L.; Jindal, R.; Bale, S.S.; Hegde, M.; McCarty, W.; Golberg, I.; Bhushan, A.; Yarmush, M.L.; Usta, O.B. Long-term maintenance of a microfluidic 3D human liver sinusoid. *Biotechnol. Bioeng.* **2016**, *113*, 241–246. [CrossRef] [PubMed]
97. Torino, S.; Corrado, B.; Iodice, M.; Coppola, G. PDMS-Based Microfluidic Devices for Cell Culture. *Inventions* **2018**, *3*, 65. [CrossRef]
98. Campbell, S.B.; Wu, Q.; Yazbeck, J.; Liu, C.; Okhovatian, S.; Radisic, M. Beyond Polydimethylsiloxane: Alternative Materials for Fabrication of Organ-on-a-Chip Devices and Microphysiological Systems. *ACS Biomater. Sci. Eng.* **2021**, *7*, 2880–2899. [CrossRef]
99. Khetani, S.R.; Bhatia, S.N. Microscale culture of human liver cells for drug development. *Nat. Biotechnol.* **2008**, *26*, 120–126. [CrossRef]
100. Farooqi, H.; Kang, B.; Khalid, M.; Salih, A.; Hyun, K.; Park, S. Real-time monitoring of liver fibrosis through embedded sensors in a microphysiological system. *Nano Converg.* **2021**, *8*, 1–12. [CrossRef]
101. de Hoyos-Vega, J.M.; Hong, H.J.; Stybayeva, G.; Revzin, A. Hepatocyte cultures: From collagen gel sandwiches to microfluidic devices with integrated biosensors. *APL Bioeng.* **2021**, *5*, 041504. [CrossRef]
102. Choe, A.; Ha, S.K.; Choi, I.; Choi, N.; Sung, J.H. Microfluidic Gut-liver chip for reproducing the first pass metabolism. *Biomed. Microdevices* **2017**, *19*, 4. [CrossRef]
103. Esch, M.B.; Mahler, G.J.; Stokol, T.; Shuler, M.L. Body-on-a-chip simulation with gastrointestinal tract and liver tissues suggests that ingested nanoparticles have the potential to cause liver injury. *Lab Chip* **2014**, *14*, 3081–3092. [CrossRef]
104. Maschmeyer, I.; Lorenz, A.K.; Schimek, K.; Hasenberg, T.; Ramme, A.P.; Hübner, J.; Lindner, M.; Drewell, C.; Bauer, S.; Thomas, A.; et al. A four-organ-chip for interconnected long-term co-culture of human intestine, liver, skin and kidney equivalents. *Lab Chip* **2015**, *15*, 2688–2699. [CrossRef]
105. Oleaga, C.; Bernabini, C.; Smith, A.S.; Srinivasan, B.; Jackson, M.; McLamb, W.; Platt, V.; Bridges, R.; Cai, Y.; Santhanam, N.; et al. Multi-Organ toxicity demonstration in a functional human in vitro system composed of four organs. *Sci. Rep.* **2016**, *6*, 20030. [CrossRef]
106. Ewart, L.; Apostolou, A.; Briggs, S.A.; Carman, C.V.; Chaff, J.T.; Heng, A.R.; Jadalannagari, S.; Janardhanan, J.; Jang, K.J.; Joshipura, S.R.; et al. Qualifying a human Liver-Chip for predictive toxicology: Performance assessment and economic implications. *bioRxiv* **2021**. [CrossRef]
107. Starokozhko, V.; Groothuis, G.M. Judging the value of 'liver-on-a-chip' devices for prediction of toxicity. *Expert Opin. Drug Metab. Toxicol.* **2017**, *13*, 125–128. [CrossRef]
108. Sung, J.H.; Kam, C.; Shuler, M.L. A microfluidic device for a pharmacokinetic-pharmacodynamic (PK-PD) model on a chip. *Lab Chip* **2010**, *10*, 446–455. [CrossRef]

*Article*

# Single Red Blood Cell Hydrodynamic Traps via the Generative Design

Georgii V. Grigorev [1,2], Nikolay O. Nikitin [3], Alexander Hvatov [3], Anna V. Kalyuzhnaya [3], Alexander V. Lebedev [4], Xiaohao Wang [5,*], Xiang Qian [5,*], Georgii V. Maksimov [6,7] and Liwei Lin [2]

1   Data Science and Information Technology Research Center, Tsinghua Berkeley Shenzhen Institute, Tsinghua University, Shenzhen 518055, China; georgii@berkeley.edu
2   Mechanical Department, University of California in Berkeley, Berkeley, CA 94703, USA; lwlin@berkeley.edu
3   NSS Lab, ITMO University, 197101 Saint-Petersburg, Russia; nnikitin@itmo.ru (N.O.N.); alex_hvatov@itmo.ru (A.H.); anna.kalyuzhnaya@itmo.ru (A.V.K.)
4   Machine Building Department, Bauman Moscow State Technical University, 105005 Moscow, Russia; mr.aleksanderl@yandex.ru
5   Shenzhen International Graduate School, Tsinghua University, Shenzhen 518055, China
6   Biophysics Lab, Biology Department, Moscow State University, 119192 Moscow, Russia; gmaksimov@mail.ru
7   Physical metallurgy Department, Federal State Autonomous Educational Institution of Higher Education, National Research Technological University, MISiS, 119049 Moscow, Russia
*   Correspondence: wang.xiaohao@sz.tsinghua.edu.cn (X.W.); qian.xiang@sz.tsinghua.edu.cn (X.Q.); Tel.: +86-136-001-864-22 (X.W.); +86-755-2603-67-55 (X.Q.)

Citation: Grigorev, G.V.; Nikitin, N.O.; Hvatov, A.; Kalyuzhnaya, A.V.; Lebedev, A.V.; Wang, X.; Qian, X.; Maksimov, G.V.; Lin, L. Single Red Blood Cell Hydrodynamic Traps via the Generative Design. *Micromachines* 2022, 13, 367. https://doi.org/10.3390/mi13030367

Academic Editors: Violeta Carvalho, Senhorinha Teixeira and João Eduardo P. Castro Ribeiro

Received: 12 January 2022
Accepted: 26 January 2022
Published: 26 February 2022

Publisher's Note: MDPI stays neutral with regard to jurisdictional claims in published maps and institutional affiliations.

**Abstract:** This paper describes a generative design methodology for a micro hydrodynamic single-RBC (red blood cell) trap for applications in microfluidics-based single-cell analysis. One key challenge in single-cell microfluidic traps is to achieve desired through-slit flowrates to trap cells under implicit constraints. In this work, the cell-trapping design with validation from experimental data has been developed by the generative design methodology with an evolutionary algorithm. L-shaped trapping slits have been generated iteratively for the optimal geometries to trap living-cells suspended in flow channels. Without using the generative design, the slits have low flow velocities incapable of trapping single cells. After a search with 30,000 solutions, the optimized geometry was found to increase the through-slit velocities by 49%. Fabricated and experimentally tested prototypes have achieved 4 out of 4 trapping efficiency of RBCs. This evolutionary algorithm and trapping design can be applied to cells of various sizes.

**Keywords:** microfluidics; cell trap; RBC; evolutionary algorithm; generative design; artificial intelligence

## 1. Introduction

Microfluidic devices are indispensable for studying behaviors of single living cells, such as cytological, mechanical, and electrical responses for potential applications ranging from early disease diagnosis to drug testing. To study single-cell activities, microfluidics trapping systems are important platforms; several approaches have been reported previously, such as acoustic [1,2], dielectrophoretic [3,4], hydrodynamic [5–8], magnetic [9,10], and optical trapping schemes [11,12]. Hydrodynamic trapping of single cells happens within a microfluidic channel, where the channels' geometries and flowrates require careful study and optimization. Hydrodynamic trapping utilizes mechanical barriers or arrays for the separation of target particles from the main flow. Separated particles are retained within the hydrodynamic traps for further analyses by employing various principles, such as vortices-based trapping (centrifugation assisted, cavitation microstreaming, hydrodynamic tweezers), cross streamed (viscoelastic focusing, inertial migration, dean flow and deformability selective cell separation), and external controlled approaches (pneumatic valves, PID controllers, eddy currents, electro-magnetic fields, acoustics) [6]. The advantages of hydrodynamic processing are the ease to implement the inertial focusing of enhanced cell separation and sorting with narrowed sheathed flows. As a disadvantage, the hydrodynamic single cell platform may produce stress on cell samples, and is reported to alter molecular mechanisms, and inhomogeneity issues [13–16]. The geometric design of hydrodynamic single cell trapping belongs to the category of "wet fluid-structure interaction (FSI)"; a standard approach is the topology optimization through gradient-based methods [17,18].

In this work, we approached the FSI design problem by using the evolutionary algorithm. It allowed us to extend the 'classical' statement of the topology optimization and explore more comprehensive designs with extended variability [19]. In general, the topology generative design has no a priori assumptions on the form of the initial design, i.e., optimization starts 'from scratch'. The idea of evolutionary generative design allows the algorithm to extend the possible design from the creation and improvement of the digital twins of real-world objects [20]. Researchers have also successfully applied evolutionary algorithms and other AI methods in different areas, including studies for coastal structures [21], mathematical models [22], architecture [23], and drug designs [24]. However, the adaptation of this approach to the hydrodynamic cell trappings has specific factors that should be taken into consideration. As an example, we have developed task-specific evolutionary operators, validation rules, objective functions, and post-processing procedures in this work. In addition, the parallelization of computations has been implemented due to the high computational cost of the hydrodynamic simulations. Here, the micro hydrodynamic traps were designed with a unique feature to trap RBCs (red blood cells) in channels while keeping it suspended to allow fluidic flows. The trapping chamber was designed with a specific implementation evolutionary algorithm. The prototype devices have been fabricated and tested using frog RBCs as the living cells for validations.

## 2. Materials and Methods

### 2.1. Single Cell Traps

The goal of this work is to create single-cell traps by an evolutionary-based generative design. Figure 1a shows the design principle for the single RBC trapping scheme and Figure 1b is an example for the chamber design with the trapping structure of the zero trapping chance due to the low through-trap flow. Figure 1c demonstrates the result from the evolutionary algorithm, which generated the necessary flow obstacles to obtain the high trapping probability. The COMSOL simulation results are displayed in Figure 1d, showing the velocity gradient in the final design with inflow velocity of 0.01 m/s and the no slip boundary condition. Figure 1e shows the flows schematics in the trapping chamber and Figure 1f presents the experimental results of the prototype system following the evolutionary algorithm design to trap RBCs.

The properties and constraints of the microfluidic single-RBC traps are targeted to have 4 erythrocytes trapped within one FOV (field of vision) while keeping the flow streams around RBCs. This is different to erythrocytes trapped in cavity, pocket, and well-like structures with little or no fluid flows [8,25,26]. The basic assumptions include: a single phase Navier–Stokes flow, steady-state, and no-slip boundary condition. The rheological parameters used in this work include: the kinematic viscosity of 3.3 mm$^2$/s; dynamic viscosity of 0.0035 kg/ms; and the fluid density of 1060 kg/m$^3$ (blood density). PDMS (polydimethylsiloxane) was chosen as the material with built-in properties in the COMSOL Multiphysics library. For all the simulations in COMSOL Multiphysics, the following built-in meshing parameters were used. We set sequence type as physics-controlled mesh and element size as finer, respectively.

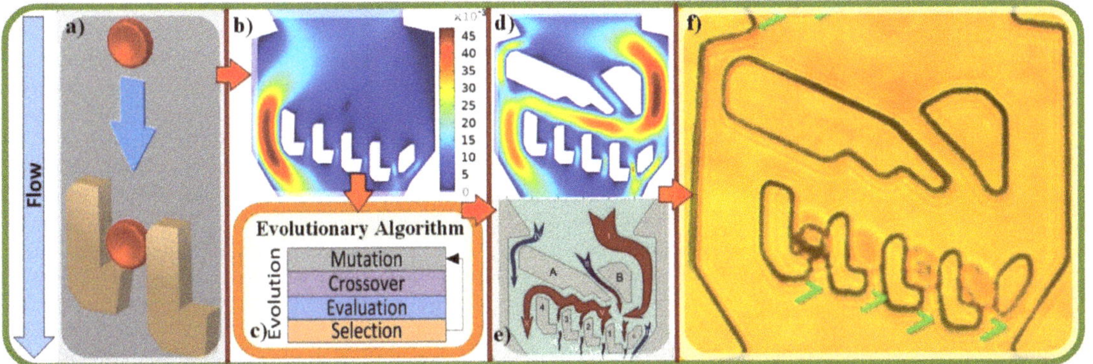



**Figure 1.** The overall device design sequence. Blue arrow indicates the flow direction; (**a**) the single RBC trapping principle; (**b**) an example of the chamber with trapping structure deign with zero trapping chance due to the low through-trap flow rate; color bar corresponds to the flow velocity distribution (**c**) the evolutionary algorithm which generated necessary flow obstacles to obtain high RBC trapping chances; (**d**) COMSOL simulation of the velocity gradient in the final geometry of the trapping chamber (inflow velocity = 0.01 m/s, no slip boundary condition); (**e**) flow schematics in the trapping chamber; (**f**) experimental results of the prototype system following the evolutionary algorithm design to trap RBCs; green "V"s mark the successfully trapped cells in each slot.

Frog RBCs were chosen as the trapping object based on their good availability and bigger nucleated RBCs were chosen for good trapping visualizations. To implicitly account for the number of cells trapped, three physical parameters were calculated for each design. The *CRV* and *CRL* parameters were used as the constraints to avoid the cases where the cells are destroyed by the flow. The *CRV* (curvature of rotated vector fields) reflects the flow curvature level. It takes the cross-section of a channel flow and cuts it into multiple pieces with probed values for each piece, and then performs a summation of all velocities from all the pieces of the cross-section as:

$$CRV = \iint_{\Omega} abs\left[\frac{u^2 * \frac{\partial v}{\partial x} - v^2 \frac{\partial u}{\partial y} + uv * \left(\frac{\partial v}{\partial y} - \frac{\partial u}{\partial x}\right)}{(u^2 + v^2)^{3/2}}\right] dxdy \tag{1}$$

where $u$ is the horizontal velocity and $v$ is the vertical velocity. Second, the *CRL* reflects rapid changes in the flow channel and calculates the maximum of velocity and velocity gradient in the designated area. It takes the sum of the maximum values of the y derivative of $u$ and $x$ derivative of $v$ as:

$$CRL = \max_{\Omega}\left(\frac{\partial u}{\partial y}\right) + \max_{\Omega}\left(\frac{\partial v}{\partial x}\right) \tag{2}$$

Third, *TVR* is used as an objective function and maximized during the optimization process. *TVR* is the flow ratio in trapping slits and those of non-trapping channels to ensure the cells can be trapped as the objective of topology optimization:

$$TVR = \frac{\sum_{k=1}^{4} v_k}{v_{PD} + v_{main}} \tag{3}$$

where $v_k$ is flow velocity (m/s) in the trapping slit, $k = 1, 2, 3, 4$; $v_{PD}$ is the flow velocity in the pressure dropping channel, and $v_{main}$ is the flow velocity in the main output channel.

Designs from Figure 2a to Figure 2b has improved efficiency by increasing the number of trapping slits to 3 with the new L-shape structure where the opening width of the slit is close to the width of the cells. Furthermore, the slits were placed at 90-degrees against the flow direction with an increased height, as shown to allow cell trappings. The flow simulation results show poor flow gradient to trap cells. Figure 2c shows the L-shape traps placed in the same direction of the flow to trap cells, while the flow velocities in the slits are nearly zero, due to the high flow resistance. To further improve the cell trapping efficiency, a more advanced version of L-shaped traps with geometric alternations were designed to increase the flow velocity in the slits, as shown in Figure 2d. In the later experimental sections, the typical length and breadth of the nucleated erythrocytes of frog were 19.8 ± 1.5 µm and 8.6 ± 0.3 µm, respectively. As such, the slit width was chosen as 11 µm. Qualitatively, the L-shaped traps possessed the built-in 90-degree microfluidic obstacle to create weak flows for cells to flow into the trap and remain there. The stagnation-point flows and Moffatt-type vortices have been studied in the trapping slits as part of the design considerations.

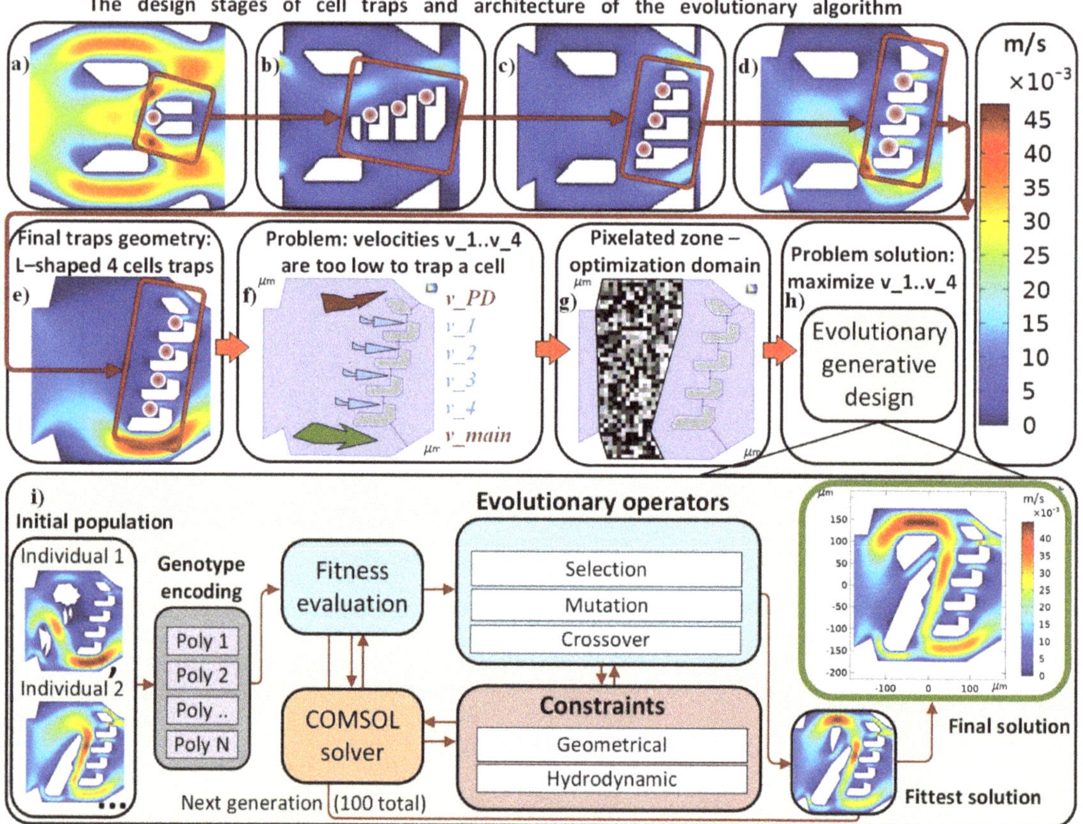


**Figure 2.** The design stages of cell traps and the architecture of the evolutionary algorithm. Flow direction from left to right. Red circles symbolize the RBCs. (**a–e**) are stages of cell trapping structures: each L-shape trap traps one cell; red rectangle highlights the design changes of the trapping slits; (**f**)—problem statement: velocities v_1 … v_4 through slits have near-zero values, and almost zero flux pass through these slits; thus it is incapable of luring a cell by this configuration. Topology optimization is needed (TVR, Equation (2)) to increase the flow velocity in the slits to insure cell trappings; Red, blue, and green colors indicate the flow (**g**)—topology optimization domain (marked pixilated zones); (**h,i**)—architecture of the evolutionary algorithm with the final design of the trapping chamber.

Figure 2e shows the final design for the L-trap with 4 traps within 1 FOV. The top of the chamber has a 21 μm-wide pressure dropping channel (red arrow in Figure 2f), which helps to slow down the flow along the traps, while an erythrocyte can pass through easily. The design also removes the two upstream rectangular trapezoidal shape structures and becomes the initial design for the topology optimization process. Figure 2f shows the specific flow velocity in the trapping slits used in Equation (3). The pixelated zone in Figure 2g is the optimization domain where the evolutionary algorithm was used by placing polygons to optimize flow velocities under the constraints of Equations (1)–(3).

Figure 2h,i are the architecture of the evolutionary algorithm with major components and stages, including initial population examples, evolutionary loops and the final solution which satisfies the objective function and constraints. The optimization pipeline presented in Figure 2i is organized accordingly. First, the initial population was set in a random way. Each design in this population was represented as a list of polygons; the values of fitness for the designs were evaluated using the COMSOL simulator. The evolutionary operators were used to modify the designs. Next, modified designs were used in the selection stage; the solution obtained after the last iteration represented the best design.

### 2.2. Evolutionary Algorithm

The initial trapping slit designs from the previous section have flow velocities of less than 0.009 m/s (simulated with COMSOL) with zero trapping probability. The generative design creates obstacles in the chamber to increase the flow velocities in slits. Our design approach refers to the alternative scheme of the topology optimization—the intelligent field of generative design [22], by using artificial intelligence (AI) and machine learning (ML) methods to create a diversity of variants that have adequate values of the objective function, while satisfying all limitations.

The evolutionary optimization of the cell trapping topology has been implemented using the concept of genetic algorithms with continuous numerical genotype encoding [27]. The pipeline of the algorithm's implementation is presented in Figure 2. As can be seen, the evaluation of the fitness function was based on the COMSOL model. The selection stage was based on the binary tournament algorithm. The pseudocode of the evolutionary algorithm is presented in Algorithm 1. It consists in a procedure for generating initial populations, evolutionary operators, constraint validations, and connectors to the hydrodynamic model. The software implementation of the algorithm has been done in Python 3.8 and is available as a part of the GEFEST framework (https://github.com/ITMO-NSS-team/GEFEST; accessed on 23 December 2021).

The convergence of the evolutionary search for the best topology is presented in Figure 3, where the convergence of the fitness function was calculated by comparing a given design solution to the specified aim for the 100 generations (iterations of evolution). We show that the diversity was successfully preserved even in the late generations of the algorithm. In addition, the optimization was not converging to local minima and the final solution was constantly improving without long stagnation.

The genotype of the designs was represented in the vector form. It consisted of N polygons with $K_i \in [3, \ldots, N_{vert}]$ vertices, where $i = 1, \ldots, N$. The objective function was based on inverted values of the flow ratio. Several custom operators for the mutation, crossover, and initialization of the initial population were implemented and the competition selection was used to preserve diversity. There are different mutations: rotation, rescaling or movement of entire geometric polygons, or single nodes in it. The crossover was implemented at the polygon level to allow for the combination of promising solutions to obtain a more effective one.

---

**Algorithm 1:** *Evolutionary algorithm for cell trap design*

input: *params* = set of hyperparameters for evolutionary algorithm (population size, number of populations, etc)

*constraints* = set of constraints for cell trap

output: Best found cell trap design

▶*Generate random initial population*
*pop*← InitPopulation(*params.pop_size, constraints*)
while not IsFinished(*params.num_pop*) do
*offsprings*← Reproduce(*pop, constraints*)
*pop.fitness*← Fitness(*pop, constraints*)
*pop* ← TournamentSelection(*offsprings*)
return Best(*pop*)

procedure Reproduce:
input: *pop, constraints*

output: *offsprings*

while not Validate(*constraints*)
*modify cell traps designs*
*offsprings*← Crossover (*pop*)
*offsprings*← Mutation (*pop*)
*return offsprings*

procedure Fitness:
input: *pop, constraints*

output: fitness values for each individual

*fitnesses={}*
for *ind* in *pop*:
if Validate(*individual, constraints*)
*run sim for cell trap described in genotype*
*fitnesses[ind]*← COMSOL_Sim(*ind*)
else
*fitnesses[ind]*← 0
*return fitnesses*

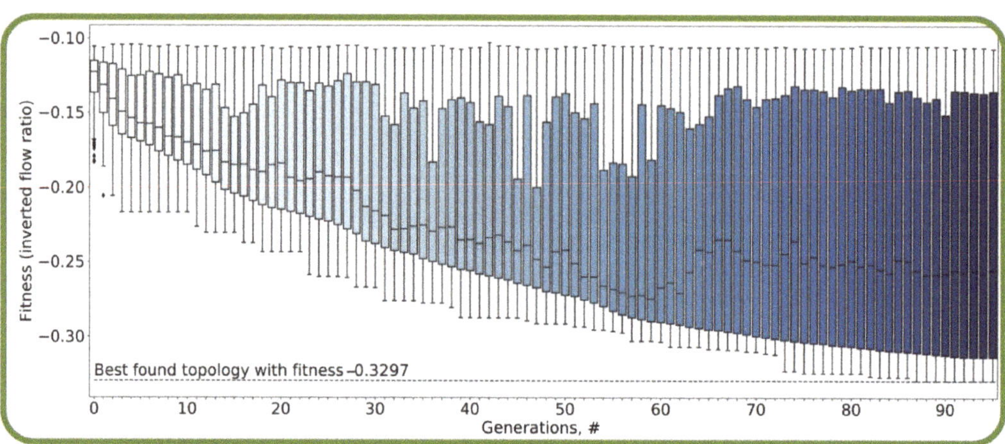


**Figure 3.** The convergence of the fitness function values (the function that is used to estimate how close a given design solution is to the specified aim) during the evolutionary optimization of cell traps for the 100 generations (#, iterations of evolution). The boxplots represent the diversity of the solutions in each population: the centerline of each box represents the median of the fitness distribution in each population, the boundaries of the box—25 and 75 percentiles of the same distribution, and the additional lines, represent the minimum and maximal values of fitness. The solution with the best fitness in all populations is highlighted with the dashed line. It can be seen that the quality of solutions improved steadily during the optimization. The shade of the boxplots' color depends on the number of a generation.

### 2.3. Fabrication

PDMS microfluidics chips were fabricated in a clean room by a typical soft lithography method and were glued to a glass substrate. The photolithography and molding were made with a negative photoresist SU-8 2025 (25 μm height). We used a Sylgard 186 Silicone Elastomer Kit for the PDMS slab fabrication.

### 2.4. Bio-Samples Preparation

Samples of frog blood were purchased from YuanXieShengWu, Shanghai, PRC (LOT: H13J9Q65425,) in accordance with the Tsinghua University (Graduate School in Shenzhen) Ethics Committee.

## 3. Results and Discussion

### 3.1. Optimal Designed Micfluidic Traps

The trapping geometry was designed in Autocad (CAD) and flow patterns were simulated in COMSOL Multiphysics. They are redesigned and re-simulated until all key flow parameters and constraints were satisfied. Two criteria were set: velocities in the slits to be close to 0.015 m/s, and a TVR value of 1.93 (chosen after analyzing the first 100 of the simulations as the minimum values). The final trapping chamber design is shown in Figure 4, where A, B, and C were the flow breaking structures to form main chamber streams; bodies 1, 2, 3, and 4 were L-shaped elements for capturing a single cell and holding it during the analyses of single RBC responses to a solution/object/stimuli; I-IX are the flow streams. IV is the top by-passing collateral flow. II is the local turbulence flow that creates the inertia moment pushing the cells towards the traps. III and IV are the valve-like streams that dump excessive pressure and I is the main inflow (large red arrow). VI, VII, VIII, and IX are weak flows to ensure single cell will flow into the trap and stay there. X is the main flow that passes the traps and meets collateral flow IV and creates the main flow that goes into the next chamber.

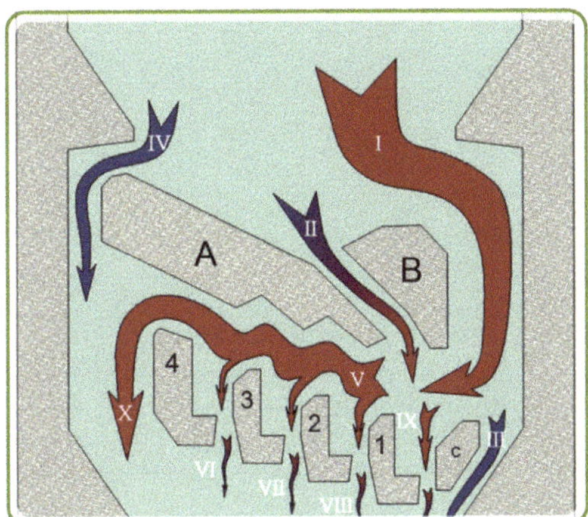


**Figure 4.** The trapping chamber flows schematics of the final design: (**A–C**)—flow breaking structures to form main streams; 1, 2, 3, & 4—L-shaped elements for capturing a single cell and holding it during the analyses of single RBC responses to a solution/object/stimuli; I–IX—flow streams. IV—the top by-passing collateral flow, II—the local turbulence flow that creates inertia moment pushing the cells towards the traps; III and IV—valve-like streams which dump excessive pressure; I—main inflow (red arrow); VI, VII, VIII, IX—weak flows to insure single cell will flow into the trap and stay there; X—main flow that passes traps and meets collateral flow IV and creates the main flow that goes into the next chamber.

To ensure the correctness of the obtained solutions, the geometry-based and flow-based constraints were involved in the optimization. The geometry-based constraints include self-intersection, minimal inter-polygon distance, and other simple checks. The flow-based constraints validate the different parameters of liquid solution in COMSOL. The value of CRL was calculated directly and with an upper limit CRL < 30,000 (chosen after analyzing the first 100 of the simulations as the maximum value). Beyond this value, the flow had multiple turns with sharp angles causing cells to pile up around the pivotal points to block the passage. The value of CVR was calculated directly and with an upper limit CVR < 7 × 10$^7$. The flow constraint 1.22 < TVR < 2 was calculated from the COMSOL liquid velocity magnitude field U (Equation (3)). The values for the constraints were chosen after analyzing the first 100 of the simulations as the minimum/maximum values to achieve the optimal slit through the flow vales of 0.015 m/s (COMSOL integral probe); designs that violate the constraints were rejected during the application of evolutionary operators.

The additional improvement of the computational performance is achieved using the parallelization of implemented operators. The probability of mutation and crossover was selected as 0.6 and 0.4. The size of the population was set as 300 and the maximum number of generations was defined as 100. Hyper parameters of evolutionary algorithm are chosen based on the best practices in the field of the geometrical structures design [28]. The through-slit velocities and parameters for the resulting geometries are listed in Table 1. The initial values of the "empty" (no obstacles) design in Figure 1e of vi, CRL, CVR, and TVR are obtained during simulations. Lines connecting the vertices at the end of the slits served as boundary edges.

**Table 1.** Values of velocities, optimization objectives, and obtained gains for the initial, and optimized solutions.

| Parameter | Units | Initial | Target Values | Optimized | Gain, % |
|---|---|---|---|---|---|
| vl_1 | m/s | 0.012038 | determined | 0.02308 | 92 |
| vl_2 | m/s | 0.009443 | | 0.01579 | 67 |
| vl_3 | m/s | 0.009478 | by | 0.012701 | 34 |
| vl_4 | m/s | 0.009544 | | 0.010092 | 6 |
| vl_PD | m/s | 0.005998 | TVR ratio | 0.012438 | 107 |
| vl_main | m/s | 0.027247 | | 0.019577 | −28 |
| CVR | 1/m | 70,769,000 | <7 × 10$^7$ | 17,113,000 | −76 |
| CRL | 1/s | 12,717 | <30,000 | 20,615 | 62 |
| TVR (target) | - | **1.22** | **1.22 < TVR < 2** | **1.93** | **58** |

The optimized values of Table 1 were obtained in the same way at the same locations but after the optimized obstacles were generated by the evolutionary algorithm and placed before the cell traps. Gain (%) is the percentage increase of vi, CRL, and TVR values compared with optimal values, where corresponding initial values were set as 100%.

### 3.2. Experimental Results

Figure 5 shows the fabricated device designed by the evolutionary algorithm successfully trapping nucleated RBCs within one FOV (objective ×20 with NA 0.45) made with camera Opax A3514DU3 (Opax; Berkeley, CA, USA) and Motic MHG-100B microscope (Motic, Berkeley, CA, USA). Figure 5a–e show the trapping sequence of three living cells. Figure 5a–c,e,a1–d1 show a single RBC approaching and being trapped. Figure 5e1 has the trapping results, where all 4 slots were occupied by a single erythrocyte. Green "V"s mark the successfully trapped cell in each slot. Time elapsed images from Figure 5a to Figure 5e take 11 s and Figure 5a1 to Figure 5e1 take 5 s.

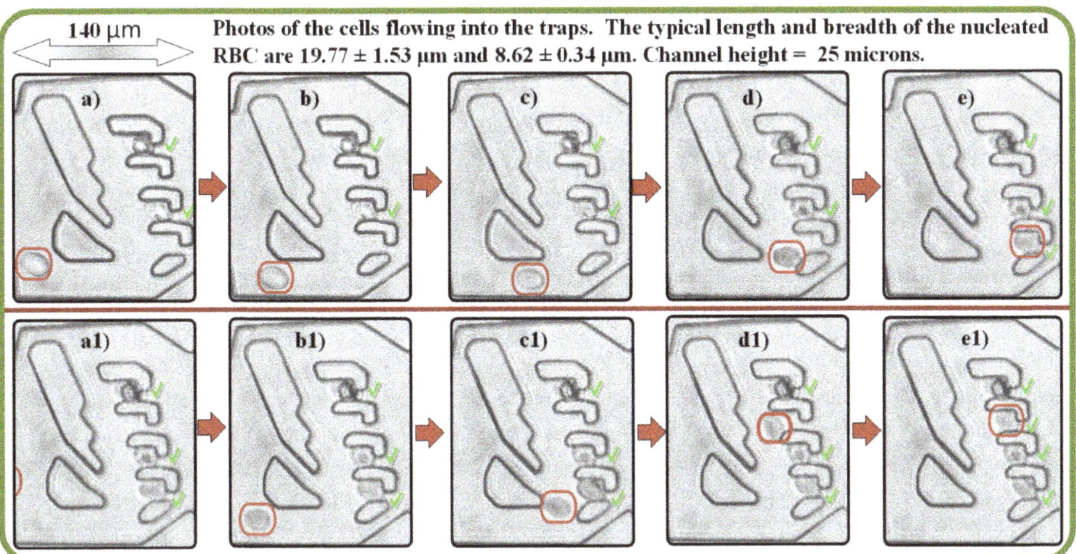



**Figure 5.** Experimental results of the prototype system following the evolutionary algorithm design to trap RBCs. The RBC cell approaching a trap is highlighted with the red oval. (**a–e**) Trapping of three living cells with one empty slot; (**a–d**) and (**a1–d1**) A single RBC is approaching the trap; (**e1**) the trapping is completed with all 4 slots occupied by erythrocytes. Green "V"s mark the successfully trapped cells in each slot. Time elapsed images from (**a**) to (**e**) take 11 s and it takes 3 s from (**a1**) to (**e1**); photos made with objective ×20 with NA 0.45, camera Opax A3514DU3, and Motic MHG-100B microscope.

Figure 6 shows the experimental results of the prototype system following the evolutionary algorithm design to trap RBCs with three consecutive flow chambers. Green "V"s mark the successfully trapped cells.

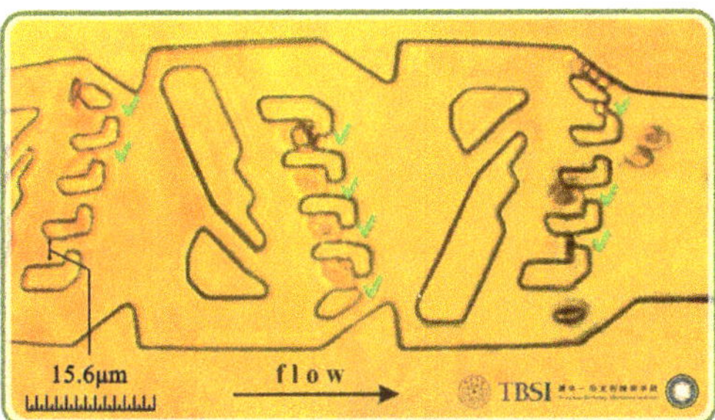



**Figure 6.** Experimental results of the prototype system following the evolutionary algorithm design to trap RBCs. Three consecutive chambers are designed to trap RBCs. Green "V"s mark the successfully trapped cells in each slot.

The experiment was conducted under the following parameters: the width of the chamber inlet of 118 μm for 4 traps, and the height of the channels of 25 μm (less than the 2 × RBC width,) with the height/width ratio of 0.211 (which is within the recommend domain of PDMS manufacturer's guidelines). Flow rates from 0.0046 to 621.4 μL/min were applied to the chamber inlet (max and min of the pump output). Applying this flow rate range is equivalent to applying a flowrate range of 0.046 μL/min to 6214 μL/min to a chip comprised of 10 branch channels. During the experiments, the viability of the cells was observed to be high enough for the RBCs to squeeze through the slits half their size and return to their original biconcave shape without turning into any damaged morphology under the high pressure.

The chamber inflow velocities are from 0.026 μm/s to 3510.6 μm/s (max and min of the pump output). The evolutionary algorithm parameters achieved for the best optimized solution are: CRL = 20,615 (62% increase compared to initial zero trapped cells design) in Figure 2e; CVR = $1.7 \times 10^7$ (initial value $7.07 \times 10^7$); TVR = 1.93 (initial, zero trapped cells design = 1.22, goal value = 1.93 fully achieved).

The evolutionary algorithm results applied to microfluidic trap design can be used for various design goals as well. For example, these results can help researchers to achieve the required flowrate parameters through the slits/channel designs. The evolutionary algorithm reduced the time required for microfluidics design and allowed for achieving desired flows with tight constraints. The setup time for the evolutionary algorithm was 40 human-hours and the computational cost for a 6-core Intel CPU was 60 h (100 generations). Furthermore, from a bio-physical viewpoint, the proposed method enabled the investigation and control of the transfer processes by RBCs in a dynamic mode.

## 4. Conclusions

In this paper, we presented a hydrodynamic trap with a unique feature to trap single RBCs in the flow channels by allowing through-slit flows. The single RBC trapping slits were designed with an evolutionary algorithm for the topology optimizations of the system. After adequate training, our system achieved desired flow parameters and met all constraints. Experimentally, we have built microfluidic devices and tested their trapping capabilities using frog RBCs for validations. Experimental results showed 4 out of 4 nucleated RBCs were trapped within one FOV.

We envision this evolutionary algorithm method can be applied to other microfluidics designs. In the future, the convergence speed of the algorithm can be further improved with expert-generated initial assumptions. In addition, the deep learning model can be applied to build the hybrid algorithms that can be used to achieve the better effectiveness of the whole system [29]. The designed cell-trapping system can be used in the microscopic studies of single cells in blood plasma flows, such as erythrocytes and lymphocytes. The automated design approach made it possible to fine-tune existing configurations and produce entirely new setups for each specific task.

## 5. Code and Data Availability

The scripts and data for the described implementation of generative design for single red blood cell traps were available in the open repository https://github.com/ITMO-NSS-team/rbc-traps-generative-design (accessed on 23 December 2021). The algorithmic implementation of evolutionary optimization of geometrically-encoded structures were added to the self-developed GEFEST framework: https://github.com/ITMO-NSS-team/GEFEST (accessed on 23 December 2021).

**Author Contributions:** Investigation, G.V.G., N.O.N., A.H., A.V.K.; Experiment, G.V.G.; Writing-original draft, G.V.G., N.O.N., A.H., A.V.K.; Methodology, G.V.G., N.O.N., A.H., A.V.K., A.V.L.; Conceptualization G.V.G., L.L., X.W., X.Q., G.V.M.; Formal analysis, G.V.G., N.O.N., A.H., A.V.K., A.V.L., L.L., X.W., X.Q., G.V.M.; Visualization, G.V.G., A.V.L., N.O.N., A.H., A.V.K.; Funding acquisition and Writing-review and editing, L.L., X.W., X.Q., A.V.K., G.V.M. All authors have read and agreed to the published version of the manuscript.

**Funding:** This research is partially supported by the Ministry of Science and Higher Education of Russian Federation, Agreement FSER-2021-0012 and by Shenzhen Fundamental Research Funding (Grant No. JSGG20201102656602006).

**Institutional Review Board Statement:** Not applicable.

**Informed Consent Statement:** Not applicable.

**Data Availability Statement:** Not applicable.

**Conflicts of Interest:** The authors declare no conflict of interest.

# References


1. Kim, M.G.; Park, J.; Lim, H.G.; Yoon, S.; Lee, C.; Chang, J.H.; Kirk Shung, K. Label-free analysis of the characteristics of a single cell trapped by acoustic tweezers. *Sci. Rep.* **2017**, *7*, 14092. [CrossRef] [PubMed]
2. Link, A.; Franke, T. Acoustic erythrocytometer for mechanically probing cell viscoelasticity. *Lab Chip* **2020**, *20*, 1991–1998. [CrossRef] [PubMed]
3. Du, E.; Dao, M.; Suresh, S. Quantitative biomechanics of healthy and diseased human red blood cells using dielectrophoresis in a microfluidic system. *Extrem. Mech. Lett.* **2014**, *1*, 35–41. [CrossRef] [PubMed]
4. Jones, P.V.; Staton, S.J.R.; Hayes, M.A. Blood cell capture in a sawtooth dielectrophoretic microchannel. *Anal. Bioanal. Chem.* **2011**, *401*, 2103–2111. [CrossRef]
5. Mohammadi, M.; Madadi, H.; Casals-Terré, J.; Sellarès, J. Hydrodynamic and direct-current insulator-based dielectrophoresis (H-DC-iDEP) microfluidic blood plasma separation. *Anal. Bioanal. Chem.* **2015**, *407*, 4733–4744. [CrossRef]
6. Narayanamurthy, V.; Nagarajan, S.; Firus Khan, A.Y.; Samsuri, F.; Sridhar, T.M. Microfluidic hydrodynamic trapping for single cell analysis: Mechanisms, methods and applications. *Anal. Methods* **2017**, *9*, 3751–3772. [CrossRef]
7. Iliescu, C.; Xu, G.; Tong, W.H.; Yu, F.; Bălan, C.M.; Tresset, G.; Yu, H. Cell patterning using a dielectrophoretic–hydrodynamic trap. *Microfluid. Nanofluid.* **2015**, *19*, 363–373. [CrossRef]
8. Tan, W.-H.; Takeuchi, S. A trap-and-release integrated microfluidic system for dynamic microarray applications. *Proc. Natl. Acad. Sci. USA* **2007**, *104*, 1146–1151. [CrossRef]
9. Watarai, H.; Namba, M. Capillary magnetophoresis of human blood cells and their magnetophoretic trapping in a flow system. *J. Chromatogr. A* **2002**, *961*, 3–8. [CrossRef]
10. Zborowski, M.; Ostera, G.R.; Moore, L.R.; Milliron, S.; Chalmers, J.J.; Schechter, A.N. Red blood cell magnetophoresis. *Biophys. J.* **2003**, *84*, 2638–2645. [CrossRef]
11. Khokhlova, M.D. Normal and system lupus erythematosus red blood cell interactions studied by double trap optical tweezers: Direct measurements of aggregation forces. *J. Biomed. Opt.* **2012**, *17*, 25001. [CrossRef] [PubMed]
12. Zhong, M.-C.; Gong, L.; Zhou, J.-H.; Wang, Z.-Q.; Li, Y.-M. Optical trapping of red blood cells in living animals with a water immersion objective. *Opt. Lett.* **2013**, *38*, 5134. [CrossRef] [PubMed]
13. Yu, Z.T.F.; Aw Yong, K.M.; Fu, J. Microfluidic blood cell sorting: Now and beyond. *Small* **2014**, *10*, 1687–1703. [CrossRef] [PubMed]
14. Henry, E.; Holm, S.H.; Zhang, Z.; Beech, J.P.; Tegenfeldt, J.O.; Fedosov, D.A.; Gompper, G. Sorting cells by their dynamical properties. *Sci. Rep.* **2016**, *6*, 1–11. [CrossRef] [PubMed]
15. Cheng, Y.; Ye, X.; Ma, Z.; Xie, S.; Wang, W. High-throughput and clogging-free microfluidic filtration platform for on-chip cell separation from undiluted whole blood. *Biomicrofluidics* **2016**, *10*, 014118. [CrossRef] [PubMed]
16. Yang, Q.; Ju, D.; Liu, Y.; Lv, X.; Xiao, Z.; Gao, B.; Song, F.; Xu, F. Design of organ-on-a-chip to improve cell capture efficiency. *Int. J. Mech. Sci.* **2021**, *209*, 106705. [CrossRef]
17. Raissi, M.; Yazdani, A.; Karniadakis, G.E. Hidden fluid mechanics: Learning velocity and pressure fields from flow visualizations. *Science* **2020**, *367*, 1026–1030. [CrossRef]
18. Man, Y.; Kucukal, E.; An, R.; Watson, Q.D.; Bosch, J.; Zimmerman, P.A.; Little, J.A.; Gurkan, U.A.; Gurkan, U.A. Microfluidic assessment of red blood cell mediated microvascular occlusion. *Lab Chip* **2020**, *20*, 2086–2099. [CrossRef]
19. Wu, J.; Qian, X.; Wang, M.Y. Advances in generative design. *Comput.-Aided Des.* **2019**, *116*, 102733. [CrossRef]
20. Oh, S.; Jung, Y.; Kim, S.; Lee, I.; Kang, N. Deep generative design: Integration of topology optimization and generative models. *J. Mech. Des. Trans. ASME* **2019**, *141*, 111405. [CrossRef]
21. Nikitin, N.O.; Polonskaia, I.S.; Kalyuzhnaya, A.V.; Boukhanovsky, A.V. The multi-objective optimisation of breakwaters using evolutionary approach. In *Developments in Maritime Technology and Engineering*; CRC Press: Boca Raton, FL, USA, 2021; pp. 767–774. [CrossRef]
22. Kalyuzhnaya, A.V.; Nikitin, N.O.; Hvatov, A.; Maslyaev, M.; Yachmenkov, M.; Boukhanovsky, A. Towards generative design of computationally efficient mathematical models with evolutionary learning. *Entropy* **2021**, *23*, 28. [CrossRef] [PubMed]
23. Toussi, H.E. The Application of Evolutionary, Generative, and Hybrid Approaches in Architecture Design Optimization. *NEU J. Fac. Archit.* **2020**, *2*, 1–20.
24. Le, T.C.; Winkler, D.A. A Bright Future for Evolutionary Methods in Drug Design. *Chem. Med. Chem.* **2015**, *10*, 1296–1300. [CrossRef] [PubMed]
25. Arita, T.; Fukata, M.; Maruyama, T.; Odashiro, K.; Fujino, T.; Wakana, C.; Sato, A.; Takahashi, K.; Iida, Y.; Mawatari, S.; et al. Comparison of Human Erythrocyte Filterability with Trapping Rate Obtained by Nickel Mesh Filtration Technique: Two Independent Parameters of Erythrocyte Deformability. *Int. Blood Res. Rev.* **2018**, *8*, 1–8. [CrossRef]
26. Carey, T.R.; Cotner, K.L.; Li, B.; Sohn, L.L. Developments in label-free microfluidic methods for single-cell analysis and sorting. *Wiley Interdiscip. Rev. Nanomed. Nanobiotechnol.* **2019**, *11*, e1529. [CrossRef]
27. Yu, X.; Gen, M. *Introduction to Evolutionary Algorithms*; Springer: Berlin/Heidelberg, Germany, 2010.

28. Nikitin, N.O.; Hvatov, A.; Polonskaia, I.S.; Kalyuzhnaya, A.V.; Grigorev, G.V.; Wang, X.; Qian, X. Generative Design of Microfluidic Channel Geometry Using Evolutionary Approach; Generative Design of Microfluidic Channel Geometry Using Evolutionary Approach. In Proceedings of the Genetic and Evolutionary Computation Conference Companion, Lille, France, 10–14 July 2021. [CrossRef]
29. Kallioras, N.A.; Lagaros, N.D. DzAIN: Deep learning based generative design. *Procedia Manuf.* **2020**, *44*, 591–598. [CrossRef]

*micromachines*

# Microfluidic Device Using Mouse Small Intestinal Tissue for the Observation of Fluidic Behavior in the Lumen

Satoru Kuriu [1,*], Naoyuki Yamamoto [2] and Tadashi Ishida [1,*]

1   Department of Mechanical Engineering, School of Engineering, Tokyo Institute of Technology, Kanagawa 226-8503, Japan
2   Department of Life Science and Technology, School of Life Science and Technology, Tokyo Institute of Technology, Kanagawa 226-8503, Japan; n-yamamoto@bio.titech.ac.jp
*   Correspondence: kuriu.s.aa@m.titech.ac.jp (S.K.); ishida.t.ai@m.titech.ac.jp (T.I.); Tel.: +81-45-924-5468 (S.K.)

**Abstract:** The small intestine has the majority of a host's immune cells, and it controls immune responses. Immune responses are induced by a gut bacteria sampling process in the small intestine. The mechanism of immune responses in the small intestine is studied by genomic or histological techniques after in vivo experiments. While the distribution of gut bacteria, which can be decided by the fluid flow field in the small intestinal tract, is important for immune responses, the fluid flow field has not been studied due to limits in experimental methods. Here, we propose a microfluidic device with chemically fixed small intestinal tissue as a channel. A fluid flow field in the small intestinal tract with villi was observed and analyzed by particle image velocimetry. After the experiment, the distribution of microparticles on the small intestinal tissue was histologically analyzed. The result suggests that the fluid flow field supports the settlement of microparticles on the villi.

**Keywords:** microfluidic device; small intestine; ex vivo; histology; embedded resin; sectioning

**Citation:** Kuriu, S.; Yamamoto, N.; Ishida, T. Microfluidic Device Using Mouse Small Intestinal Tissue for the Observation of Fluidic Behavior in the Lumen. *Micromachines* **2021**, *12*, 692. https://doi.org/10.3390/mi12060692

Academic Editors: Violeta Carvalho, Senhorinha Teixeira and João Eduardo P. Castro Ribeiro

Received: 2 May 2021
Accepted: 9 June 2021
Published: 13 June 2021

**Publisher's Note:** MDPI stays neutral with regard to jurisdictional claims in published maps and institutional affiliations.

## 1. Introduction

The mammalian gastrointestinal (GI) tract harbors various types and tremendous amounts of gut bacteria (GB) [1–3]. Different types of GB distribute in the GI tract [4,5] and maintain their host's health condition [6–8]. Among GI tissues, the small intestine (SI) controls immune responses, as over 60% of the host's immune cells inhabit it. Immune responses are induced by a GB sampling process, the transportation of GB to GI tissue by fluid flow and responses with microfold cells (M cells) [6] or dendrites being extended by dendritic cells (DCs) [9,10]. In addition to the gut-associated lymphoid tissues (e.g., Peyer's patch, which expresses M cells surrounded by SI's unique finger-like structures, called villi [11]) [6], villi also have M cells and induce immune responses through a GB sampling process [12]. The mechanisms of the gut immune system were obtained from dissected specimens after an in vivo experiment that mainly used experimental mice [13–15]. This is because in situ visualization of interactions of the GB and the M cells/dendrites being extended by DCs in the SI tracts are difficult in in vivo studies. However, GB behaviors in the SI tracts are important for immune responses because they are triggered by the encounters of GB and M cells/DCs. In the SI tracts, GB distribution is decided by fluid flow, which controls the encounters of the GB and the M cells/DCs. Therefore, fluid flow fields inside the SI tracts are important. Therefore, technologies enabling the observation of the fluid flow field in the SI tract are required at the microscopic scale.

Microfluidic technologies have high potential for the direct and microscopic observation of the fluid flow field in luminal circumstances. This is because they can reconstruct biological systems on chips and build elaborate fluidic systems with biological samples, such as cells and tissues, under microscopic observation. An in vitro biomimetic "gut-on-a-chip" was proposed to mimic intestinal biological functions and enable microscopic observations [16–22]. Those chips commonly have two microchannels in parallel, separated by a flexible and a porous membrane. On both sides of the membrane, single or different kinds of cell (e.g., human colorectal adenocarcinoma cell (Caco-2 cell) and human umbilical vein endothelial cell (HUVEC)) are cultured over the long term (>7 days). To apply shear stress on the epithelial cells, continuous flow is perfused in both microchannels. This culture method mirrors intestinal biological features. With the gut-on-a-chip as a simulation of disease, drug tests were conducted [18,21]. However, the configurations were different in size and structure from the surface of the SI tract, which were assumed to be important for the fluid flow field in the SI tracts. On the other hand, microfluidic devices utilizing the natural features of full thickness intestinal tissues ex vivo were proposed [23–26]. Those devices have a sheet of a dissected intestinal tissue in the channel. They enable tissue culture for 3 days to monitor some functions in living intestinal tissues. In addition, microfluidic devices enable the observation of the intestinal tissue ex vivo [26]. However, the observations are only at the macro scale, and the analyses are performed only by the measurement of effluent from the outlet of the device. Although microfluidic technologies can study biological aspects of the gut, they do not focus on the studies of fluid flow in the gut at a microscopic scale.

For the observations of fluid flow field in the SI tract, especially around the villi at the microscopic level, a microfluidic device with the geometry of the SI tract is necessary. Here, we propose a microfluidic intestinal channel device (MIC) to reproduce and visualize the flow field inside an SI tract. MIC has a chemically fixed SI tissue dissected from a mouse. The SI tissue is used as microfluidic channel to visualize the flow field around its villi, although it is not an in vivo model but an in vitro one. In the fabrication and assembly of MIC, the handling of the SI tissue is the critical issue due to its small, soft and crooked characteristics. Modules to hold or to set the SI tissue were designed, and stable handling was achieved. With this device, we demonstrated the observation of fluid flow field around the villi at the microscopic scale by particle image velocimetry (PIV) and histological analysis of SI tissue.

## 2. Materials and Methods

### 2.1. Material and Reagents

Materials for device fabrication and assembly were obtained from the following suppliers: polydimethylsiloxane (PDMS) (Silpot 184 W/C, Dow Corning Toray, Tokyo, Japan), epoxy-based embedding resin (EPOX) (Technovit® EPOX, Kulzer, Hanau, Germany), agar (Bacto™ Agar, Wako, Osaka, Japan) and liquid bandage (Mentholatum, Rohto, Osaka, Japan).

Reagents for experiments were obtained from the following suppliers: 4% paraformaldehyde (4% PFU) (Nacalai Tesque, Inc., Kyoto, Japan), fluorescent beads (Fluorescent Polystyrene Microspheres, 1.00 μm, Dragon Green, Bangs Laboratories, Inc., Fishers, IN, USA), anti-integrin antibodies conjugated to AF555 (Anti-integrin $\alpha v \beta 5$ Antibody, Alexa Fluor® 555 Conjugated, Bioss, Woburn, MA, USA).

### 2.2. Tissue Preparation

Balb/c mice (female) aged 9 weeks were obtained from Charles River Laboratory, Japan (Yokohama, Japan), and housed in the animal facilities at Tokyo Institute of Technology. Animal experiments were approved by the Animal Experiment Committees of Tokyo Institute of Technology (authorization numbers D2019006-2), and carried out in accordance with relevant guidelines. Full thickness SI tissues taken from a part between duodenum and jejunum were dissected from mice, and immediately fixed in 4% *w/v* paraformaldehyde solution. The fixation prevented the degradation caused by digestive enzymes (e.g., trypsin, chymotrypsin, and many amino-/carboxy-peptidases) released from the SI tissues themselves at 37 °C, which is a suitable temperature for GB. The fixation preserved the geometry of the villi, which was important for the flow field inside the SI. While it might change the surface condition of charges, the flow field was not affected at a micro scale.

Although the density of GB and immunologic activity are different for each part of SI tissue (e.g., duodenum, jejunum and ileum), they are mainly covered by villi and have similar surface geometries of villi.

### 2.3. Image Acquisition and Processing

Bright-field and confocal fluorescent images were captured by an inverted microscope (IX83, Olympus, Tokyo, Japan) equipped with a high-speed scanner (CSU10, Yokogawa, Tokyo, Japan) and an electron multiplying charge coupled device (EMCCD) camera (iXon Ultra, Andor, Tokyo, Japan). For the excitation light source used for confocal microscopy, solid-state lasers (wavelength: 488 and 561 nm, TAC, Saitama, Japan) were used. The confocal fluorescence images were saved in 16-bit TIFF format using a capture software (iQ3, Andor, Tokyo, Japan). The images were processed and quantified by Image J (National Institute of Health, Bronx, NY, USA).

### 2.4. PIV

Suspension of tracer particles (fluorescent microbeads of 1.0 μm in diameter) were perfused in a microchannel. The tracer particles in flow were captured at an interval of 1 s. The images were captured in 1280 pixels × 720 pixels. Two serial images were compared by dividing them into interrogation areas of 64 pixels × 64 pixels. The interrogation areas were cross-correlated with each other, pixel by pixel, using a function of Image J, resulting in instantaneous velocity fields.

### 2.5. MIC

#### 2.5.1. Design of the MIC

A schematic illustration of the MIC is shown in Figure 1. The MIC consisted of three modules: container, connector and channel. The container module worked as the guide of the SI tissue. A SI tissue was placed on the guide in the container module, whose tube-like structure worked as a microfluidic channel (SI channel). The container module, including the SI tissue, could be sectioned after experiments. The container module also held the position of the connector module. The connector module worked as an inlet and an outlet of the SI channel. Both ends of the SI tissue were inserted into the connector module for liquid perfusion and glued to prevent liquid from the leakage. The channel module was the bottom part of the MIC. The channel module worked as the substitute of the sliced part of the SI tissue for observations inside. To keep the transparency for observation, all the modules were made of PDMS or EPOX.

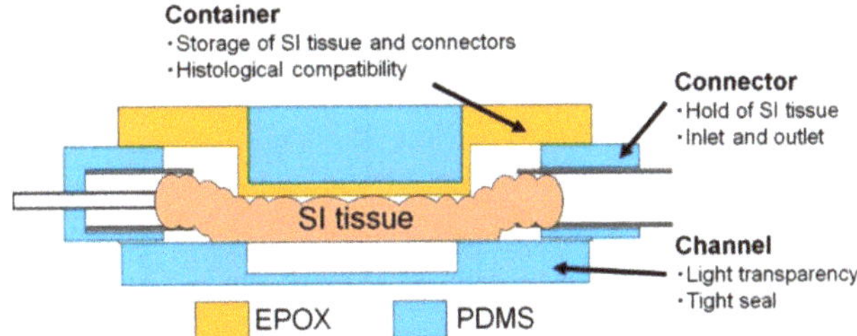


**Figure 1.** Schematic illustration of the MIC. The container module stores the SI tissue with an opening for microscopic observation. The connector module holds both edges of the SI tissue. The channel module is the bottom part of the MIC, a substitute of the SI tissue.

To achieve the MIC, the container module, connector module and channel module were separately designed. The typical dimensions of mouse SI tissue were 10 mm in length and 3 mm in diameter in our experiments. The guide for the SI tissue was 3.5 mm in width, which was slightly larger than typical diameter of the SI tissues. The height of the guide was 1.5 mm, half of the diameter, because the half of the SI tissue was removed to make an opening for the observation of the inner structure of the SI channel. Furthermore, histological analysis was necessary for the SI tissue after experiments. The SI tissue needed to be sliced inside the container module to suppress any changes and damages in the sectioning process. For the sectioning of the SI tissue set on the guide, the container module was made of EPOX, which is a suitable material for sectioning and microfluidic devices. Mechanical strength of the guide was weak due to the thin structures, especially the ceiling thickness of 0.1 mm (Figure 2A). This is because the guide for the SI tissue needed to be isolated from the container module and transferred to the sectioning process. To reinforce the thin structures, PDMS was cured on the thin structure considering its transparency and easy removal from EPOX.

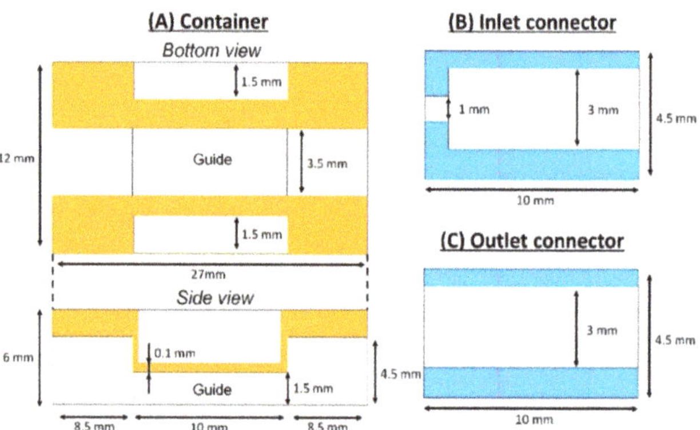


**Figure 2.** Schematic illustrations of the container and connectors with dimensions: (**A**) Container; bottom view (upper) and side view (lower). (**B**) Inlet connector. (**C**) Outlet connector.

To use the SI tissue as a microfluidic channel, inlet and outlet were the key modules for the stable liquid perfusion. The connection with a tight seal between the SI tissue and accessories (i.e., tubes or stainless pipes) was important for the stable liquid perfusion. Therefore, we designed the connector module which tightly hold both ends of the SI tissue (Figure 2B,C). The inlet connector had a double pipe (i.e., coaxial stainless pipes of 1 mm and 3 mm in diameter) (Figure 3). The stainless pipe was 1 mm in diameter and had a spacer of 2 mm in diameter at the tip. The pipe that was 1 mm in diameter was inserted into the SI tissue, and the SI tissue was covered by the pipe 3 mm in diameter. The outlet connector only had a single stainless pipe of 3 mm in diameter. This is because the stainless pipe of 3 mm in diameter could reduce fluid resistance, leading to good drainage (i.e., the fluid resistance in the case of the pipe that was 3 mm in diameter was 27 times lower than that of the inlet pipe that was 1 mm in diameter). The pipes and SI tissue were tightly sealed by a liquid bandage.

The channel module was the bottom part of the MIC. The dimensions of the channel should be close to the sliced SI tissue and designed to be a rectangular cross section of 10 mm (length) × 3 mm (width) × 500 μm (height). This is because, for the microscopic observation of the inner structures of the SI channel, the bottom of the channel module should be flat. Further, the channel module should be transparent. Although PDMS was a good choice from the point of transparency, the conventional permanent bonding of PDMS microfluidic devices was not applicable for sealing. This is because the SI tissue is weak against dryness or heat. Therefore, the components of MIC should be mechanically assembled by sealing with screws. For the tight mechanical sealing, PDMS was a good material from the point of softness and adhesion to EPOX structures [27].

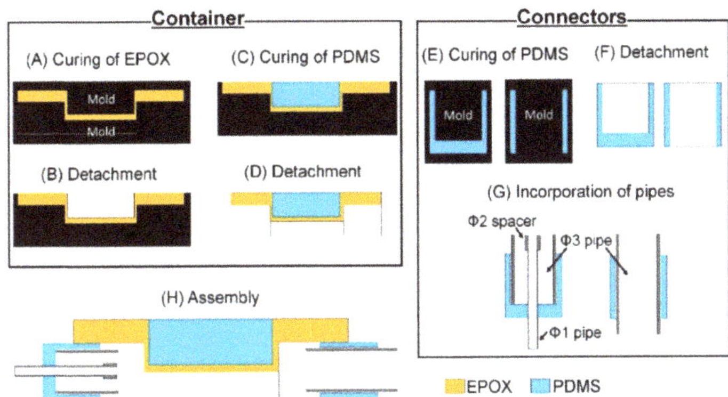

**Figure 3.** Fabrication procedure of the container and connectors. Fabrication procedure of the container: (**A**) Curing of EPOX. (**B**) Detachment. Upper mold is detached. (**C**) Curing of PDMS. PDMS is filled in the hollow of the cured EPOX and cured. (**D**) Detachment. Lower mold is detached. Fabrication procedure of the connectors. (**E**) Curing of PDMS. (**F**) Detachment. (**G**) Incorporation of pipes. Stainless pipes (diameter: 1 or 3 mm) are incorporated. (**H**) Assembly.

### 2.5.2. Fabrication Procedure of the MIC

The fabrication procedures of the container module and connector module are described in Figure 3. The container module and connector module were fabricated by conventional soft lithography. As for the fabrication process of the container module, EPOX (base EPOX to curing agent ratio was 2:1 by weight) was casted on molds and cured for 3 h at 100 °C in an air-vented oven (Figure 3A). The upper mold was detached (Figure 3B) and PDMS (base PDMS to curing agent ratio was 10:1 by weight) was casted on a thin structure of the container and cured for 3 h at 100 °C in an air-vented oven (Figure 3C). The container module was detached from the mold (Figure 3D). As for the fabrication process of the connector module, PDMS was casted on molds for an inlet and an outlet and cured for 3 h at 100 °C in an air-vented oven (Figure 3E). The cured PDMS were detached from the molds (Figure 3F). Stainless pipes were incorporated in each connector (Figure 3G). Finally, the container and connectors were assembled (Figure 3H). As for the fabrication of the aforementioned molds and holders, polyacetal plates were made by a milling machine. The molds were polished with abrasive compound (PiKAL NIHON MARYO-KOGYO) to remove residual traces of the milling. They were designed by three-dimensional computer aided design software.

### 2.5.3. Assembly Procedure of the MIC

The assembly procedure is described in Figure 4. A SI tissue was set on the guide (Figure 4A). The SI tissue and connectors were glued by liquid bandage. The SI tissue was inflated by air. The inflation by air made the SI tissue straight and smooth to avoid difficulties in the slicing process of the SI tissue due to its softness and crooked shape, such as wrinkles, waves and curves (Figure 4B and Video S1). The inflated SI tissue was embedded with pre-gel solution (agar to deionized (DI) water 5% $w/v$). The solution was gelated within a few seconds at room temperature (Figure 4C). Here, the concentration of the pre-gel solution was determined by the performance of holding and slicing the SI tissue in the following step. The agar also prevented the SI tissue from drying. The excessive volume of the agar and SI tissue were sliced by a blade (Figure 4D). The agar embedding made the handle of the SI tissue easy. The container module and connecter module including the SI tissue were assembled by holders, and they were screwed (Figure 4E). Note that the outlet connector was plugged to avoid the influx of liquid bandage and pre-gel solution into the SI tissue in the processes of tissue setting and agar embedding.

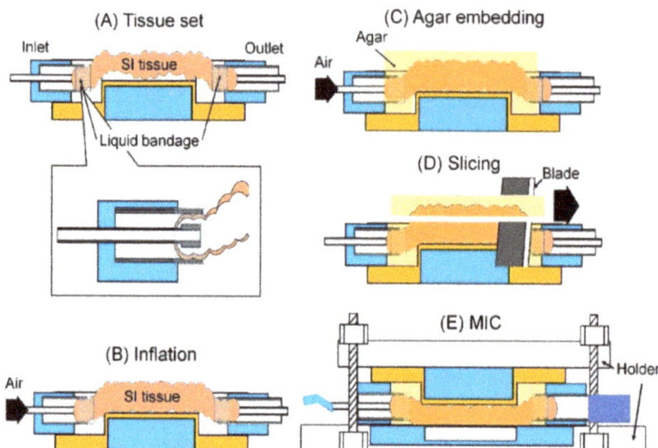



**Figure 4.** Assembly procedure of the MIC: (**A**) Tissue set. (**B**) Inflation. Air is pumped into the SI tissue. (**C**) Agar embedding. (**D**) Slicing. Excessive volume of the agar including SI tissue is sliced. (**E**) MIC.

### 2.5.4. Experimental Procedure Using MIC

The suspension of fluorescent microbeads ($1.0 \times 10^7$ particle/mL) was manually introduced into the SI tissue with a syringe. The MIC was placed in the manner of Figure 5A for 1.5 h for the settlement of the microbeads on the surface of the villi. DI water was introduced into the SI channel at a flow rate of 7 μL/min (Figure 5B). The flow rate was sufficient for the wash-out of the microbeads, which was calculated by flow rate and cross-sectional area [28]. Ethanol in a 95% concentration was perfused in the SI channel for 1 h at a flow rate of 5 μL/min for dehydration (Figure 5C). A hole of 3 mm in diameter was punched with a biopsy punch on the channel module. EPOX was introduced into the SI channel via the hole. The SI channel was embedded by EPOX and cured at room temperature overnight (Figure 5D). The channel module and PDMS reinforcement of the container module were removed from the MIC, and the guide was manually taken apart from the container module (Figure 5E). The EPOX-embedded SI tissue was sectioned by a thickness of 30 μm for histological observations (Figure 5F).

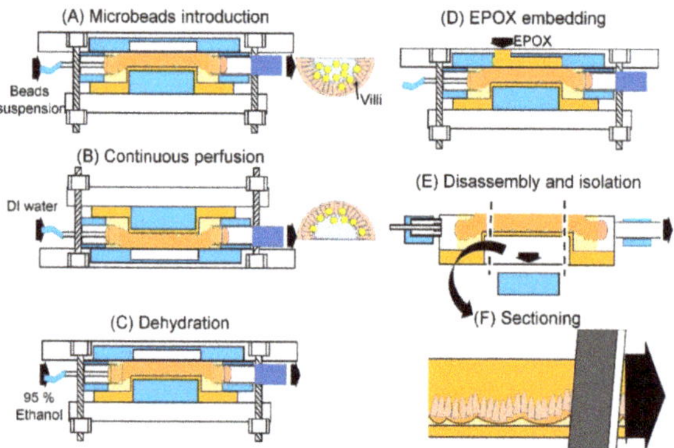



**Figure 5.** Experimental procedure using the MIC. (**A**) Microbeads introduction. (**B**) Continuous perfusion. DI water is perfused in the SI channel at the flow rate of 7 μL/min. (**C**) Dehydration. Ethanol at a 95% concentration is perfused in the SI channel for 1 h at a flow of 5 μL/min. (**D**) EPOX embedding. (**E**) Disassembly and isolation. (**F**) Sectioning. The guide including embedded SI tissue is sectioned with a thickness of 30 μm.

## 3. Results and Discussion

### 3.1. Fabricated MIC

The fabricated container module, connector module and MIC are shown in Figure 6A–C, respectively. The villi of the SI tissue in MIC were visualized through an opening of the SI tissue and the channel module with a microscope. The villi were stained with red-fluorescent labeled anti-integrin antibody, and visualized at different heights (i.e., bottom, middle and top. Figure 6D). The structures of the villi were well maintained in the MIC.

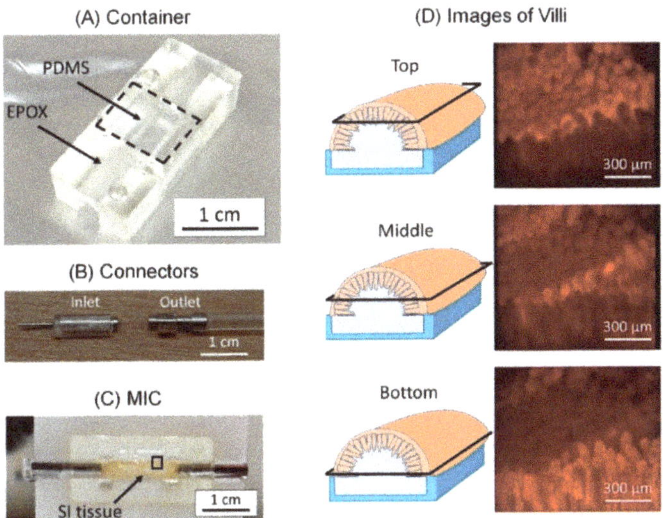



**Figure 6.** Fabricated modules and the MIC. (**A**) Fabricated container. Location of PDMS reinforcement is indicated by black dashed line. (**B**) Fabricated connectors for the inlet (left) and outlet (right). (**C**) MIC. (**D**) Images of villi in the black square in (**C**). The villi were captured at the top, middle and bottom. The villi were stained by red fluorescent-labeled anti-integrin antibodies.

### 3.2. Analysis of the Microbeads Distribution in SI Channel

The settled fluorescent microbeads on the surface of the villi were observed after the wash-out by the flow of DI water (Figure 7, Video S2). The distributions of microbeads (Figure 7A) and villi (Figure 7B) were observed. We found that the microbeads did not cover the villi themselves but the base of the villi (Figure 7C).

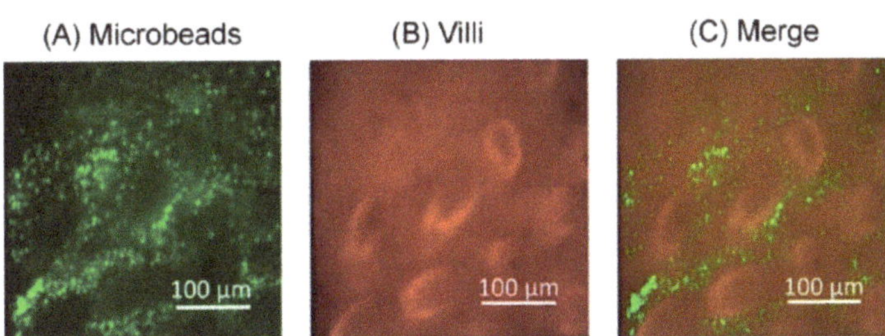



**Figure 7.** Observation of inner structures of the SI tissue in the MIC. (**A**) Microbeads. They are indicated by green fluorescent. (**B**) Villi. They are indicated by red fluorescent. (**C**) Merge.

### 3.3. Analysis of Flow Field in SI Channel

To analyze the flow field in the SI channel under the continuous flow, DI water was perfused into the SI channel at a constant flow rate as a simple model after the settlement of microbeads. The size of the microbeads was comparable to that of GB. The motions of the microbeads inside the SI channel were observed at the bottom, middle and top (Figure 8). PIV analysis was conducted against the images. The velocity fields in the planes at the bottom and middle were fast (26.3–48.8 µm/s) and stable (Figure 8A,B), while the velocity field of the top plane was slow (1.3–6.3 µm/s) (Figure 8C). From the defecation time in mouse, the flow speed of GB was empirically estimated around 2.8–14 µm/s. The speed of the microbeads flowing in the SI channel was comparable to that in the GI tract.

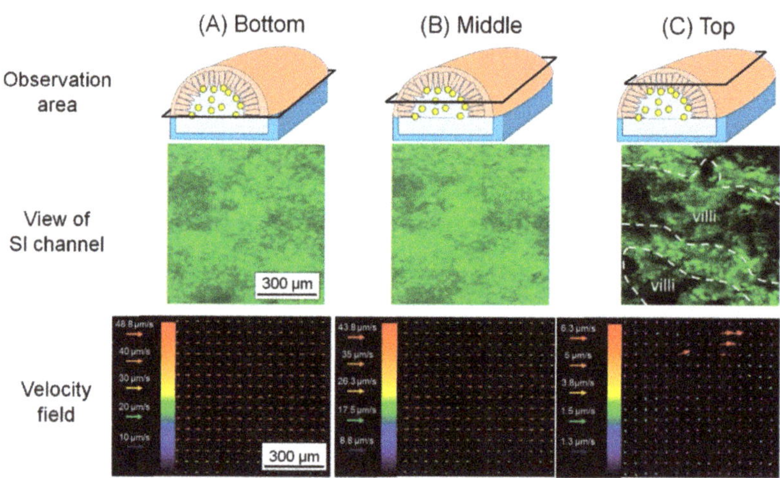


**Figure 8.** PIV analysis in the SI tissue. The fluorescent microbeads were tracked to calculate the velocity vector in the SI channel. Top column: the schematic of the observation area at the (**A**) bottom, (**B**) middle and (**C**) top. Middle column: view of the microbeads in the SI channel at the (**A**) bottom, (**B**) middle and (**C**) top. Lower column: the velocity fields at the (**A**) bottom, (**B**) middle and (**C**) top. In the image at the top, the villi are surrounded by white dash line.

The flow velocities at different planes follow the basic trend of three-dimensional velocity field of the flow in a pipe; that is, the velocity was maximal at the center of the pipe and decreased near the pipe wall. Furthermore, the microbeads flew around the villi at the top, which was different from the flows at the bottom and middle. The flow between villi was slow but some of them headed for villi, while the velocity over villi was ~1 µm/s. This result suggests that flow around the villi may transport and settle GB on the surface of the villi.

### 3.4. Transition of the Fluorescent Beads Distribution around the Villi against Time

We evaluated the temporal transition of the fluorescent intensity in the SI channel. The change in the fluorescent intensity was caused by the density change in the fluorescent microbeads around the villi, namely at the top in Figure 8C (Figure 9). Fluorescent intensity was measured every 5 min. Fluorescent intensity gradually decreased to 71 and 52% in 15 and 30 min, respectively (Figure 9B,C), compared to that in the initial state (Figure 9A). However, the fluorescent beads should not move a great deal, at least according to the PIV analysis in Figure 8C. The fluorescent microbeads on the surface of the villi could gradually fall as a result of gravity rather than the flow.

### 3.5. Histological Analysis of the Microbeads Distribution

The MIC was disassembled and the guide, including the SI tissue, was manually isolated (Figure 10A), as explained in Section of 2.5.1. The guide channel was sectioned at a thickness of 30 µm, and a series of slices was acquired (Figure 10B). The SI tissue in the MIC was histologically analyzed after the experiments. The observation of the series of slices revealed that the fluorescent microbeads were attached on the side surface of the villi, and they were especially trapped in the pit-like structures (Figure 11). This result suggests that the geometrical features of the villi may affect the trapping performance of microparticles for the settlement of GB.

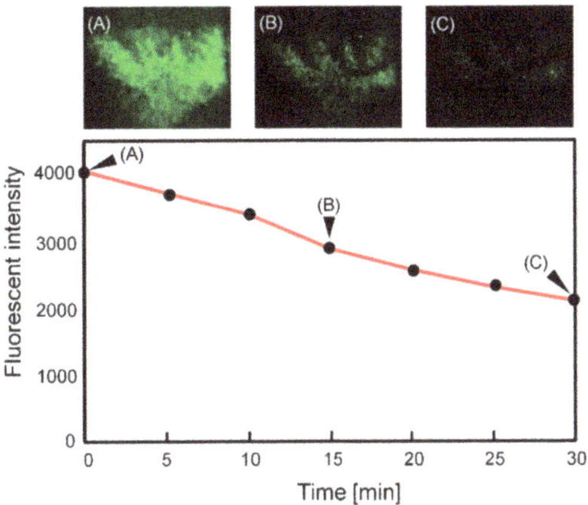

**Figure 9.** Transition of fluorescent intensity around the villi as a function of time. Fluorescent images were captured every 5 min: (**A**) Fluorescent image at 0 min, (**B**) 15 min and (**C**) 30 min. Black arrows in the graph indicate that the data acquired from fluorescent images (**A**), (**B**) and (**C**).

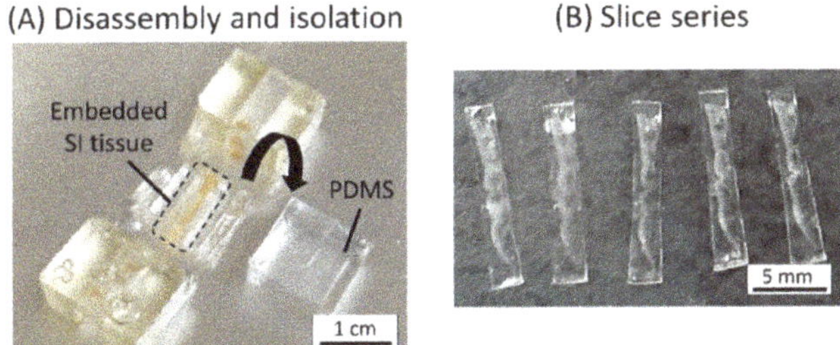


**Figure 10.** Sectioning of the guide channel including embedded SI tissue by EPOX: (**A**) Disassembly and isolation. The PDMS reinforcement was removed, and the SI tissue embedded in the container by EPOX was isolated. (**B**) Slice series. The embedded SI tissue was sectioned by the thickness of 30 μm.

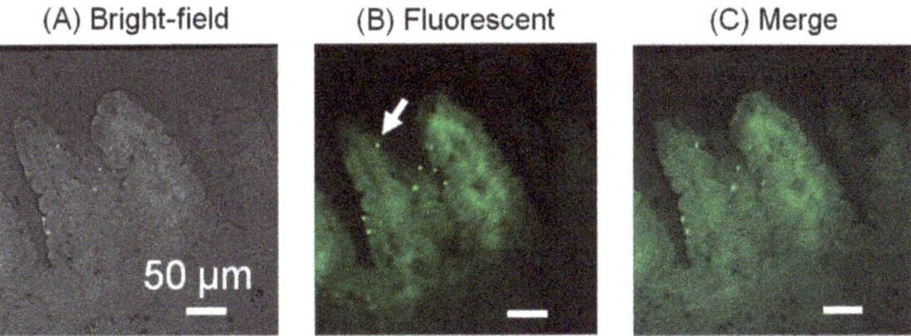


**Figure 11.** Histological analysis of the SI tissue: (**A**) Bright-field image, (**B**) fluorescent and (**C**) merged images of the cross-sections of the villi. The bright dots represent the fluorescent microbeads attached to the villi. One of the fluorescent microbeads is indicated by the white arrow in (**B**). Villi are indicated by autofluorescence.

## 4. Conclusions

We developed a microfluidic device with mouse SI tissue as a microfluidic channel. The microfluidic device consisted of a container, connectors and channel modules, and it achieved easy handling in the mouse SI tissue. With this device, the villi of the SI tissue and fluorescent microbeads were observed clearly under liquid flow at the microscopic level. We perfused the steady flow in the SI channel and analyzed the flow velocity field inside the SI tissue by PIV analysis. Some of the microbeads headed to the villi and attached to the side surface of the villi. This result suggested that the flow field may play an important role in the settlement of GB on the villi.

For future work, we will perfuse mucus-like liquid [29,30] or isotonic saline suspending living GB to understand the roles of the flow field in forming GB distribution in the SI tissue using this microfluidic device, and we will apply an inconstant flow as a simulation of peristatic motion. Furthermore, although the SI tissue used in our study was fixed and lost its biological activities, the superficial molecules on mucosal tissues and the shape of villi were properly maintained. We plan to study the intermolecular affinity between the intestinal microbial community and different parts of the SI tissue (e.g., duodenum, jejunum and ileum) with an MIC, as the different geometry of the villi may affect the fluid's behavior.

**Supplementary Materials:** The following are available online at https://www.mdpi.com/article/10.3390/mi12060692/s1, Video S1: Inflation of SI tissue. Video S2: Fluorescent beads flow inside SI channel.

**Author Contributions:** S.K. conceived, designed and performed the experiments, S.K., N.Y. and T.I. analyzed the data and wrote the paper; T.I. supervised the research. All authors have read and agreed to the published version of the manuscript.

**Funding:** This research received no external funding.

**Institutional Review Board Statement:** The present study was conducted according to the guidelines of the Declaration of Helsinki, and approved by the Animal Experiment Committees of Tokyo Institute of Technology (authorization numbers D2019006-2 on 1 September 2020), and carried out in accordance with relevant guidelines.

**Conflicts of Interest:** The authors declare no conflict of interest.

## References


1. Huttenhower, C.; Gevers, D.; Knight, R.; Abubucker, S.; Badger, J.H.; Chinwalla, A.T. The Human Microbiome Project Consortium. Structure, function and diversity of the healthy microbiome. *Nature* **2012**, *486*, 207.
2. Geva-Zatorsky, N.; Sefik, E.; Kua, L.; Mathis, D.; Benoist, C.; Kasper, D.L. Mining the Human Gut Microbiota for Immunomodulatory Organisms. *Cell* **2017**, *168*, 928–943. [CrossRef] [PubMed]
3. Marteau, P.; Pochart, P.; Dore, J.; Bera-Maillet, C.; Bernalier, A.; Corthier, G. Comparative Study of Bacterial Groups within the Human Cecal and Fecal self. *Appl. Environ. Microbiol.* **2001**, *67*, 4939–4942. [CrossRef] [PubMed]
4. Espey, M.G. Role of oxygen gradients in shaping redox relationships between the human intestine and its microbiota. *Free Radic. Biol. Med.* **2013**, *55*, 130–140. [CrossRef] [PubMed]
5. Albenberg, L.; Esipova, T.V.; Judge, C.P.; Bittinger, K.; Chen, J.; Laughlin, A.; Grunberg, S.; Baldassano, R.N.; Lewis, J.D.; Li, H.; et al. Correlation Between Intraluminal Oxygen Gradient and Radial Partitioning of Intestinal Microbiota in Humans and Mice. *Gastroenterology* **2014**, *147*, 1055–1063. [CrossRef]
6. Silva, C.D.; Wagner, C.; Bonnardel, J.; Gorvel, J.P.; Lelouard, H. The Peyer's Patch Mononuclear Phagocyte System at Steady State and during Infection. *Front. Immunol.* **2017**, *8*, 1254. [CrossRef]
7. Takiishi, T.; Fenero, C.I.M.; Câmara, N.O.S. Intestinal barrier and gut microbiota: Shaping our immune responses throughout life. *Tissue Barriers* **2017**, *5*, e13732208. [CrossRef]
8. Atarashi, K.; Tanoue, T.; Ando, M.; Hattori, M.; Umesaki, Y.; Honda, K. Th17 Cell Induction by Adhesion of Microbes to Intestinal Epithelial Cells. *Cell* **2015**, *163*, 367–380. [CrossRef] [PubMed]
9. Chieppa, A.; Rescigno, M.; Huang, A.Y.C.; Germain, R.N. Dynamic imaging of dendritic cell extension into the small bowel lumen in response to epithelial cell TLR engagement. *J. Exp. Med.* **2006**, *203*, 2841–2852. [CrossRef] [PubMed]
10. Chang, S.Y.; Ko, H.J.; Kweon, M.N. Mucosal dendritic cells shape mucosal immunity. *Exp. Mol. Med.* **2014**, *46*, e84. [CrossRef]
11. Suh, S.H.; Choe, K.; Hong, S.P.; Jeong, S.H.; Mäkinen, T.; Kim, K.S.; Alitalo, K.; Surh, C.D.; Koh, G.Y.; Song, J.H. Gut microbiota regulates lacteal integrity by inducing VEGF-C in intestinal villus macrophages. *EMBO Rep.* **2019**, *20*, e46927. [CrossRef] [PubMed]

12. Jang, M.H.; Kweon, M.N.; Iwatani, K.; Yamamoto, M.; Terahara, K.; Sasakawa, C.; Suzuki, T.; Nochi, T.; Yokota, Y.; Rennert, P.D.; et al. Intestinal villous M cells: An antigen entry site in the mucosal epithelium. *Proc. Natl. Acad. Sci. USA* **2004**, *101*, 6110–6115. [CrossRef] [PubMed]

13. Torii, A.; Torii, S.; Fujiwara, S.; Tanaka, H.; Inagaki, N.; Nagai, H. *Lactobacillus Acidophilus* Strain L-92 Regulates the Production of Th1 Cytokine as well as Th2 Cytokines. *Allergol. Int.* **2007**, *56*, 293–301. [CrossRef] [PubMed]

14. Yanagihara, S.; Kanaya, T.; Fukuda, S.; Nakato, G.; Hanazato, M.; Wu, X.R.; Yamamoto, N.; Ohno, H. Uromodulin-SlpA binding dictates *Lactobacillus acidophilus* uptake by intestinal epithelial M cells. *Int. Immunol.* **2017**, *29*, 357–363. [CrossRef]

15. Shreedhar, V.K.; Kelsall, B.L.; Neutra, M.R. Cholera Toxin Induces Migration of Dendritic Cells from the Subepithelial Dome Region to T- and B-Cell Areas of Peyer's Patches. *Infect. Immun.* **2003**, *71*, 504–509. [CrossRef] [PubMed]

16. Kim, H.J.; Huh, D.; Hamilton, G.; Ingber, D.E. Human gur-on-a-chip inhabited by microbial flora that experiences intestinal peristalsis-like motions and flow. *Lab Chip* **2012**, *12*, 2165–2174. [CrossRef]

17. Donkers, J.M.; Amirabadi, H.E.; Steeg, E.V.D. Intestine-on-a-chip: Next level in vitro research model of the human intestine. *Curr. Opin. Toxicol.* **2021**, *25*, 6–14. [CrossRef]

18. Shah, P.; Fritz, J.V.; Glaab, E.; Desai, M.S.; Greenhalgh, K.; Frachet, A.; Niegowska, M.; Estes, M.; Jäger, C.; Seguin-Devaux, C.; et al. A microfluidics-based in vitro model of the gastrointestinal human-microbe interface. *Nat. Commun.* **2016**, *7*, 11535. [CrossRef] [PubMed]

19. Jing, B.; Wang, Z.A.; Zhang, C.; Deng, Q.; Wei, J.; Luo, Y.; Zhang, X.; Li, J.; Du, Y. Establishment and Application of Peristaltic Human Gut-Vessel Microsystem for Studying Host-Microbial Interaction. *Front. Bioeng. Biotechnol.* **2020**, *8*, 272. [CrossRef]

20. Bein, A.; Shin, W.; Jalili-Firoozinezhad, S.; Park, M.H.; Sontheimer-Phelps, A.; Tovaglieri, A.; Chalkidaki, A.; Kim, H.J.; Ingber, D.E. Microfluidic Organ-on-a-Chip Models of Human Intestine. *CMGH Cell. Mol. Gastroenterol. Hepatol.* **2017**, *5*, 659–668. [CrossRef]

21. Kim, H.J.; Lee, J.; Choi, J.H.; Bahinski, A.; Ingber, D.E. Co-culture of Living Microbiome with Microengineered Human Intestinal Villi in a Gut-on-a-Chip Microfluidic Device. *J. Vis. Exp.* **2016**, *114*, e54344. [CrossRef] [PubMed]

22. Kim, H.J.; Ingber, D.E. Gut-on-a-Chip microenvironment induces human intestinal cells to undergo villus differentiation. *Integr. Biol.* **2013**, *5*, 1130–1140. [CrossRef]

23. Costa, M.O.; Nosach, R.; Harding, J.C.S. Development of a 3D printed device to support long term intestinal culture as an alternative to hyperoxic chamber methods. *3D Print. Med.* **2017**, *3*, 9. [CrossRef] [PubMed]

24. Baydoun, M.; Treizeibré, A.; Follet, J.; Vanneste, S.B.; Creusy, C.; Dercourt, L.; Delaire, B.; Mouray, A.; Viscogliosi, E.; Certad, G.; et al. An interphase Microfluidic Culture System for the Study of Ex Vivo Intestinal Tissue. *Micromachiens* **2020**, *11*, 150. [CrossRef] [PubMed]

25. Dawson, A.; Dyer, C.; Macfie, J.; Davise, J.; Karsai, L.; Greenman, J.; Jacobsen, M. A microfluidic chip based model for the study of full thickness human intestinal tissue using dual flow. *Biomicrofluidics* **2016**, *10*, 064101. [CrossRef] [PubMed]

26. Richardson, A.; Schwerdtfeger, L.A.; Eaton, D.; Mclean, I.; Henry, C.S.; Tobet, S.A. A microfluidic organotypic device for culture of mammalian intestines ex vivo. *Anal. Method.* **2020**, *12*, 297–303. [CrossRef]

27. Kuriu, S.; Kadonosono, T.; Kizaka-Kondoh, S.; Ishida, T. Slicing Spheroid in Microfluidic Devices for Morphological and Immunohistochemical Analysis. *Micromachines* **2020**, *11*, 480. [CrossRef]

28. Cremer, J.; Segota, I.; Yang, C.Y.; Arnoldini, M.; Sauls, J.T.; Zhang, Z.; Gutierrez, E.; Groisman, A.; Hwa, T. Effect of flow and peristaltic mixing on bacterial growth in a gut-like channel. *Proc. Natl. Acad. Sci. USA* **2016**, *113*, 11414–11419. [CrossRef]

29. Demouveaux, B.; Gouyer, V.; Robbe-Masselot, C.; Gottrand, F.; Narita, T.; Desseyn, J.L. Mucin CYS domain stiffens the mucus gel hindering bacteria and spermatozoa. *Sci. Rep.* **2019**, *9*, 16993. [CrossRef]

30. Lieleg, O.; Vladescu, I.; Ribbeck, K. Characterization of Particle Translocation through Mucin Hydrogels. *Biophys. J.* **2010**, *98*, 1782–1789. [CrossRef] [PubMed]

*Review*

# Computational Simulations in Advanced Microfluidic Devices: A Review

Violeta Carvalho [1,2,*], Raquel O. Rodrigues [3], Rui A. Lima [1,4] and Senhorinha Teixeira [2]

1   MEtRICs, Campus de Azurém, University of Minho, 4800-058 Guimarães, Portugal; rl@dem.uminho.pt
2   ALGORITMI, Campus de Azurém, University of Minho, 4800-058 Guimarães, Portugal; st@dps.uminho.pt
3   Center for MicroElectromechanical Systems (CMEMS-UMinho), Campus de Azurém, University of Minho, 4800-058 Guimarães, Portugal; raquel.rodrigues@dei.uminho.pt
4   CEFT, R. Dr. Roberto Frias, Faculty of Engineering of the University of Porto (FEUP), 4200-465 Porto, Portugal
*   Correspondence: violeta.carvalho@dem.uminho.pt

**Abstract:** Numerical simulations have revolutionized research in several engineering areas by contributing to the understanding and improvement of several processes, being biomedical engineering one of them. Due to their potential, computational tools have gained visibility and have been increasingly used by several research groups as a supporting tool for the development of preclinical platforms as they allow studying, in a more detailed and faster way, phenomena that are difficult to study experimentally due to the complexity of biological processes present in these models—namely, heat transfer, shear stresses, diffusion processes, velocity fields, etc. There are several contributions already in the literature, and significant advances have been made in this field of research. This review provides the most recent progress in numerical studies on advanced microfluidic devices, such as organ-on-a-chip (OoC) devices, and how these studies can be helpful in enhancing our insight into the physical processes involved and in developing more effective OoC platforms. In general, it has been noticed that in some cases, the numerical studies performed have limitations that need to be improved, and in the majority of the studies, it is extremely difficult to replicate the data due to the lack of detail around the simulations carried out.

**Keywords:** computational simulations; drug discovery; organ-on-a-chip; microfluidic devices; preclinical models; numerical simulations

**Citation:** Carvalho, V.; Rodrigues, R.O.; Lima, R.A.; Teixeira, S. Computational Simulations in Advanced Microfluidic Devices: A Review. *Micromachines* **2021**, *12*, 1149. https://doi.org/10.3390/mi12101149

Academic Editor: Aleksander Skardal

Received: 9 August 2021
Accepted: 21 September 2021
Published: 23 September 2021

**Publisher's Note:** MDPI stays neutral with regard to jurisdictional claims in published maps and institutional affiliations.

## 1. Introduction

Over the years, the demand for more reliable and effective preclinical models has grown as we try to reduce and, where possible, replace the utilization of animal models for assessing drug efficacy and safety [1]. Although animal models provide a physiological structure similar to that of human cells and organs, they fail to adequately forecast the behavior of and represent human metabolism. These experiments also require laborious protocols, present ethical issues, and are expensive [2–5]. Due to the low throughput of in vivo testing, researchers have been using advanced microfluidic devices that integrate biomodels to replicate the physiology and function of cells and/or organs. A microfluidic device is the building block of the microfluidic chip family, defined as a set of technologies embedded in a chip that contains microchannels with a dimension of tens to hundreds of micrometers to process a small volume of fluids. These microdevices first emerged mainly for analytical purposes, and due to the small amount of reagents and sample requirements, they attracted increasing interest thanks to their several advantages over other standard technologies, such as the short time needed for analysis, reduction in fabrication and reagent costs, miniaturization, selectivity, sensitivity, portability, and biocompatibility. In time, the ability to integrate multiple processes in a single device led to the concept of lab-on-a-chip or micro total analysis systems (μTASs), which were first introduced by Andreas Manz in 1990 [6,7]. Their versatility enables their application in a variety of fields, from environmental monitoring and the food industry and microelectronics to biomedical or clinical applications.

Their biocompatibility and ability to precisely study, evaluate, or diagnose individual or multiple cells in a dynamic and accessible way brought forth the concept of cells-on-a-chip. In these biodevices, cells are inserted (mixed with an organic fluid) into the microfluidic chip, which allows assessing their unique function and behavior outside of the body, maintaining their integrity and biological functionality.

Lastly, and more recently, another concept has been introduced: organ-on-a-chip (OoC), also known as microphysiological systems. In these advanced microfluidic devices, 3D cell structures containing mono or multiple cell lines are embedded in an extracellular matrix to emulate the complexity and architecture of the mimicked living organ [8,9]. These devices allow more complex studies and analyses which are physiologically closer to what we can find in a human body system. In these organ-on-a-chip devices, studies can be devoted to single organs and functions, but also to the interaction and response of multiple organs, where the ultimate goal is a body-on-a-chip which is able to replicate the entire human body in a single platform [10,11]. Among the several attributes of these models, their ability to accelerate research, accurately reproduce the micro-environment of human cells, and be mass-produced in a cost-effective way should be stressed [12,13]. Nevertheless, there are also certain drawbacks associated with them, which have to be overcome over time. Frequently, polydimethylsiloxane (PDMS) is used in the fabrication of biomedical and advanced microfluidic devices owing to its outstanding properties, including optical transparency, permeability to gases, and biocompatibility [14–22]. However, small hydrophobic molecules, such as biomolecules, proteins, and drugs, can be adsorbed to the surface and consequently compromise the quality of the experiments' analysis in OoC platforms [23–25]. Nonetheless, as the advantages of these devices exceed their limitations, these models continue to be a valuable tool in the study of new health treatments. Many researchers have developed several OoCs to simulate the function of the heart [26], lungs [27], liver [28], kidney [29], brain [30], bone [31], etc., or a combination of several organs, known as body-on-a-chip [32]. The last one allows a better understanding of not only the efficacy of a drug, but also its potential toxicity in other organs, which in turn guarantees a better validity of the results and extrapolation to reality [33]. In addition to the previous models, recently, Si et al. [34] used a human-airway-on-a-chip to investigate the antimalarial drug amodiaquine as a powerful inhibitor of infection with SARS-CoV-2. The research team modeled a human bronchial airway epithelium and pulmonary endothelium infected with pseudotyped severe acute respiratory syndrome coronavirus 2 and verified that the antimalarial drug amodiaquine has the potential to inhibit infection. Hence, the use of OoC helped in the rapid advances witnessed in the testing of new drugs to fight the new coronavirus, which has become an urgent need for the fight against the COVID-19 pandemic.

Along with the progress made in the development of new microphysiological systems, computational tools have played an important role. These allow complementing experimental studies, from an engineering point of view, allowing us to study physical phenomena, assess the viability of a device, and optimize the design at a lower cost [35–39]. Furthermore, their capability to determine critical parameters which are difficult to measure via experimental methods, such as pressure, velocity, shear rate, and temperature, constitute other benefits of using numerical methods. For these reasons, researchers have complemented their studies with numerical methods [40–42]. Although a substantial number of review papers exist in the field of microfluidic OoC-based systems, only a small number have addressed the different approaches using numerical simulations as an auxiliary tool in the development of these devices [11,35,37,43]. Hence, the present work aims to provide an overview of the most recent progress in numerical studies in advanced microfluidic devices and how researchers have applied computational tools in their investigations.

## 2. Applications of Numerical Simulations in Advanced Microfluidic Devices

In this section, the usefulness of conducting numerical simulations in parallel with experimental studies is presented, namely, in the study of fluid flow and mass transfer, nanoparticle development, and also in the optimization of advanced microfluidic devices, i.e., OoC and cell-on-a-chip.

### 2.1. Fluid Flow and Mass Transfer

A key factor that affects the functionality of OoC platforms is fluid flow, which includes the evaluation of several parameters, such as velocity fields, shear stress, and the concentration of vital parameters for cellular function, such as oxygen, glucose, and temperature.

The flow of the culture medium is of great importance to maintain an adequate amount of nutrients for the cell culture as well as the shear stresses exposed to cells, and therefore, understanding flow behavior is vital in the design of OoC microfluidic devices [44]. Kocal et al. [45] simulated flow behavior in a cancer microenvironment to understand the uniformity of fluid shear stress along with microfluidic devices. The results showed that adherent cancer cells experienced a homogenous wall shear stress of up to 25 μPa in the microchannels (Figure 1a). Another interesting study was conducted by Lo et al. [46]. They investigated the influence of shear stress and antioxidant concentrations on the production of reactive oxygen species in lung cancer cells by combining experimental and numerical simulations, and analogous results were obtained for both methods. They verified that an increase in shear stress leads to an increase in the production of reactive oxygen species (Figure 1b). Additionally, two antioxidants, α-tocopherol and ferulic acid, were tested to verify their ability to reduce reactive oxygen species. It was found that a high dose of α-tocopherol was not able to eliminate reactive oxygen species, but a lower dose could. The same outcome was not observed for ferulic acid. This antioxidant could eliminate the production of reactive oxygen species in a concentration-independent way. Kou et al. [47], on the other hand, developed a microfluidic system with the ability to create four different shear flows in only one device to study the cytosolic calcium concentration dynamics of osteoblasts. They observed that cytosolic calcium concentration in the osteoblasts increased proportionally to the intensity of shear stress from 0.03 to 0.30 Pa (Figure 1c). Komen et al. [48] developed a microfluidic device with the ability to expose cancer cells to an in-vivo-like concentration profile of a drug and measure the effectiveness on-chip. This device comprised a cell culture chamber and a drug-dosing channel separated by a transparent membrane so that the drug is not exposed to shear stresses and also to allow a label-free growth quantification. They simulated cell exposure and verified that it followed the blood concentrations measured in vivo.

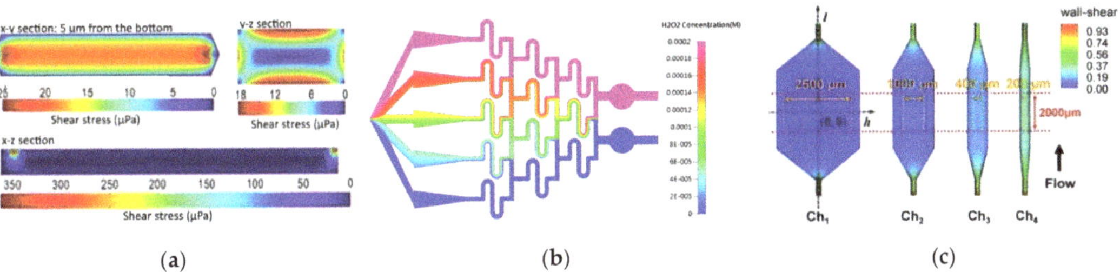


**Figure 1.** (**a**) CFD simulation of shear stress on the XY, YZ, and ZX planes. Adapted from [45]; (**b**) numerical simulation of H₂O₂ concentration inside the microfluidic chip. Adapted from [46]; (**c**) contours of wall shear stress distribution in designed chambers. From left to right, central areas of each chamber, with different horizontal distances. The orange and red dashed lines indicate the uniform wall shear stress area boundary. Adapted from [47].

The diffusion of species is also of utmost importance for cell maintenance, and porous zones are commonly modeled to study these phenomena. Wong et al. [36] created a computational model of a microfluidic device that incorporates flow-induced shear stress and monitors cell-secreted biomolecules. For that, they modeled a bilayer device in which cells are cultured on a porous membrane support. This way, researchers can adjust the device to operate with physiological shear stresses, while ensuring the rapid transport of secreted biomolecules to the biosensor (Figure 2a). In another study [49], Wong et al. used the same model but to simulate the concentration of electroactive tracers, which they validated with experiments. A similar model was proposed by Chen et al. [50], consisting of a porous membrane-separated and a microfluidic chip channel with two layers. In their study, they evaluated the dependences of flow features on the chip geometry and membrane permeability using immersed boundary methods—IBM. The results presented elucidated the dependences of flow flux, wall shear stress, and transmembrane pressure difference on the chip geometry (channel height and length) and membrane permeability, which were consistent with the experimental results. For a membrane with lower permeability, a lengthier channel was required to achieve a permeability-independent stage. Moreover, it was also observed that wall shear stress was dependent on channel length and membrane permeability. Mosavati et al. [51] developed a numerical model of a three-dimensional placenta-on-a-chip model with a porous medium to model the membrane separating the two channels (Figure 2b). The researchers aimed to analyze the unsteady flow into microchannels and glucose concentration profiles in different locations of the microfluidic chip. After validation of the model, Mosavati et al. studied the effects of flow rate and membrane porosity on glucose diffusion through the placental barrier. Similarly, in a recent study [4], Banaeiyan et al. explored the diffusion of glucose, since this is the main element present in the culture medium, through diffusion channels. By contrast, Hu and Li [52] numerically investigated oxygen concentration in the interior of a three-dimensional microchannel containing tumor spheroids and tested two different velocities, 50 $\mu$m/s and 100 $\mu$m/s. They observed that with lower velocity, oxygen concentration downstream of the spheroid surface is not enough for spheroid growth. By raising the perfusion velocity, oxygen concentration increases and may consequently enhance the growth of the spheroids. Another model comprising a permeable layer was presented by Zahorodny-Burke and Elias [53]. Briefly, they developed a rectangular channel with a polymer layer permeable to oxygen linked to a glass substrate seeded with one layer of oxygen-consuming cells. In contrast, Bhise and coworkers [54] did not simulate a porous membrane where oxygen passes through it. Instead, they modeled hydrogels as a porous medium with a homogenous volumetric oxygen consumption rate related to the total number of encapsulated cells. Other researchers have studied not only the diffusion of species but also the convection of carbon dioxide—$CO_2$ [55]. A computational model was developed to investigate $CO_2$ transport in a cylindrical microfluidic device. The level of $CO_2$ at the base of the water chamber for different $CO_2$ supplying concentrations was investigated, and it was found that by increasing the level of $CO_2$ in the feeding gas to above 10%, $CO_2$ concentration at the bottom of the chamber reaches the desired level of 5%.

The study of the effects of radiation treatment on microcirculation was recently addressed by Cicchetti et al. [56]. The authors developed a numerical model to understand the effects of radiation therapy on normal tissues surrounding a tumor. They modeled the vasodilation and variation of membrane permeability as inputs with consequent fluid extravasation and the deposition of fibrin or the formation of edema. In a second simulation, the researchers modeled the impact of variations in vessel wall elasticity on the distribution of blood flow velocity.

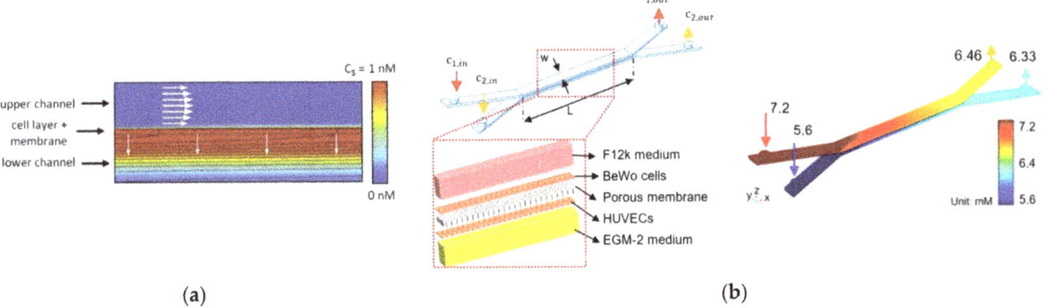

**Figure 2.** (**a**) Computational modeling of fluid flow and analyte transport in microchannels and porous membrane. White arrows indicate flow direction in the membrane and the upper microfluidic channel. The upper surface of the cell layer has a fixed concentration equal to 1 nM. Analyte flux streamlines are shown in black. Adapted from [36]; (**b**) chip design and glucose concentration profiles at a flow rate of 50 μL/h in the bare membrane, respectively. Adapted from [51].

Another important parameter in these types of studies is temperature because it affects not only the environment but also the behavior and function of cells [57]. Hence, its monitoring is of great importance for cell culture. Peng et al. [58] conducted a multiphysics simulation (fluid flow and heat transfer) in a microfluidic device to obtain an accurate and rapid control of on-chip local temperature control. According to the results, the device was able to operate with temperatures ranging between 2 °C and 37 °C using cooling and heating components (Figure 3a). A similar study was conducted by Das and coworkers [59], in which a microfluidic device to study the viability and activity of the cells under a temperature gradient was developed. Through numerical simulations, the temperature gradient that can be reached in the proposed chip was assessed, and it was found that the proposed chip is capable of creating temperature conditions that realistically mimic physiological/biological conditions (Figure 3b).

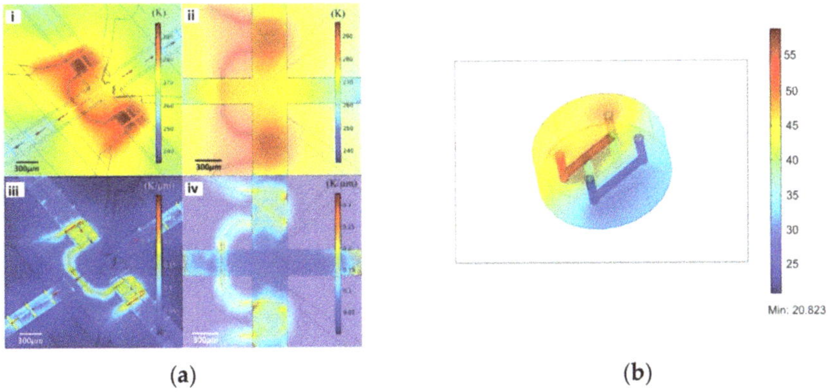

**Figure 3.** (**a**) Numerical simulation of on-chip temperature profiles including spatial temperature distribution (i), planar temperature distribution (ii), spatial temperature gradient distribution (iii), and planar temperature gradient distribution (iv). Adapted from [58]; (**b**) temperature distribution in the chip in the incubator environment (370 K ambient and bath at 900 K). Adapted from [59].

Although single-phase simulations are mostly used, the use of multiphase models is a valuable option to improve the precision of simulations and obtain more detailed information. To study the behavior of circulating tumor cells in a 3D bioprinted vasculature chip, Hynes et al. [60] performed numerical simulations using both the lattice

Boltzmann method (LBM) and finite element method (Figure 4a). The particles were treated as either rigid or deformable spheres, and the results matched closely to the experimental findings, proving that the numerical model used was appropriate. By contrast, Ye et al. [61] used a smoothed dissipative particle dynamics (SDPD) method to model the fluid flow and an IBM for the fluid–cell interaction to study the RBC mechanics in a microfluidic chip (Figure 4b). Likewise, Zhang et al. [62], in order to simulate mechanical interactions between flow and particles (cells) for a cells-on-a-chip device, used a two-way Euler/Lagrange multiphase model. Another example of the use of multiphase models in this type of study was presented by Tanaka et al. [63]. The authors developed a hybrid pump driven by cardiomyocytes that self-organize into bridges between elastic micro-pier structures (Figure 4c). In addition, they performed numerical simulations to understand the flow pattern and distribution of flow velocity, and the results were in good agreement with the experimental results. They used the volume of fluid (VOF) model to better understand and verify the flow generation, flow pattern, and velocity distribution of the ultra-small fluid oscillation unit.

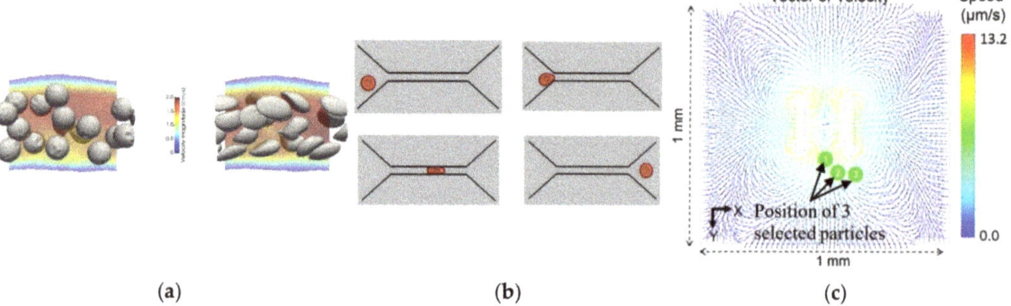



**Figure 4.** (**a**) Simulation of relatively stiff beads with a shear elastic modulus, Gs, of 10−3 N/m, compared to deformable cells Gs = 10−5 N/m. Adapted from [60]; (**b**) deformation of an RBC passing through a narrow microchannel. Adapted from [61]; (**c**) vector of velocity on the XY plane and position of the selected particles in experimental tests. Adapted from [63].

Another key parameter that affects the function of a device is the material. As previously mentioned, PDMS is preferred due to its suitable properties and low price [5,23,64]. Moreover, this material can be obtained with different air permeability, which is interesting for improving the diffusion of species in advanced microfluidic devices. Lamberti et al. [65] demonstrated that by increasing the PDMS mixing ratio, air permeability can be improved up to 300% when compared to the standard 10:1 ratio. However, this material also has certain limitations that encourage researchers to seek alternative materials, one of them being the possibility of absorption of small hydrophobic species [23,64]. Two alternative materials are poly(-methylmethacrylate) (PMMA) [66–69] and cyclic olefin copolymer (COC) [70]. Zahorodny-Burke et al. [53] studied the nature of oxygen transport within microfluidic cell culture devices considering PDMS, COC, and PMMA through a two-dimensional convection–diffusion mass transfer model. To this end, flow rate, diffusive layer thickness, and material selection were investigated in order to determine their relation to ensure adequate supplies of oxygen for cell growth. However, special attention must be paid when using COC or PMMA. Due to the low oxygen diffusion of these materials, flow rate must be optimized to ensure a sufficient supply of oxygen to the cells by applying high flow rates, at an order of $10^{-2}$ m/s (Figure 5). By contrast, Ochs and coworkers [71] investigated not only PDMS and COC, but also a different type of material, polymethylpentene (PMP). They experimentally and numerically studied the influence of PMP on oxygen concentration in cell culture and verified that this is a possible alternative for PDMS. PMP was shown to be able to deliver sufficient oxygen for various types of cells, namely, hepatocytes with a high oxygen consumption rate.

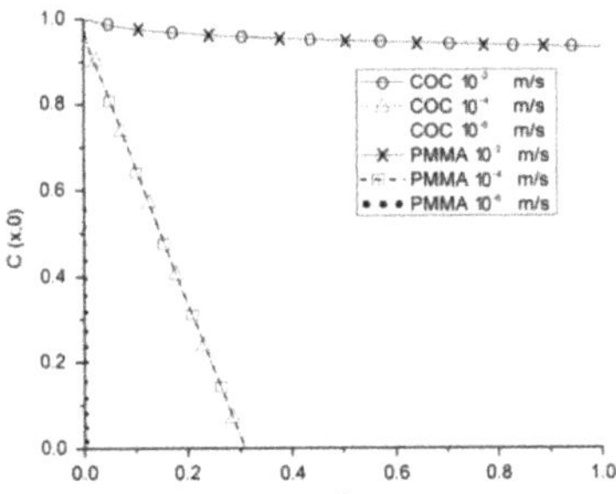


**Figure 5.** COC and PMMA at various average flow rates with a polymer capping layer with a thickness of 1.9 mm. At this thickness, for corresponding speeds, both polymers have virtually identical profiles. Adapted from [53].

*2.2. Nanoparticle Simulations*

Nanoparticles (NPs) have been attracting interest for their potential use as drug delivery systems, but the optimization of their physical properties, such as their size and shape, is a point of interest to accomplish appropriate cell responses [72–74]. Due to the multidisciplinarity of computational software, several researchers have taken advantage of its capabilities to broaden the understanding of the transport of NPs and drugs in OoC [75–79]. Soheili et al. [75] performed a numerical and experimental study on the microfluidic preparation and production of chitosan NPs. To develop NPs with the desired size, they implemented a model that relates NPs' diameter to the flow rate, and sodium tripolyphosphate (TPP) concentration as a critical factor that affects cell flow behavior. Further, the researchers found the optimal mixing efficiency of tripolyphosphate conditions by associating TPP concentration with the length of the channels and flow rate. Kwak and coworkers [79] developed a tumor-microenvironment-on-chip to recapitulate the key features of complex transport of drugs and NPs within a tumor microenvironment. By combining computational simulations and experimental tests, they obtained more detailed information about the dynamic transport behavior of NPs and concluded that NPs should be conceived considering their interactions with the tumor microenvironment.

Other authors have been motivated by the "catch and release" of biomolecules observed in physiological processes and used numerical simulations to design synthetic systems that can catch, transport, and release targeted species inside the surrounding solution, which can be useful for effective separation processes within microfluidic devices [77]. Their setup allows the determination of the conditions where the oscillating fins of the device can selectively bind and "catch" target NPs in the upper channel, and then release them in the lower channel. Their findings have provided insights into fabrication devices to detect and separate target molecules from complex fluids. This topic has been addressed by various researchers [80–83]. Maleki and coworkers [78] developed a hybrid drug carrier of doxorubicin (DOX) due to the high consumption rate of riboflavin in cancerous cells. In parallel, molecular simulations to study the effects of the microfluidic environment and shift metal dichalcogenide nanolayers on the stability, size, and self-assembly interaction energies of the nanocarriers were conducted (Figure 6a). The results showed that among the studied configurations, the most suitable ones for the adsorption of DOX molecules are $MoSe_2$, MoSSe, $MoO_3$, and $MoS_2$, respectively. Furthermore, the appearance of the new coronavirus has triggered several scientific investigations, and the study conducted by Arefi and colleagues [76] was one of them. They studied the effect of airborne NPs on human health through both numerical simulations and experimental work. Transport and adsorption in a microfluidic lung-on-a-chip device were investigated, and two different multiphase models were used. For the fine NPs, particle transport was estimated using an Eulerian advection-diffusion approach, while for coarse NPs, the Lagrangian particle tracking approach was used (Figure 6b). One of their findings showed that during physical exercise, particle deposition increases, and thus, outdoor exercise on days with poor air quality should be reduced.

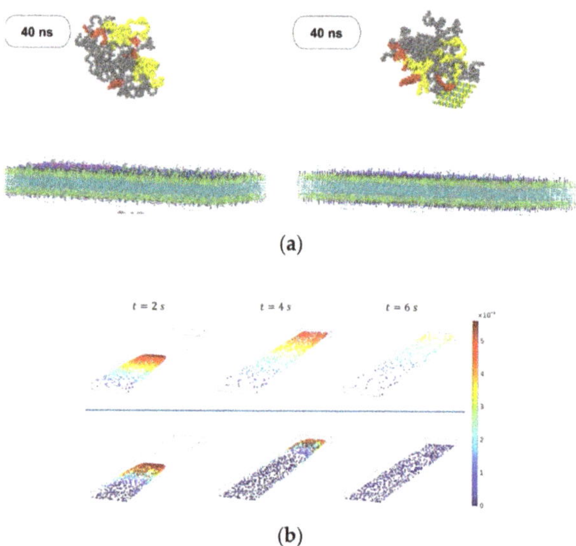


**Figure 6.** (a) Nanocarrier penetration into the cancer cell membrane without $MoSe_2$ and with $MoSe_2$, respectively. Adapted from [78]; (b) Distribution and instantaneous velocity of non-Brownian particles at t = 2, 4 and 6 s for particle diameter d = 100 nm and 500 nm. Adapted from [76].

## 3. Numerical Optimization

The use of numerical simulations to optimize devices is one of the main steps that must be carried out during the design process. This helps researchers to improve and accelerate the development of more efficient and representative models for a given problem by subjecting the chip to design changes and improvements. Experimental conditions can also be optimized in accordance with numerical simulations. For instance, Karakas and coworkers [84] developed a microfluidic chip for screening individual cancer cells and performed numerical simulations in order to optimize the exposure time for autophagy, a cellular mechanism where proteins are consumed and recycled to give another source of energy to cells (Figure 7a). This numerical study allowed investigators to determine the minimum exposure time required to ensure the viability of the experiments, saving time and resources during experimental tests. On the other hand, Zhang et al. [81] studied the effect of varying the inlet flow rate of blood and buffer on the separation performance of circulating tumor cells in a microfluidic chip and optimized the operating conditions to increase separation efficiency. Another way to improve a device's accuracy was presented by Jun-Shan et al. [85], who developed a microfluidic chip with micropillar arrays for 3D cell culture, and using numerical simulations, the space between micropillars was optimized, which allowed the nutrients in the medium to quickly diffuse into the chamber, while cell metabolites could diffuse out of the chamber in a timely manner (Figure 7b). A similar study with micropillars was performed by Chen and coworkers [86]. However, in this case, the authors studied pillars with circular, elliptical, and square cross-sections assembled in both aligned and staggered patterns. Through numerical simulations, the researchers found that in the latter case, fluid flows through both the center of the array and around the pillars. These observations led the research team to address this approach since in staggered arrangements, fluid is better distributed throughout the device. Zhang et al. [62] used a two-way Euler/Lagrange multiphase model to simulate mechanical interactions between flow and particles (cells) for cells-on-a-chip devices. The authors developed three different designs and studied the effect of using different cell densities, inlet flow velocities, and inlet cell numbers. The results showed that regions of lower strain rates had high cell concentrations, and when lower inlet velocities (10 and 20 $\mu$m/s) were used, high cell concentrations were observed. However, they observed that cells were not able to reach the outlet, while with a velocity of 40 $\mu$m/s, some cells were.

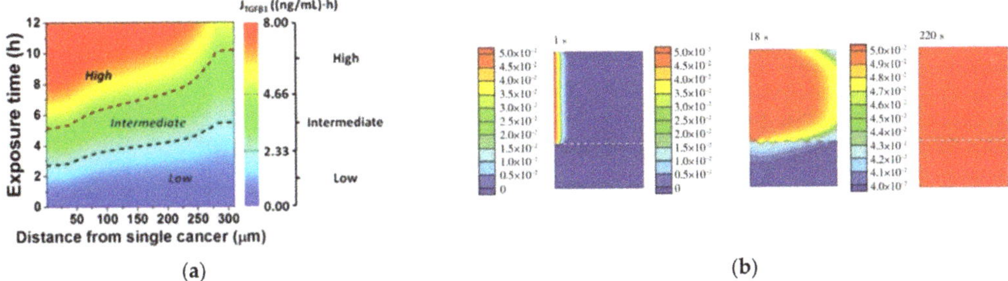

**(a)**

**(b)**

**Figure 7.** (**a**) Contour plot of the TGF$\beta$1 exposure ($J_{TGF\beta1}$) in (distance, exposure time)-space, where the numerical simulation was conducted. Adapted from [84]; (**b**) glucose concentration profile at different instants (1, 18, and 220 s) in the chip. Adapted from [85].

Optimization methods are also of great importance for discovering the relationship between several variables. For instance, Huang and Nguyen [87] simulated a cell-stretching device and optimized important dimensional parameters. They performed a parametrical optimization to determine the relationship between the lateral displacement of the membrane and wall height, wall thickness, membrane thickness, and vacuum strength in order to maximize the output strain of the stretched membrane.

In Table 1, a summary of the numerical considerations used in the previously reviewed papers is presented, including the software used, the flow behavior, flow type, fluid rheology, mesh, flow analysis, and also the validation.

**Table 1.** Summary of the considerations made in the numerical studies.

| Software | Flow Regime | Phases Number | Fluid Rheology | Mesh | Time-Dependence | Validation | Ref |
|---|---|---|---|---|---|---|---|
| COMSOL | Laminar and turbulent | Single-phase | Newtonian | n/d | Steady | × | [45] |
| CFD-ACE + | Laminar | Single-phase | n/d | n/d | Steady | ✓ | [46] |
| Ansys | Laminar | Single-phase | Newtonian | 78,000 nodes | Steady | × | [47] |
| COMSOL | n/d | Single-phase | n/d | n/d | Transient | ✓ | [48] |
| COMSOL | Laminar | Single-phase | Newtonian | 182,900 elements | Transient | × | [36] |
| n/d | Laminar | Single-phase | n/d | n/d | Transient | ✓ | [49] |
| IBM, MATLAB and Ansys | Laminar | Multiphase | Newtonian | Mesh independence test (1 $\mu$m) | Transient | × | [50] |
| COMSOL | Laminar | Single-phase | Newtonian | Mesh independence test (n/d) | Transient | ✓ | [51] |
| COMSOL | Laminar | Single-phase | Newtonian | Fine mesh (n/d) | Transient | × | [4] |
| COMSOL | n/d | Single-phase | n/d | n/d | Transient | ✓ | [52] |
| COMSOL | Laminar | Single-phase | n/d | 7000 elements | Steady | ✓ | [53] |
| COMSOL | n/d | Single-phase | n/d | Mesh independence test (800,000 nodes) | Transient | × | [54] |
| COMSOL | n/d | Multiphase | Newtonian | 24,300 elements | Transient | ✓ | [55] |
| User-Defined Software | Laminar | Single-phase | n/d | n/d | Steady | × | [56] |
| COMSOL | Laminar | Single-phase | n/d | n/d | Steady | × | [58] |
| COMSOL | Laminar | Single-phase | n/d | 123,334 elements | Steady | ✓ | [59] |
| User-Defined Software | n/d | Multiphase | Newtonian | n/d | Transient | ✓ | [60] |
| SDPD and IBM | n/d | Multiphase | n/d | n/d | Transient | ✓ | [61] |
| Ansys | Laminar | Multiphase | Newtonian | Mesh independence test (220,000 elements) | Transient | ✓ | [62] |
| Ansys | Laminar | Multiphase | Newtonian | n/d | Steady | ✓ | [63] |

**Table 1.** *Cont.*

| Software | Flow Regime | Phases Number | Fluid Rheology | Mesh | Time-Dependence | Validation | Ref |
|---|---|---|---|---|---|---|---|
| COMSOL | n/d | Single-phase | n/d | $1.8 \times 10^6$ elements | Transient | ✓ | [71] |
| COMSOL | Laminar | Multiphase | n/d | Mesh independence test (6000 elements) | Transient | × | [75] |
| n/d | n/d | Multiphase | n/d | n/d | Transient | ✓ | [79] |
| n/d | n/d | Multiphase | n/d | n/d | Transient | × | [77] |
| GROMACS | n/d | Multiphase | Newtonian | n/d | Transient | ✓ | [78] |
| COMSOL | Laminar | Multiphase | n/d | Mesh independence test (168,000 elements) | Steady and Transient | ✓ | [76] |
| COMSOL | n/d | Single-phase | n/d | n/d | Transient | × | [84] |
| n/d | n/d | Multiphase | n/d | n/d | Transient | × | [81] |
| COMSOL and Ansys | n/d | Single-phase | n/d | n/d | Transient | ✓ | [85] |
| Ansys | n/d | Single-phase | n/d | n/d | Steady | × | [86] |
| Ansys | n/d | Single-phase | n/d | 2–8 $\mu$m element size | Transient | ✓ | [87] |

n/d—non-defined.

A brief analysis of the results reveals several blank spaces which represent the absence of information regarding the assumptions made when performing the numerical works. Additionally, the preferred software for these types of studies was found to be COMSOL Multiphysics® software, followed by Ansys software®, although researchers have begun to develop their own simulation software to overcome some of the limitations of the existing ones [56,60]. Moreover, a fundamental step in numerical simulations is mesh development. Without guaranteeing that the mesh is appropriate, one cannot obtain truthful and accurate results. Hence, both mesh quality and independence have to be ensured, and this is a step that is commonly missing in the studies presented in this review article. Some authors conducted mesh independence studies, but the majority only indicate the number of elements or do not present any information. One way to carry out such studies with more accuracy is through the calculation of the grid convergence index, as represented in the study conducted by Soheili et al. [75], increasing, in this way, the reliability of the presented results. Moreover, it can be seen that the validation of these types of studies is usually disregarded, but this is an extremely important step that should be addressed whenever possible.

## 4. Final Remarks and Future Directions

Advanced microfluidic devices have revolutionized drug development processes and in vitro tests by allowing to save time and money but also, more importantly, to present a preclinical platform that can accurately mimic human tissues/organs and thus diminish the use of animals in clinical testing. These devices have become research accelerators, therefore rapidly achieving the status of a promising technology, which has now been proven to be useful for advances in emerging diseases, such as research related to the new coronavirus, Sars-Cov-2, and cancer treatment. Nevertheless, although this technology has been shown to be extremely valuable, it is still far from being able to precisely mimic a proper human organ, which has hindered the passage of these devices to the market. Therefore, research and development of more complex and reliable advanced microfluidic devices will continue to progress for several years, and numerical simulations are expected to boost this process by not only allowing mass transport insights, but also reducing the costs and saving time in research. Hence, computational methods are envisioned to become a fundamental tool for the advance of research in advanced microfluidic platforms.

The present review intended to encourage current and future researchers to include this tool in their studies, and it has shown the importance of using numerical simulations in parallel with experimental works to improve the performance and quality of advanced microfluidic devices. Nevertheless, it was found that most numerical studies did not include details about the simulations used, which makes it unfeasible for other researchers to recreate such simulations and understand what methods were actually used to obtain certain outputs.

In addition to the advantages of using numerical tools, there are also certain challenges and limitations associated with numerical modeling platforms. As biomedical devices become more complicated, mathematical modeling platforms need further improvement and validation to become more precise, and this is a lengthy and difficult process. Moreover, in these biological studies, there are a lot of physical and chemical processes occurring both at a micro- and macro-scale, and it becomes difficult to assess and represent these in one numerical model, making simulations extremely complex and time-consuming. However, given the advancements in this type of technology, the field is expected to grow by leaps and bounds, and more biological features will be included in the existing software. For instance, molecular dynamics simulations may be one good alternative to perform more complex simulations.

**Author Contributions:** Conceptualization, V.C., R.O.R., R.A.L., and S.T.; methodology, V.C.; validation, R.O.R., R.A.L., and S.T.; formal analysis, R.O.R., R.A.L., and S.T.; writing—original draft preparation, V.C.; writing—review and editing, V.C., R.O.R., R.A.L., and S.T.; supervision, R.O.R., R.A.L., and S.T.; funding acquisition, R.O.R., R.A.L., and S.T. All authors have read and agreed to the published version of the manuscript.

**Funding:** The authors are grateful for the funding of FCT through projects NORTE-01-0145-FEDER-029394, NORTE-01-0145-FEDER-030171 funded by COMPETE2020, NORTE2020, PORTUGAL2020, and FEDER. This work was also supported by Fundação para a Ciência e a Tecnologia (FCT) under the strategic grants UIDB/04077/2020, UIDB/00319/2020, UIDB/04436/2020, and UIDB/00532/2020.

**Acknowledgments:** Violeta Carvalho acknowledges the PhD scholarship UI/BD/151028/2021 attributed by FCT.

**Conflicts of Interest:** The authors declare no conflict of interest.

# References

1. Filipovic, N.; Nikolic, M.; Sustersic, T. Simulation of organ-on-a-chip systems. In *Biomaterials for Organ and Tissue Regeneration*; Vrana, N.E., Knopf-Marques, H., Barthes, J., Eds.; Woodhead Publishing Series in Biomaterial; Woodhead Publishing: Sawston, UK, 2020.
2. Soker, S.; Carolina, N. A reductionist metastasis-on-a-chip platform for in vitro tumor progression modeling and drug screening. *Biotechnol. Bioeng.* **2016**, *113*, 2020–2032. [CrossRef]
3. Metiner, P.S.; Gulce-Iz, S.; Biray-Avci, C. Bioengineering-inspired three-dimensional culture systems: Organoids to create tumor microenvironment. *Gene* **2018**, *686*, 203–212. [CrossRef]
4. Banaeiyan, A.A.; Theobald, J.; Paukstyte, J.; Wölfl, S.; Adiels, C.B.; Goksör, M. Design and fabrication of a scalable liver-lobule-on-a-chip microphysiological platform. *Biofabrication* **2017**, *9*, 015014. [CrossRef] [PubMed]
5. Carvalho, V.; Maia, I.; Souza, A.; Ribeiro, J.; Costa, P.; Puga, H.; Teixeira, S.; Lima, R.A. In vitro biomodels in stenotic arteries to perform blood analogues flow visualizations and measurements: A review. *Open Biomed. Eng. J.* **2020**, *14*, 87–102. [CrossRef]
6. Rodrigues, R.O.; Lima, R.; Gomes, H.T.; Silva, A.M.T. Polymer microfluidic devices: An overview of fabrication methods. *U. Porto. J. Eng.* **2017**, *1*, 67–79. [CrossRef]
7. Manz, A.; Graber, N.; Widmer, H. Miniaturized total chemical analysis systems: A novel concept for chemical sensing. *Sens. Actuators B Chem.* **1990**, *1*, 244–248. [CrossRef]
8. Huh, D.; Kim, H.J.; Fraser, J.P.; E Shea, D.; Khan, M.; Bahinski, A.; Hamilton, G.A.; Ingber, D.E. Microfabrication of human organs-on-chips. *Nat. Protoc.* **2013**, *8*, 2135–2157. [CrossRef]
9. Fabre, K.M.; Livingston, C.; Tagle, D.A. Organs-on-chips (microphysiological systems): Tools to expedite efficacy and toxicity testing in human tissue. *Exp. Biol. Med.* **2014**, *239*, 1073–1077. [CrossRef]
10. Wu, Q.; Liu, J.; Wang, X.; Feng, L.; Wu, J.; Zhu, X.; Wen, W.; Gong, X. Organ-on-a-chip: Recent breakthroughs and future prospects. *Biomed. Eng. Online* **2020**, *19*, 9. [CrossRef]
11. Carvalho, V.; Gonçalves, I.; Lage, T.; Rodrigues, R.; Minas, G.; Teixeira, S.; Moita, A.; Hori, T.; Kaji, H.; Lima, R. 3D printing techniques and their applications to organ-on-a-chip platforms: A systematic review. *Sensors* **2021**, *21*, 3304. [CrossRef]
12. Huh, D.; Hamilton, G.A.; Ingber, D.E. From 3D cell culture to organs-on-chips. *Trends Cell Biol.* **2011**, *21*, 745–754. [CrossRef] [PubMed]
13. Polini, A.; Prodanov, L.; Bhise, N.S.; Manoharan, V.; Dokmeci, M.R.; Khademhosseini, A. Organs-on-a-chip: A new tool for drug discovery. *Expert Opin. Drug Discov.* **2014**, *9*, 335–352. [CrossRef] [PubMed]
14. Pinho, D.; Carvalho, V.; Gonçalves, I.M.; Teixeira, S.; Lima, R. Visualization and measurements of blood cells flowing in microfluidic systems and blood rheology: A personalized medicine perspective. *J. Pers. Med.* **2020**, *10*, 249. [CrossRef] [PubMed]

15. Faustino, V.; Rodrigues, R.O.; Pinho, D.; Costa, E.; Santos-Silva, A.; Miranda, V.; Amaral, J.S.; Lima, R. A microfluidic deformability assessment of pathological red blood cells flowing in a hyperbolic converging microchannel. *Micromachines* **2019**, *10*, 645. [CrossRef] [PubMed]
16. Catarino, S.O.; Rodrigues, R.O.; Pinho, D.; Miranda, J.M.; Minas, G.; Lima, R. Blood cells separation and sorting techniques of passive microfluidic devices: From fabrication to applications. *Micromachines* **2019**, *10*, 593. [CrossRef]
17. Quirós-Solano, W.F.; Gaio, N.; Stassen, O.M.J.A.; Arik, Y.B.; Silvestri, C.; Van Engeland, N.C.A.; van der Meer, A.; Passier, R.; Sahlgren, C.; Bouten, C.; et al. Microfabricated tuneable and transferable porous PDMS membranes for organs-on-chips. *Sci. Rep.* **2018**, *8*, 13524. [CrossRef]
18. Boas, L.V.; Faustino, V.; Lima, R.; Miranda, J.M.; Minas, G.; Fernandes, C.S.V.; Catarino, S.O. Assessment of the deformability and velocity of healthy and artificially impaired red blood cells in narrow polydimethylsiloxane (PDMS) microchannels. *Micromachines* **2018**, *9*, 384. [CrossRef]
19. Lima, R.; Wada, S.; Tanaka, S.; Takeda, M.; Ishikawa, T.; Tsubota, K.-I.; Imai, Y.; Yamaguchi, T. In vitro blood flow in a rectangular PDMS microchannel: Experimental observations using a confocal micro-PIV system. *Biomed. Microdevices* **2008**, *10*, 153–167. [CrossRef]
20. Pinho, D.; Muñoz-Sánchez, B.N.; Anes, C.F.; Vega, E.J.; Lima, R. Flexible PDMS microparticles to mimic RBCs in blood particulate analogue fluids. *Mech. Res. Commun.* **2019**, *100*, 103399. [CrossRef]
21. Mata, A.; Fleischman, A.J.; Roy, S. Characterization of polydimethylsiloxane (PDMS) properties for biomedical micro/nanosystems. *Biomed. Microdevices* **2005**, *7*, 281–293. [CrossRef]
22. Leclerc, E.; Sakai, Y.; Fujii, T. Cell culture in 3-dimensional microfluidic structure of PDMS (polydimethylsiloxane). *Biomed. Microdevices* **2003**, *5*, 109–114. [CrossRef]
23. Halldorsson, S.; Lucumi, E.; Gomez-Sjoberg, R.; Fleming, R.M. Advantages and challenges of microfluidic cell culture in polydimethylsiloxane devices. *Biosens. Bioelectron.* **2015**, *63*, 218–231. [CrossRef] [PubMed]
24. Wang, Y.; Chen, S.; Sun, H.; Li, W.; Hu, C.; Ren, K. Recent progresses in microfabricating perfluorinated polymers (Teflons) and the associated new applications in microfluidics. *Microphysiol. Syst.* **2018**, *2*, 6. [CrossRef]
25. Sosa-Hernández, J.E.; Villalba-Rodríguez, A.M.; Romero-Castillo, K.D.; Aguilar-Aguila-Isaías, M.A.; García-Reyes, I.E.; Hernández-Antonio, A.; Ahmed, I.; Sharma, A.; Parra-Saldívar, R.; Iqbal, H.M.N. Organs-on-a-chip module: A review from the development and applications perspective. *Micromachines* **2018**, *9*, 536. [CrossRef]
26. Sakamiya, M.; Fang, Y.; Mo, X.; Shen, J.; Zhang, T. A heart-on-a-chip platform for online monitoring of contractile behavior via digital image processing and piezoelectric sensing technique. *Med. Eng. Phys.* **2019**, *75*, 36–44. [CrossRef]
27. Zamprogno, P.; Wüthrich, S.; Achenbach, S.; Thoma, G.; Stucki, J.D.; Hobi, N.; Schneider-Daum, N.; Lehr, C.-M.; Huwer, H.; Geiser, T.; et al. Second-generation lung-on-a-chip with an array of stretchable alveoli made with a biological membrane. *Commun. Biol.* **2021**, *4*, 168. [CrossRef]
28. Deng, J.; Wei, W.; Chen, Z.; Lin, B.; Zhao, W.; Luo, Y.; Zhang, X. Engineered liver-on-a-chip platform to mimic liver functions and its biomedical applications: A review. *Micromachines* **2019**, *10*, 676. [CrossRef]
29. Benam, K.H.; Dauth, S.; Hassell, B.; Herland, A.; Jain, A.; Jang, K.-J.; Karalis, K.; Kim, H.J.; MacQueen, L.; Mahmoodian, R.; et al. Engineered in vitro disease models. *Annu. Rev. Pathol. Mech. Dis.* **2015**, *10*, 195–262. [CrossRef]
30. Ndyabawe, K.; Cipriano, M.; Zhao, W.; Haidekker, M.; Yao, K.; Mao, L.; Kisaalita, W.S. Brain-on-a-chip device for modeling multiregional networks. *ACS Biomater. Sci. Eng.* **2021**, *7*, 350–359. [CrossRef]
31. Hao, S.; Ha, L.; Cheng, G.; Wan, Y.; Xia, Y.; Sosnoski, D.M.; Mastro, A.M.; Zheng, S.-Y. A spontaneous 3D bone-on-a-chip for bone metastasis study of breast cancer cells. *Small* **2018**, *14*, e1702787. [CrossRef]
32. Sung, J.H.; Wang, Y.I.; Sriram, N.N.; Jackson, M.; Long, C.; Hickman, J.J.; Shuler, M.L. Recent advances in body-on-a-chip systems. *Anal. Chem.* **2019**, *91*, 330–351. [CrossRef] [PubMed]
33. Shuler, M.L. Organ-, body- and disease-on-a-chip systems. *Lab Chip* **2017**, *17*, 2345–2346. [CrossRef] [PubMed]
34. Si, L.; Bai, H.; Rodas, M.; Cao, W.; Oh, C.Y.; Jiang, A.; Moller, R.; Hoagland, D.; Oishi, K.; Horiuchi, S.; et al. A human-airway-on-a-chip for the rapid identification of candidate antiviral therapeutics and prophylactics. *Nat. Biomed. Eng.* **2021**, *5*, 815–829. [CrossRef]
35. Sung, J.H.; Wang, Y.; Shuler, M.L. Strategies for using mathematical modeling approaches to design and interpret multi-organ microphysiological systems (MPS). *APL Bioeng.* **2019**, *3*, 021501. [CrossRef]
36. Wong, J.F.; Young, E.W.K.; Simmons, C.A. Computational analysis of integrated biosensing and shear flow in a microfluidic vascular model. *AIP Adv.* **2017**, *7*, 115116. [CrossRef]
37. Sheidaei, Z.; Akbarzadeh, P.; Kashaninejad, N. Advances in numerical approaches for microfluidic cell analysis platforms. *J. Sci. Adv. Mater. Devices* **2020**, *5*, 295–307. [CrossRef]
38. Carvalho, V.; Pinho, D.; Lima, R.A.; Teixeira, J.C.; Teixeira, S. Blood flow modeling in coronary arteries: A review. *Fluids* **2021**, *6*, 53. [CrossRef]
39. Lopes, D.; Agujetas, R.; Puga, H.; Teixeira, J.; Lima, R.; Alejo, J.; Ferrera, C. Analysis of finite element and finite volume methods for fluid-structure interaction simulation of blood flow in a real stenosed artery. *Int. J. Mech. Sci.* **2021**, *207*, 106650. [CrossRef]
40. Carvalho, V.; Rodrigues, N.; Ribeiro, R.; Costa, P.F.; Teixeira, J.C.F.; Lima, R.A.; Teixeira, S.F.C.F. Hemodynamic study in 3D printed stenotic coronary artery models: Experimental validation and transient simulation. *Comput. Methods Biomech. Biomed. Eng.* **2020**, *24*, 623–636. [CrossRef]

41. Carvalho, V.; Rodrigues, N.; Ribeiro, R.; Costa, P.F.; Lima, R.A.; Teixeira, S.F. 3D printed biomodels for flow visualization in stenotic vessels: An experimental and numerical study. *Micromachines* **2020**, *11*, 549. [CrossRef]

42. Doutel, E.; Viriato, N.; Carneiro, J.; Campos, J.B.; Miranda, J.M. Geometrical effects in the hemodynamics of stenotic and non-stenotic left coronary arteries—Numerical and in vitro approaches. *Int. J. Numer. Methods Biomed. Eng.* **2019**, *35*, e3207. [CrossRef]

43. Rodrigues, R.O.; Sousa, P.C.; Gaspar, J.; Bañobre-López, M.; Lima, R.; Minas, G. Organ-on-a-chip: A preclinical microfluidic platform for the progress of nanomedicine. *Small* **2020**, *16*, e2003517. [CrossRef]

44. Nguyen, N.T.; Wereley, S.T.; Shaegh, S.A.M. *Fundamentals and Applications of Microfluidics*, 3rd ed.; Artech House: Norwood, MA, USA, 2019; Volume 110.

45. Kocal, G.C.; Guven, S.; Foygel, K.; Goldman, A.; Chen, P.; Sengupta, S.; Paulmurugan, R.; Baskin, Y.; Demirci, U. Dynamic microenvironment induces phenotypic plasticity of esophageal cancer cells under flow. *Sci. Rep.* **2016**, *6*, 38221. [CrossRef]

46. Lo, K.-Y.; Zhu, Y.; Tsai, H.-F.; Sun, Y.-S. Effects of shear stresses and antioxidant concentrations on the production of reactive oxygen species in lung cancer cells. *Biomicrofluidics* **2013**, *7*, 64108. [CrossRef]

47. Kou, S.; Pan, L.; van Noort, D.; Meng, G.; Wu, X.; Sun, H.; Xu, J.; Lee, I. A multishear microfluidic device for quantitative analysis of calcium dynamics in osteoblasts. *Biochem. Biophys. Res. Commun.* **2011**, *408*, 350–355. [CrossRef]

48. Komen, J.; Westerbeek, E.Y.; Kolkman, R.W.; Roesthuis, J.; Lievens, C.; Berg, A.V.D.; Van Der Meer, A.D. Controlled pharmacokinetic anti-cancer drug concentration profiles lead to growth inhibition of colorectal cancer cells in a microfluidic device. *Lab Chip* **2020**, *20*, 3167–3178. [CrossRef]

49. Wong, J.F.; Simmons, C.A. Microfluidic assay for the on-chip electrochemical measurement of cell monolayer permeability. *Lab Chip* **2019**, *19*, 1060–1070. [CrossRef] [PubMed]

50. Chen, S.; Xue, J.; Hu, J.; Ding, Q.; Zhou, L.; Feng, S.; Cui, Y.; Lü, S.; Long, M. Flow field analyses of a porous membrane-separated, double-layered microfluidic chip for cell co-culture. *Acta Mech. Sin.* **2020**, *36*, 754–767. [CrossRef]

51. Mosavati, B.; Oleinikov, A.V.; Du, E. Development of an organ-on-a-chip-device for study of placental pathologies. *Int. J. Mol. Sci.* **2020**, *21*, 8755. [CrossRef]

52. Hu, G.; Li, D. Three-dimensional modeling of transport of nutrients for multicellular tumor spheroid culture in a microchannel. *Biomed. Microdevices* **2007**, *9*, 315–323. [CrossRef] [PubMed]

53. Zahorodny-Burke, M.; Nearingburg, B.; Elias, A. Finite element analysis of oxygen transport in microfluidic cell culture devices with varying channel architectures, perfusion rates, and materials. *Chem. Eng. Sci.* **2011**, *66*, 6244–6253. [CrossRef]

54. Bhise, N.S.; Manoharan, V.; Massa, S.; Tamayol, A.; Ghaderi, M.; Miscuglio, M.; Lang, Q.; Zhang, Y.S.; Shin, S.R.; Calzone, G.; et al. A liver-on-a-chip platform with bioprinted hepatic spheroids. *Biofabrication* **2016**, *8*, 014101. [CrossRef] [PubMed]

55. Mäki, A.-J.; Peltokangas, M.; Kreutzer, J.; Auvinen, S.; Kallio, P. Modeling carbon dioxide transport in PDMS-based microfluidic cell culture devices. *Chem. Eng. Sci.* **2015**, *137*, 515–524. [CrossRef]

56. Cicchetti, A.; Laurino, F.; Possenti, L.; Rancati, T.; Zunino, P. In silico model of the early effects of radiation therapy on the microcirculation and the surrounding tissues. *Phys. Medica* **2020**, *73*, 125–134. [CrossRef]

57. Kim, M.; Kim, G.; Kim, D.; Yoo, J.; Kim, D.-K.; Kim, H. Numerical study on effective conditions for the induction of apoptotic temperatures for various tumor aspect ratios using a single continuous-wave laser in photothermal therapy using gold nanorods. *Cancers* **2019**, *11*, 764. [CrossRef]

58. Peng, J.; Fang, C.; Ren, S.; Pan, J.; Jia, Y.; Shu, Z.; Gao, D. Development of a microfluidic device with precise on-chip temperature control by integrated cooling and heating components for single cell-based analysis. *Int. J. Heat Mass Transf.* **2019**, *130*, 660–667. [CrossRef]

59. Das, S.K.; Chung, S.; Zervantonakis, I.; Atnafu, J.; Kamm, R.D. A microfluidic platform for studying the effects of small temperature gradients in an incubator environment. *Biomicrofluidics* **2008**, *2*, 034106. [CrossRef]

60. Hynes, W.F.; Pepona, M.; Robertson, C.; Alvarado, J.; Dubbin, K.; Triplett, M.; Adorno, J.J.; Randles, A.; Moya, M.L. Examining metastatic behavior within 3D bioprinted vasculature for the validation of a 3D computational flow model. *Sci. Adv.* **2020**, *6*, eabb3308. [CrossRef]

61. Ye, T.; Shi, H.; Phan-Thien, N.; Lim, C.T.; Li, Y. Numerical design of a microfluidic chip for probing mechanical properties of cells. *J. Biomech.* **2019**, *84*, 103–112. [CrossRef]

62. Zhang, M.; Zheng, A.; Zheng, Z.C.; Wang, M.Z. Multiphase flow experiment and simulation for cells-on-a-chip devices. *Proc. Inst. Mech. Eng. Part H J. Eng. Med.* **2019**, *233*, 432–443. [CrossRef]

63. Tanaka, N.; Yamashita, T.; Yalikun, Y.; Amaya, S.; Sato, A.; Vogel, V.; Tanaka, Y. An ultra-small fluid oscillation unit for pumping driven by self-organized three-dimensional bridging of pulsatile cardiomyocytes on elastic micro-piers. *Sens. Actuators B Chem.* **2019**, *293*, 256–264. [CrossRef]

64. Subramaniam, A.; Sethuraman, S. Biomedical applications of nondegradable polymers. In *Natural and Synthetic Biomedical Polymers*; Elsevier Science: Amsterdam, The Netherlands, 2014.

65. Lamberti, A.; Marasso, S.L.; Cocuzza, M. PDMS membranes with tunable gas permeability for microfluidic applications. *RSC Adv.* **2014**, *4*, 61415–61419. [CrossRef]

66. Li, J.M.; Liu, C.; Dai, X.D.; Chen, H.H.; Liang, Y.; Sun, H.L.; Tian, H.; Ding, X.P. PMMA microfluidic devices with three-dimensional features for blood cell filtration. *J. Micromechanics Microengineering* **2008**, *18*, 95021. [CrossRef]

67. Faustino, V.; Catarino, S.O.; Lima, R.; Minas, G. Biomedical microfluidic devices by using low-cost fabrication techniques: A review. *J. Biomech.* **2016**, *49*, 2280–2292. [CrossRef]
68. Singhal, J.; Pinho, D.; Lopes, R.; Sousa, P.C.; Garcia, V.; Schutte, H.; Lima, R.; Gassmann, S. Blood flow visualization and measurements in microfluidic devices fabricated by a micromilling technique. *Micro Nanosyst.* **2015**, *7*, 148–153. [CrossRef]
69. Lopes, R.; Rodrigues, R.O.; Pinho, D.; Garcia, V.; Schütte, H.; Lima, R.; Gassmann, S. Low cost microfluidic device for partial cell separation: Micromilling approach. In Proceedings of the 2015 IEEE International Conference on Industrial Technology (ICIT), Seville, Spain, 17–19 March 2015; pp. 3347–3350.
70. Aghvami, S.A.; Opathalage, A.; Zhang, Z.; Ludwig, M.; Heymann, M.; Norton, M.; Wilkins, N.; Fraden, S. Rapid prototyping of cyclic olefin copolymer (COC) microfluidic devices. *Sens. Actuators B Chem.* **2017**, *247*, 940–949. [CrossRef]
71. Ochs, C.J.; Kasuya, J.; Pavesi, A.; Kamm, R.D. Oxygen levels in thermoplastic microfluidic devices during cell culture. *Lab Chip* **2014**, *14*, 459–462. [CrossRef] [PubMed]
72. Santiago, G.T.-D.; Flores-Garza, B.G.; Negrete, J.T.; Lara-Mayorga, I.M.; González-Gamboa, I.; Zhang, Y.S.; Rojas-Martínez, A.; Ortiz-López, R.; Álvarez, M.M. The tumor-on-chip: Recent advances in the development of microfluidic systems to recapitulate the physiology of solid tumors. *Materials* **2019**, *12*, 2945. [CrossRef] [PubMed]
73. Hua, S.; De Matos, M.B.C.; Metselaar, J.M.; Storm, G. Current trends and challenges in the clinical translation of nanoparticulate nanomedicines: Pathways for translational development and commercialization. *Front. Pharmacol.* **2018**, *9*, 790. [CrossRef]
74. Ahn, J.; Ko, J.; Lee, S.; Yu, J.; Kim, Y.; Jeon, N.L. Microfluidics in nanoparticle drug delivery; From synthesis to pre-clinical screening. *Adv. Drug Deliv. Rev.* **2018**, *128*, 29–53. [CrossRef]
75. Soheili, S.; Mandegar, E.; Moradikhah, F.; Doosti-Telgerd, M.; Javar, H.A. Experimental and numerical studies on microfluidic preparation and engineering of chitosan nanoparticles. *J. Drug Deliv. Sci. Technol.* **2020**, *61*, 102268. [CrossRef]
76. Arefi, S.M.A.; Yang, C.W.T.; Sin, D.D.; Feng, J.J. Simulation of nanoparticle transport and adsorption in a microfluidic lung-on-a-chip device. *Biomicrofluidics* **2020**, *14*, 044117. [CrossRef] [PubMed]
77. Liu, Y.; Bhattacharya, A.; Kuksenok, O.; He, X.; Aizenberg, M.; Aizenberg, J.; Balazs, A.C. Computational modeling of oscillating fins that "catch and release" targeted nanoparticles in bilayer flows. *Soft Matter* **2016**, *12*, 1374–1384. [CrossRef] [PubMed]
78. Maleki, R.; Khedri, M.; Malekahmadi, D.; Mohaghegh, S.; Jahromi, A.M.; Shahbazi, M.A. Simultaneous doxorubicin encapsulation and in-situ microfluidic micellization of bio-targeted polymeric nanohybrids using dichalcogenide monolayers: A Molecular in-silico study. *Mater. Today Commun.* **2021**, *26*, 101948. [CrossRef]
79. Kwak, B.; Ozcelikkale, A.; Shin, C.; Park, K.; Han, B. Simulation of complex transport of nanoparticles around a tumor using tumor-microenvironment-on-chip. *J. Control. Release* **2014**, *194*, 157–167. [CrossRef] [PubMed]
80. Nasiri, R.; Shamloo, A.; Akbari, J.; Tebon, P.; Dokmeci, M.R.; Ahadian, S. Design and simulation of an integrated centrifugal microfluidic device for CTCs separation and cell lysis. *Micromachines* **2020**, *11*, 699. [CrossRef]
81. Zhang, X.; Xu, X.; Ren, Y.; Yan, Y.; Wu, A. Numerical simulation of circulating tumor cell separation in a dielectrophoresis based Y-Y shaped microfluidic device. *Sep. Purif. Technol.* **2021**, *255*, 117343. [CrossRef]
82. Ma, J.-T.; Xu, Y.-Q.; Tang, X.-Y. A numerical simulation of cell separation by simplified asymmetric pinched flow fractionation. *Comput. Math. Methods Med.* **2016**, *2016*, 2564584. [CrossRef]
83. Shamloo, A.; Boodaghi, M. Design and simulation of a microfluidic device for acoustic cell separation. *Ultrasonics* **2018**, *84*, 234–243. [CrossRef]
84. Karakas, H.E.; Kim, J.; Park, J.; Oh, J.M.; Choi, Y.; Gozuacik, D.; Cho, Y.-K. A microfluidic chip for screening individual cancer cells via eavesdropping on autophagy-inducing crosstalk in the stroma niche. *Sci. Rep.* **2017**, *7*, 2050. [CrossRef]
85. Liu, J.-S.; Zhang, Y.-Y.; Wang, Z.; Deng, J.-Y.; Ye, X.; Xue, R.-Y.; Ge, D.; Xu, Z. Design and validation of a microfluidic chip with micropillar arrays for three-dimensional cell culture. *Chin. J. Anal. Chem.* **2017**, *45*, 1109–1114. [CrossRef]
86. Chen, W.; Li, J.; Wan, X.; Zou, X.; Qi, S.; Zhang, Y.; Weng, Q.; Li, J.; Xiong, W.; Xie, C.; et al. Design of a microfluidic chip consisting of micropillars and its use for the enrichment of nasopharyngeal cancer cells. *Oncol. Lett.* **2019**, *17*, 1581–1588. [CrossRef] [PubMed]
87. Huang, Y.; Nguyen, N.-T. A polymeric cell stretching device for real-time imaging with optical microscopy. *Biomed. Microdevices* **2013**, *15*, 1043–1054. [CrossRef] [PubMed]

*Review*

# Diagnosis Methods for COVID-19: A Systematic Review

Renata Maia [1,2], Violeta Carvalho [1,2,3,4], Bernardo Faria [1,2], Inês Miranda [1,2], Susana Catarino [1,2], Senhorinha Teixeira [4], Rui Lima [3,5,6], Graça Minas [1,2] and João Ribeiro [6,7,8,9,*]

1  Microelectromechanical Systems Research Unit (CMEMS-UMinho), School of Engineering, Campus de Azurém, University of Minho, Guimarães, Portugal
2  LABBELS-Associate Laboratory, Braga/Guimarães, Portugal
3  MEtRICs, Mechanical Engineering Department, Campus de Azurém, University of Minho, 4800-058 Guimarães, Portugal
4  ALGORITMI, Production and Systems Department, School of Engineering, Campus de Azurém, University of Minho, 4800-058 Guimarães, Portugal
5  CEFT, Faculty of Engineering, University of Porto, 4200-465 Porto, Portugal
6  ALiCE, Faculty of Engineering, University of Porto, 4200-465 Porto, Portugal
7  Campus de Santa Apolónia, Instituto Politécnico de Bragança, 5300-253 Bragança, Portugal
8  Centro de Investigação de Montanha (CIMO), Campus de Santa Apolónia, Instituto Politécnico de Bragança, 5300-253 Bragança, Portugal
9  Laboratório Associado para a Sustentabilidade e Tecnologia em Regiões de Montanha (SusTEC), Campus de Santa Apolónia, Instituto Politécnico de Bragança, 5300-253 Bragança, Portugal
*  Correspondence: jribeiro@ipb.pt

**Abstract:** At the end of 2019, the coronavirus appeared and spread extremely rapidly, causing millions of infections and deaths worldwide, and becoming a global pandemic. For this reason, it became urgent and essential to find adequate tests for an accurate and fast diagnosis of this disease. In the present study, a systematic review was performed in order to provide an overview of the COVID-19 diagnosis methods and tests already available, as well as their evolution in recent months. For this purpose, the Science Direct, PubMed, and Scopus databases were used to collect the data and three authors independently screened the references, extracted the main information, and assessed the quality of the included studies. After the analysis of the collected data, 34 studies reporting new methods to diagnose COVID-19 were selected. Although RT-PCR is the gold-standard method for COVID-19 diagnosis, it cannot fulfill all the requirements of this pandemic, being limited by the need for highly specialized equipment and personnel to perform the assays, as well as the long time to get the test results. To fulfill the limitations of this method, other alternatives, including biological and imaging analysis methods, also became commonly reported. The comparison of the different diagnosis tests allowed to understand the importance and potential of combining different techniques, not only to improve diagnosis but also for a further understanding of the virus, the disease, and their implications in humans.

**Keywords:** COVID-19; diagnosis; image analysis; PCR; SARS-CoV-2

**Citation:** Maia, R.; Carvalho, V.; Faria, B.; Miranda, I.; Catarino, S.; Teixeira, S.; Lima, R.; Minas, G.; Ribeiro, J. Diagnosis Methods for COVID-19: A Systematic Review. *Micromachines* **2022**, *13*, 1349. https://doi.org/10.3390/mi13081349

Academic Editor: Nam-Trung Nguyen

Received: 2 August 2022
Accepted: 16 August 2022
Published: 19 August 2022



## 1. Introduction

Acute respiratory syndrome coronavirus 2 (SARS-CoV-2), which causes the COVID-19 disease, appeared in December 2019 in China [1] and rapidly spread around the world, being declared by the World Health Organization as a global pandemic in March 2020 [2]. SARS-CoV-2 is an enveloped, positive-sense, and single-stranded Ribonucleic acid (RNA) beta-coronavirus, capable of infecting animals and humans. Regarding its biochemical constitution, the SARS-CoV-2 genome contains around 30,000 nucleotides, encoding structured proteins such as spike (S), membrane (M), envelope (E), and hemagglutinin esterase (HE), and nucleocapsid (N) [3]. The two SARS viruses' S proteins were shown to have similar and low nanomolar binding affinities to human angiotensin-converting enzyme 2 (ACE2), the host surface receptor that the viruses exploit to enter cells (Figure 1) [4].

**SARS-COV-2**

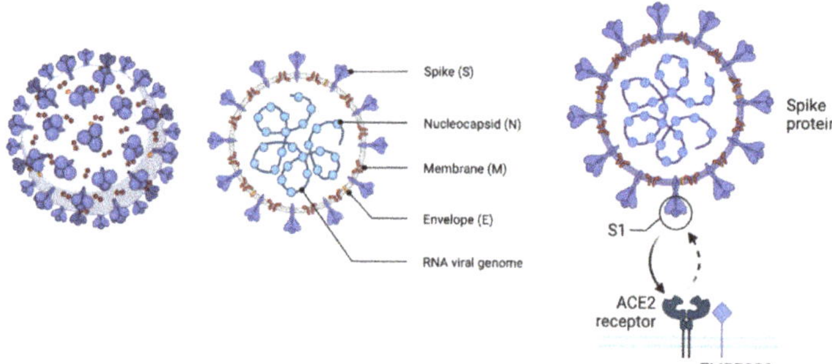


**Figure 1.** Sars-COV-2 virus representation.

This virus can lead to respiratory, gastrointestinal, and neurologic syndromes [3]. Particularly, COVID-19 early infections' most common symptoms are fever, cough, and other respiratory issues [5]. Although these are the main indicators of the disease, it is important to notice that some patients do not report any disease symptoms but may still contribute to the transmission of the virus to other human hosts [3].

The transmission of SARS-CoV-2 among humans occurs, mainly, via close contact with an infected individual that produces respiratory droplets and aerosols while coughing or sneezing, within a range up to 2 m [6]. Therefore, the scientific community recognized transmission by airborne particles as the primary route of infection, although contaminated surfaces or objects can also cause the virus to spread [6]. The main form of control and avoiding the virus transmission has been the use of personal protection equipment and social distance, as the general population has been oriented to use masks as a mechanical barrier to prevent droplet dispersion.

COVID-19 clinical diagnosis has been mainly based on signs and symptoms evaluation and confirmed by nucleic acid amplification tests (NAAT), for example, RT-PCR (Reverse Transcription Polymerase Chain Reaction) of nasopharyngeal or oropharyngeal swabs [7]. PCR-based methods are simple, highly sensitive, and highly specific and, therefore, they are routinely and reliably capable of detecting coronavirus infection in patients. These assays, widely used to amplify minimum quantities of deoxyribonucleic acid (DNA), start with the conversion of the coronavirus ribonucleic acid (RNA) into complementary DNA by reverse transcription. Subsequently, PCR is performed, and the resultant amplification of DNA is subjected to specific detection through different analytical methods. RT-PCR is a gold-standard method to detect most coronaviruses, including SARS-CoV-2. However, this method has some disadvantages, such as requiring expensive specialized equipment and highly trained analysts and technicians. Furthermore, PCR requires up to 4–8 h to process the samples and additional 1–3 days to report results, also having a high false-negative rate [8].

NAAT includes loop-mediated isothermal amplification (LAMP), multiple cross displacement amplification (MCDA), recombinase-aided amplification (RPA), CHAnicking and extension amplification reaction (NEAR), and Clustered Regularly Interspaced Short Palindromic Repeat associated (Cas) proteins (CRISPR–CasN)-based assays [9]. Serological tests can be performed as lateral flow immunochromatographic lateral flow assay (ILFA) [10], lateral flow immunochromatographic strip (LFICS) [11], chemiluminescence immunoassay (CLIA) [12], and enzyme-linked immunosorbent assay (ELISA) [13] based on antibody detection. Now, there have been some studies and commercial antigen detection kits for SARS-CoV-2 [14]. Although the molecular assay is the most common method to diagnose COVID-19, other diagnosis methods have been widely reported: chest computed tomography (CT) scan combined with the evaluation of clinical symptoms [15], potential electrochemical (EC) biosensors [16], field-effect transistor (FET)-based biosensors [17], surface plasmon resonance (SPR)-based biosensors [18] and artificial intelligence methods [19]. Figure 2 represents some of the tests. In order to diagnose various disorders, CT is a frequently employed auxiliary detection technology. The diagnosis of COVID-19 can be made on the basis of changes to lungs imaging brought on by SARS-CoV-2 infection [20].

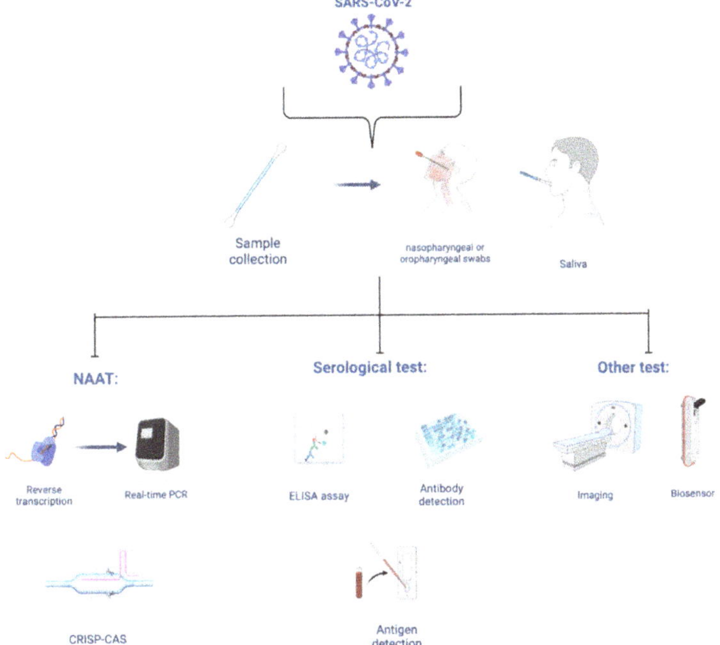


**Figure 2.** Sars-COV-2 virus diagnosis methods.

Particularly, CT scans have several limitations for COVID-19 diagnosis, since they do not allow the identification of specific viruses, and many clinics and laboratories do not have access to proper equipment [19]. Regarding ELISA, this assay is based on the optical measurement of labeled fluorescent markers and, although ELISA has been used for the detection of SARS-CoV-2, the assay takes at least a few hours and requires specific spectral analyzers [21]. Although RNA detection based on RT-qPCR and antibody detection based on ELISA and LFICS have been developed, both methods have certain practical limitations. Concerning biosensors, they have the potential to be alternative tools since they can provide fast, accurate, sensitive early detection, especially smartphone-driven biosensors [18]. The advantages of EC biosensing assays are low cost, simplicity, easily miniaturization and mass manufacture. Additionally, they feature a rapid test result and can have the potential for being point-of-care, but their mass production for distribution is limited because it is a new technique for which the mass manufacturing process is in a process of development [19].

SARS-CoV-2 is highly contagious and, to date, although vaccines were approved and disseminated, there is no effective treatment for the disease. Due to the global pandemic state and the progressive human-to-human transmission, it became essential to develop preventive methods and COVID-19 diagnosis approaches while searching for novel treatments. As the research community started looking for solutions to diagnose COVID-19 faster and more efficiently, several approaches were developed to improve RT-PCR, and biosensors. Imaging analysis was used in machine learning algorithms to analyze CT and X-ray images on a large scale [22]. Since mass testing remains imperative, the present systematic review aims to identify the new advances in COVID-19 diagnosis that were achieved during 2021.

## 2. Methods

### 2.1. Search Strategy and Selection Criteria

The aim of this systematic review was to identify new solutions to diagnose COVID-19 and to overcome the disadvantages of the PCR gold standard tests. Thus, this review intended to analyze different innovative approaches developed and under development for COVID-19 diagnosis, compare them, and observe their advantages and disadvantages.

The present systematic literature review was conducted according to the Preferred Reporting Items for Systematic Reviews and Meta-Analyses (PRISMA) guidelines. An electronic comprehensive search on ScienceDirect (SD), Scopus, and PubMed (PM) databases was performed. From database inception up to 21 January 2022, studies that included different types of tests for COVID-19, besides the real-time RT-PCR diagnosis, were searched. The search strategy was established by combining several keywords and the use of AND/OR Boolean operators. The relevant studies resulting from the database search were manually analyzed to identify other potential studies to be included. The exclusion criteria were: reviews, comments, overviews, case reports, viewpoints, and perspectives. Additionally, tests with low effectiveness and/or ambiguity of data were also excluded. Studies not written in the English language were also excluded. Figure 3 presents the PRISMA flow diagram of the conducted study.

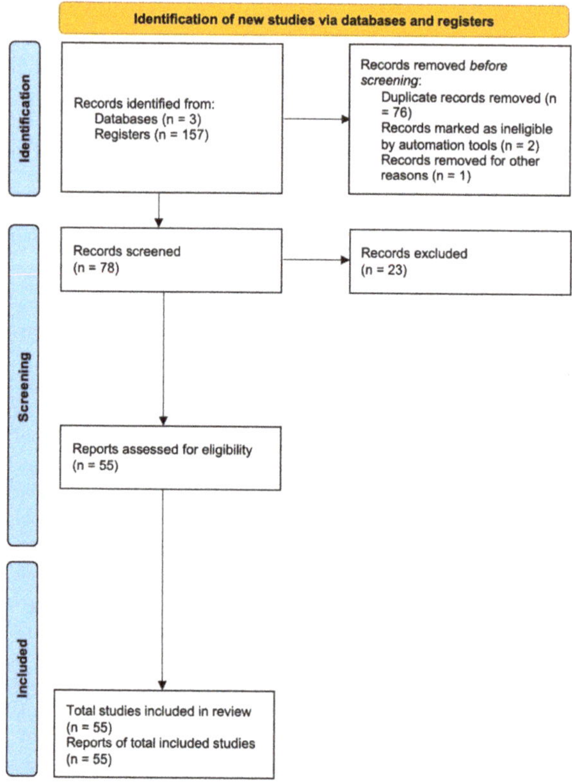


**Figure 3.** PRISMA flow diagram of search strategy conducted in the systematic review.

### 2.2. Data Collection

The database and additional manual searches provided 157 results. After removing duplicates, 78 studies were considered. First, titles and abstracts were screened. All abstracts were read and those that did not fit the purpose of this systematic review were excluded. According to the defined exclusion criteria, full texts were reviewed and, finally, data were extracted. Extracted data were: test design, experimental settings, quantitative outcomes and reported study limitations, as well as other relevant comments.

### 2.3. Outcomes

In the current systematic review, the outcomes of interest were the inclusion of a set of advantages and disadvantages/limitations of each diagnosis method. It was noticed that all the studies included in the review reported tests in few subjects, since COVID-19 is a very recent disease and, therefore, there were not a lot of samples available for research testing. Additionally, some of the studies also presented quantitative results such as sensibility and sensitivity. However, due to the low number of considered samples (around 100 samples per study), those parameters were not taken into consideration.

### 2.4. Data Analysis

All records were extracted to Mendeley software, where duplicated documentation was removed and manually checked. The titles of all documentation were searched by 3 authors, each one focused on one of the following categories of search: RT-PCR, Biosensors, and Imagology-Computational Methods. Sequentially, all abstracts and full texts were examined and evaluated by the same authors. In the occurrence of ambiguity, a consensus on the evaluation was reached by having an additional element.

## 3. Results

A total of 55 studies met all the eligibility criteria and were identified to be included in this review. Of those, 50 studies reported specific methods for diagnosis of COVID-19 and 5 studies were comparative.

Amongst the 50 studies, 38 were based on biological tests, and 12 reported computational techniques on imagological images. Some of the studies tested have their assays on clinical samples containing the SARS-CoV-2 gene, while others only reported tests in artificial samples. All the selected studies present the tests of the developed methods both on subjects infected with SARS-CoV-2 and controls (non-infected subjects). Table 1 presents the methods reported in the literature for COVID-19 detection using biological analysis assays, comprising the principle of operation, advantages, and limitations of all methods.

**Table 1.** Methods reported in the literature for COVID-19 detection through biological analysis.

| Detection Method | Principle of Operation | Advantages | Disadvantages/Limitation | Authors |
|---|---|---|---|---|
| Dual staining assay (Immunohistochemistry-IHC/in situ hybridization-ISH) | Reaction with S and N antibodies | Very precise; Useful for studying the pathogenesis of SARS-CoV-2 | No quantitative results; Was not tested on clinical samples, only on Formalin-Fixed Paraffin-Embedded (FFPE) pellets | J. Liu et al. [23] |
| Graphene-based FET biosensor | Electric response of SARS-CoV-2 coupling S antibody | No cross-reaction with Middle East respiratory syndrome coronavirus (MERS-CoV); Highly sensitive and instantaneous measurement; Low-noise detection; Clinical samples do not need preparations/pre-processing | Needs novel materials for a more accurate detection | G. Seo et al. [24] |

**Table 1.** *Cont.*

| Detection Method | Principle of Operation | Advantages | Disadvantages/Limitation | Authors |
|---|---|---|---|---|
| RT-LAMP | Auto-cycling strand displacement DNA synthesis using Orf1lab and S antibodies as target genes | Faster than RT-PCR assay; No cross-reactivity with other respiratory pathogens; Easy to handle; Does not require skilled personnel or specialized instruments; Results are easy to read | Few complete genomes available on databases; Mutation occurring with the spread of the virus | C. Yan et al. [25] |
| DNA nanoscaffold hybrid chain reaction (DNHCR) | The presence of SARS-CoV-2 triggers a cascade reaction along the DNA nanoscaffold, lighting up the structure (detected by fluorescence) | High signal gain; Short reaction time; High specificity; Room temperature response; Cost-effectiveness; Readily available reagents | Output of the fluorescence signal requires the use of specialized equipment | J. Jiao et al. [26] |
| Multiplex RT-LAMP coupled with nanoparticle-based lateral flow biosensor (LFB) assay-mRT-LAMP-LFB | LAMP amplification, reverse transcription, and multiplex analysis, allowing detection of orf1lab and N antibodies at the same time | Easy-to-use; Simple and objective; Less error-prone; Avoids the requirement of complex processes, special reagents and expensive instruments | RNA templates are sensitive to degradation by inadequate sample handling, post-mortem processes, or storage; Small number of clinical samples; Not evaluated on other clinical samples (e.g., sputum, blood, urine) | X. Zhu et al. [16] |
| COVID-19 associated ROS diagnosis (CRD) | Reactive oxygen species (ROS) released from infected cells would react with a working electrode covered by functionalized multi-wall carbon nanotubes, releasing electric charges that are posteriorly measured | Extremely rapid method; Non-invasive | Some false-negatives | Z. S. Miripour et al. [27] |
| Multiplex real-time RT-PCR (rRT-PCR) | Simultaneous detection of N and E gene | High sensitivity; Reduced reagents, costs and time required | Poor reproducibility; Maximum signal intensities were low | T. Ishige et al. [28] |
| Simplexa™ direct assay (RT-PCR) | Targeting of E and RdRp genes | Fast; Easy-to-use; Does not require extra laboratory equipment; Low train required; No cross-reactivity with other viruses | Small number of samples which can be tested in a run | L. Bordi et al. [29] |
| RT-PCR CRISPR-Cas12a | RT-PCR is used to amplify target regions from viral RNA and the resulting amplicons are transferred to the gRNA/Cas12a-based Clustered Regularly Interspaced Short Palindromic Repeats (CRISPR) system for fluorescence detection | Sensitive and robust; Readily available equipment | Small number of samples; Does not give quantitative results | Z. Huang et al. [30] |
| RT-PCR | E and RdRo detection | Very sensitive; No cross-reactivity with other viruses | Weak initial reactivity | V. M. Corman et al. [31] |
| RT-PCR | Fully automatic PCR platform for detection of E gene | No cross-reactivity with other viruses; Allows a large number of patients to be screened in a reasonable timeframe | Was not evaluated with clinical samples; Results have to be confirmed with an independent PCR | S. Pfefferle et al. [32] |
| rRT-PCR | Viral load detected in saliva samples | Low risk of transmission at the collection; Less invasive | Low number of samples; Qualitative results | L. Azzi et al. [33] |

**Table 1.** *Cont.*

| Detection Method | Principle of Operation | Advantages | Disadvantages/Limitation | Authors |
|---|---|---|---|---|
| Serological Immunochromatographic (IC) assay | Immunochromatography strip assay for detection of IgM and IgC antibodies | Ready-to-use and time-saving; High detection capacity; Blood collection less risky than nasal swab samples | Assay carried out without specificity analysis; Qualitative results | Y. Pan et al. [34] |
| Quotient MosaiQ™ (microarray-based assay) | Automated detection of antibodies directed to the spike protein | High specificity and clinical sensitivity; Rapid throughput of samples; No cross-reactivity with other viruses | Weak repeatability and reproducibility; Did not include positive samples for other coronaviruses | C. Martinaud et al. [35] |
| BioFire® Respiratory Panel 2.1 (RT-PCR) | Nucleic acid amplification platform for detection of M and S genes | Detects low levels of viral RNA; Allows the simultaneous differentiation between viruses; Easy to use | Non-specified | H. M. Creager et al. [36] |
| ACE2-based LFIA | Detect SARS-CoV-2 S1 protein using LFIA with a matched pair consisting of ACE2 and an antibody. | Detect the S1 antigen of SARS-CoV-2 | Tests performed only on two different corona-related spike antigens | Lee et al. [37] |
| Cell-based biosensors for the detection of the SARS CoV-2 spike S1 protein | Molecular Identification through Membrane Engineering | Portable, high throughput, and low-cost system | No clinical validation | Mavrikou et al. [38] |
| Detection of IgM Antibodies against the SARS-CoV-2 Virus via Colloidal Gold Nanoparticle-Based Lateral-Flow Assay | A colloidal gold nanoparticle-based lateral-flow assay to detect the IgM antibody against the SARS-CoV-2 virus through the indirect immunochromatography method | Low sample consumption | Satisfactory specificity. Weak comparison with PCR results; | Huang et al. [39] |
| Fluorescent immunochromatographic assay based on multilayer quantum dot nanobead | Two-channel fluorescent Immunochromatographic assay method for ultrasensitive and simultaneous detection of SARS-CoV-2/FluA in real biological samples | Simultaneous detection of SARS-CoV-2 antigen and influenza A virus | Missing information about the number of samples tested | Wang et al. [40] |
| Gold nanoparticle-based biosensor | Combined colorimetric and electrochemical biosensor to detect SARS-CoV-2 spike antigen | Does not require for sensor preparation and modification. Saliva samples; | Satisfactory Selectivity | Karakus et al. [41] |
| SARS-CoV-2-specific biosensor for antigen detection | Lateral flow immunoassay-based biosensor using single-chain variable fragment-crystallizable fragment (scFv-Fc) fusion antibodies. | Time-saving, good detection limit | Needs optimization. Satisfactory sensitive | Kim et al. [42] |
| Aptamer | Detection of SRAS-CoV2 N protein using DNA-based aptamers | Aptamers can be synthesized easily and the process is less expensive than antibody production. | Lack of serum samples | Chen et al., [43] |
| Point-of-care nucleic acid amplification test for diagnosis of active COVID-19 | Based on the principle of LAMP | Easy to operate and does not require skilled personnel | Some false-negatives | Deng et al. [44] |
| Point-of-care testing for SARS-CoV-2 virus nucleic acid detection | Catalytic hairpin assembly reaction-based signal amplification system coupled with a lateral flow immuno-assay strip | Highly sensitive, Fast. | Limited number of clinical samples | Zou et al. [45] |

**Table 1.** *Cont.*

| Detection Method | Principle of Operation | Advantages | Disadvantages/Limitation | Authors |
|---|---|---|---|---|
| Reverse transcription–enzymatic recombinase amplification | Detect the SARS-CoV-2 gene by applying reverse transcription–enzymatic recombinase amplification | No need of thermocyclers | Dual detection and single-copy sensitive | Xia e al. [13] |
| CLIA | Serological test for detecting SARS-CoV-2 specific IgA as well as IgM and IgG | Measure levels of the three types of antibodies in blood | Few cases of COVID-19 patients | Ma et al. [46] |
| EC impedance-based detector | Electrochemical detection of SARS-CoV-2 antibodies using a commercially available impedance sensing platform. | Rapid screening of patient samples, expanded serological surveys to assess anti-SARS-CoV-2 antibody levels in the community. | Further testing is needed to determine the limit of detection | Rashed et al. [47] |
| RT-LAMP | RT-LAMP method designed to target the nucleocapsid protein gene | High sensitivity and specificity, low cost. | False-positive single read-out and sensitivity to aerosol contaminants during assay manipulations | Baek et al. [48] |

Table 2 presents the methods reported in the literature for COVID-19 detection based on imaging analysis, detailing the implemented machine learning algorithms.

**Table 2.** Methods reported in the literature for COVID-19 detection through image analysis.

| Methodology | Architecture | Machine Learning Algorithms | Optimization | Pre-Processing/Pre-Training | Reference |
|---|---|---|---|---|---|
| X-ray | Convolutional neural network (CNN) | Support vector machines (SVM); Decision Tree (DT); k-nearest neighbors (KNN) | Bayesian algorithm | None | M. Nour et al. [49] |
| | CNN | ConvXNet | Stacking algorithm | X-ray images of COVID-19 and other cases of pneumonia | T. Mahmud et al. [50] |
| | CNN | nCOVnet | VGG-16 | X-ray images of COVID-19 true positive patients | H. Panwar et al. [51] |
| | Xception (CNN based) | CoroNet (SVM based) | Depends on the availability of the training data | ImageNet | A. I. Khan et al. [52] |
| | Residual Exemplar Local Binary Pattern (ResExLBP) and ReliefF (CNN based) | DT; Linear discriminant (LD); Subspace discriminant (SD); SVM; KNN | Local Binary Pattern (LBP); 10-fold cross-validation; | Leave-One-Out Cross-Validation (LOOCV); 10-fold cross-validation holdout validation | T. Tuncer et al. [53] |
| | CNN | SVM | Stochastic gradient descent (SGD) | Fuzzy Colour Technique MobileNetV2 SqueezeNet | M. Toğaçar et al. [54] |
| CT | Multi-Scale CNN (MSCNN) | Multi-scale spatial pyramid (MSSP) decomposition | None | 2D Images | T. Yan et al. [55] |
| | CNN | Enhanced KNN classifier (EKNN); Hybrid feature selection methodology (HFSM) | KNN optimization | Gray Level Co-occurrence Matrix (GLCM) COVID_CT | W. M. Shaban et al. [56] |

## 4. Discussion

Due to the high variety of methods on which the COVID-19 diagnosis is based, particularly those exploiting the reaction chain of human antibodies, it would be interesting to perform a comparison between all the inherent characteristics of those studies, in order to properly discuss their advantages and limitations. Particularly, the sensitivity and specificity of the assays would be a relevant topic to address. However, since not all presented studies reported such information, and those who reported it were based on a small number of tested samples, it was not possible to faithfully compare those specifications. Additionally, another important parameter to assess the quality of the methods would be the cost-effectiveness per assay, which could not be defined or inferred due to the lack of well-founded literature, i.e., only one article reported that data [16]. Therefore, the comparison expressed here is sustained only by the main advantages, disadvantages, and reliability of the methods to be applied.

### 4.1. Nucleic Acid Amplification Tests

All the selected studies demonstrated that there is an urgent need to develop better methods for the diagnosis of COVID-19. RT-PCR, the most used technique, allows the processing of DNA/RNA to search for specific genes, and it is commonly used to diagnose a variety of viruses (for example, Grapevine virus T [57] and Zaire ebolavirus [58]). Particularly, in the case of SARS-CoV-2, the genes that are normally used for its detection are the N, S, E, M, ORF1a, and ORF1b genes [59]. The target gene regions, primer, and probe sequences used in different RT-PCR setups for SARS-CoV-2 detection are described by Yuce et al., 2021 [60]. As soon as the genomic sequence of SARS-CoV-2 was discovered, it was possible to perform RT-PCR on the discovered sequence, becoming the gold-standard method for COVID-19 diagnosis [61], assuring high accuracy.

RT-PCR is recommended as the most sensitive NAAT method [29,31,32]. However, most studies report RT-PCR as an assay that cannot fulfill the urgent requirements of the COVID-19 pandemic, as its technology relies on expensive and sophisticated equipment and reagents (with all the logistics complications that arise from their high prices), and it can only be performed by qualified people working in a laboratory specified to handle pathogens, it is also time-consuming and it may lead to some false-negative results [33,62]. Based on the classic and gold-standard RT-PCR technique, different studies reported new valuable methods, improved in comparison with the standard RT-PCR assays (as the rRT-PCR, with lower reagents and time consumption [28]), that would not need any new equipment or facilities in the existent medical units [33,49]. Also, digital PCR emerged as an improvement of RT-PCR. For example, Suo et al., 2020, optimized a droplet digital PCR used for the detection of SARS-CoV-2, which showed that the limit of detection is significantly lower than that of RT-PCR [63].

In addition, various isothermal techniques were performed. Isothermal techniques are a powerful tool that do not require expensive thermocycling or professional skills [64]. These isothermal amplification assays are more sensitive and independent of a heat cycler, making them better suited to the development of quick, high-throughput, and low-cost assays. LAMP is the most well-established isothermal amplification method [48], with LAMP assays for SARS-CoV-2 detection being the most widely used isothermal amplification method [25,44]. Although LAMP has many advantages, any aerosol produced can lead to false positives. LAMP can be combined with colorimetric [65], reverse transcription (RT) [66], and others [9]. Other tests can use this technology as DNA nano-scaffold based on SARS-CoV-2 RNA triggered isothermal amplification [26].

Given their high specificity, sensitivity, simplicity, and repeatability, CRISPR-based nucleic acid detection approaches have recently demonstrated significant potential in the development of next-generation molecular diagnostic technology and have been used to diagnose COVID-19 [67]. The CRISPR-CasN-based assay is an effective gene-editing technique, and it can be analyzed with a fluorescent reader or lateral flow strips [30] or in a system based on lateral flow. For example, Broughton et al., 2020, described a CRISPR-Cas12- based lateral flow to detect SARS-COV-2, it was shown to be a rapid and easy-to-implement method [67]. There have been several Cas-protein-based assays such as CRISPR-Cas12 [67], CRISPR-Cas3 [68], and CRISPR-Cas13 [69]. Before the detection of CRISP- R–CasN-based assays, an isothermal amplification step was frequently introduced, for example, Joung et al.,2020, described a system namely SHERLOCK based on RPA and Cas13 [69]. However, CRISPR-based assay requires further testing.

Nucleic acid aptamers are short, single-stranded DNA (ssDNA) or RNA molecules that are selected for binding to a specific target [70]. Recently, aptamers have begun to be applied to detect SARS-COV-2 [43]; however, few methods have been described until now. Song et al., 2020, described an aptamer to target the receptor-binding domain of the SARS-CoV-2 spike glycoprotein [71]. Aptamer-based detection is more flexible, less costly, more stable, and easier to produce than antibody-based assays [72]. Nevertheless, it still needs further investigations and optimizations.

### 4.2. Serological Tests

Alternatively, different research teams developed serological assays to detect COVID-19. These methods test a clinical sample, combined with specific antigens, to detect SARS-CoV-2 antibodies in the human subject and it can contribute to epidemiological investigations of COVID-19. The goal of this technology is to perform a qualitative or semi-quantitative evaluation of the antibodies using different techniques, such as ELISA, which consists of the antigen protein immobilized on the surface of microplate wells, binds to the target antibody [73]. CLIA combines chemiluminescence techniques with immunochemical reactions [74], and lateral flow immunoassays (LFIA) [75]. Currently, and to the best of the authors' knowledge, the most promissory test seems to be LFIA, because it has the advantages of decreased technical requirement, affordability, lower sampling, and specimen preparation risk, higher detection sensitivity and specificity, and it can deliver results in 15 min [76].

Serological tests can be divided into antigen-based tests [37,40] and antibody-based tests [34,35,39,46]. Antigen-based diagnostics detect protein fragments on or within the virus. This type of testing can detect active infections within 15 min compared to hours with RT-PCR [77]. The widely available SARS-CoV-2 antigen kits employ two methods: (1) the ICT assay, which uses colloid gold conjugated antibodies to produce visible colored bands to indicate positivity, and (2) the FIA, which uses an automated immunofluorescence reader to provide results [78]. An antibody is a protein produced by the immune system in response to an antigen. Antibody-based diagnostic measures the presence/concentration of IgG and IgM levels in the blood/serum/plasma samples to determine if the body is fighting with a pathogen, for example, a contagious virus [79]. ELISA, LFIA, and CLIA are widely used in the detection of anti-SARS-CoV-2 antibodies [73]. These methods are not as specific as the tests recognizing RNA sequences in the virus.

This systematic review includes 3 articles that compare and analyze different serological tests for COVID-19 diagnosis [76,80]. While some of the tests were performed with only a specific antibody (IgG/IgM/IgA), others are performed with a combination of two of them. For each of the compared tests, their specificity and sensibility were presented. The main outcome of these works was the reporting that the results of serological immunoassays highly depend on the course of the COVID-19 disease, since the number of antibodies varies at different stages of infection [73,76,80]. Contrarily to RT-PCR, serological tests are less time-consuming, have a lower risk associated with the manipulation of specimens, and need fewer technical requirements. Serological tests can be useful as a complement to RT-PCR, especially in asymptomatic individuals, allowing the improvement of the epidemiological studies and the clinical diagnosis of COVID-19.

### 4.3. Biosensors

In addition to RT-PCR and serological testing, there is a lot of interest in developing new COVID-19 biosensors that are fast, reliable, and sensitive [81]. Some biosensors have been developed to achieve SARS-COV-2 detection. For example, some covid biosensors that have been developed were COVID-19 biosensors based on surface nucleoproteins that attach to the ACE-2 receptor [37], gold-nanoparticle biosensors [41], biosensors to detect antigen [42], FET [21], and ROS biosensors [27].

Biosensors based on FET technology are very promising for COVID-19 diagnosis due to their advantages, namely fast and ultra-sensible response [21,24]. Detection based on ROS levels is also promising, since these biosensors assure easy, fast, and cost-effective detection, with high sensitivity [27]. However, the large-scale production of these biosensors (based both on FET technology and ROS levels) and their distribution in medical units is yet a drawback that disables the possibility of currently using them. The electrochemical biosensor is another interesting form of biosensor for highly sensitive point-of-care SARS-CoV-2 detection [47]. In diluted human serum samples, Djaileb et al., 2020, based on surface plasmon resonance (SPR) using gold chips modified by viral nucleocapsid proteins, detected antibodies in 15 min with nanomolar sensitivity [82]. Biosensors have various advantages, including high sensitivity and specificity, low analysis costs, quick execution time, the low limit of detection, and the ability to construct tiny platforms that can be utilized directly at the point of care, but more research and testing are still required [83].

*4.4. Imaging*

Based on Table 2, it is possible to verify that there are several studies reporting machine learning algorithms in complement to radiological imaging techniques for COVID-19 diagnosis. It is possible to observe, from the reported studies, that there is a considerable preference for chest X-rays over CT scans for COVID-19 diagnosis. There are two main reasons for that [84]: (1) X-ray machines have a higher availability in the hospital environment, with lower inherent economical cost than CT scan machines; (2) X-ray scans involve lower ionizing radiations than CT scans. Besides the physical techniques usually applied in the acquisition of chest imaging, research teams worldwide have studied artificial intelligence [85], such as machine learning algorithms and deep learning to analyze those acquired images [86].

Machine learning algorithms, based on neural network architectures, such as CNN, combined with advanced artificial intelligence techniques embedded in radiological imaging, are useful for accurate COVID-19 detection or as an auxiliary tool for distinguishing COVID-19 from pneumonia [87]. Due to the higher accuracy and sensitivity of CNN and CNN-based model architectures, these architectures are applied more often than deep neural network (DNN) architectures [88]. By applying machine learning algorithms and optimization methods, it is possible to significantly diminish the processing time of a radiological test, highly reducing the time interval needed to acquire a COVID-19 diagnosis [89].

Many studies have described CT manifestations in COVID-19 patients; however, they have some differences because CT findings are strongly related to the stage of infection after symptoms begin [90]. Additionally, through pre-processing or pre-training methods, it is also possible to enhance the methods, improving their capacity to distinguish COVID-19 from pneumonia [91]. These methods also allow for a remote diagnosis, which avoids physical contact between the radiologist and the infected patient. However, these methods carry the necessity for large databases of medical images for pre-training and pre-processing, as well as an intrinsic cost for software programs [92]. Nevertheless, radiological tests, including CT imaging, are believed to be the assistive diagnostic method in which the presence of consolidation lesions in CT images indicates the onset of the COVID-19 disease [93].

*4.5. Microfluidic Approach*

Microfluid systems provide a platform for many diagnostic tests, including RT-PCR, RT-LAMP, nested-PCR, nucleic acid hybridization, ELISA, among others [94]. Microfluidic devices are made up of interconnected miniaturized compartments that can perform multiple experimental tasks, either individually or in parallel, in an integrated way. Usually, there are two distinct types of microfluidic devices, namely, paper-based and channel-based. The channel-based is manufactured by four main methods which are 3D-printing, molding, laminate, and nanofabrication. This type of microfluidic devices requires channels to create a reservoir for the integration of reagents. The paper-based microfluidic device is made of a series of nitrocellulose fibers or hydrophilic cellulose which guide liquid in paper by absorption [94]. Before the manufacturing process, these devices can be previously analyzed and developed using numerical simulations [95] to improve the performance of the sensor. The well-known advantages of the microfluidic systems, including the need for small volumes of samples, portability, and fast detection, are gaining increasing popularity as a tool to help to improve the detection and diagnosis of several diseases, such as malaria and diabetes [96–100] and also to evaluate the potential of novel therapies [101–104]. Additional microfluidic devices have been used commercially to detect and diagnose COVID-19. For example, during the second wave of COVID-19 in Italy, a microfluidic antigen was used in emergency rooms. The test showed high sensitivity, proving the potential of microfluidics as a tool for COVID-19 point-of-care tests [105]. The closest systems that use technology based on several microfluidic phenomena are the COVID-19 rapid test kits [9].

The use of microfluiddic point-of-care systems in the serological test is one of the research avenues being investigated for the diagnosis of COVID-19. Small sample volume, miniaturization, portability, multiplexed analysis, quick detection, and signal amplification techniques are advantages of microfluidic systems. They can also increase the sensitivity of analyte detection [106]. Additionally, new technologies have been developed to enhance the transfer rate of antigen and accelerate the reaction processes [107].

Recently, some researchers, like Torrente et al. [108], reported new approaches in which is possible to determine simultaneously both the viral and serologic status of an individual. They reported a novel multiplexed electrochemical platform ultra-rapid detection of COVID-19. Within physiologically relevant ranges, that platform quantitatively detects COVID-19-specific biomarkers in blood and saliva, including SARS-CoV-2 nucleocapsid protein (NP), specific immunoglobulins (Igs) against SARS-CoV-2 spike protein (S1) (S1-IgM and S1-IgG), and CRP [108].

Organs-on-a-chip have also been used in the research of alternative therapies for SARS-CoV-2 [109]. For instance, Si et al. used a human-airway-on-a-chip to investigate the antimalarial drug amodiaquine as a powerful inhibitor of infection with SARS-CoV-2 [110]. The research team modeled a human bronchial airway epithelium and pulmonary endothelium infected with pseudotyped severe acute respiratory syndrome coronavirus [110,111].

## 5. Conclusions

With the exponential growth of the pandemic situation caused by the coronavirus, a fast and effective method is needed to diagnose COVID-19, allowing mass testing of the population in a short period for greater control of the transmission. This systematic review presented the methods and assays, reported in 2021, for COVID-19 diagnosis, focusing on both biological and image analysis methodologies. The advantages, limitations, and main characteristics of each method were reported. Basically, PCR-based techniques are the gold standard for a reliable diagnosis but require highly trained personnel, and they are time-consuming. Diagnoses methods based on biosensors seem a viable option for a fast and sensitive response. However, this technology still needs to be produced on a large scale. Computational techniques in conjunction with the previous ones are very useful as a complementary diagnosis tool, but their processing takes a long time. In addition, while a plethora of methods and techniques are already available for an accurate COVID-19 diagnosis that can be individually tested, or as a complement to other methods, the limitations of the current methods and the instability of the SARS-CoV-2 coronavirus enhance the necessity of new developments in this field. Particularly, as SARS-CoV-2 variations are found, new improvements in the detection are required in the search for faster, more accurate, more sensitive, and simpler tests that are able to detect the virus as soon as possible and limit its transmission between humans.

**Author Contributions:** Conceptualization, J.R., R.L., S.T., S.C., V.C. and G.M.; methodology, R.M., V.C., B.F. and I.M.; validation, J.R., R.L., S.T., S.C., V.C. and G.M.; formal analysis, J.R., R.L., S.T. and G.M.; investigation, R.M. and V.C.; data curation R.M., B.F. and I.M.; writing—original draft preparation, R.M.,V.C., B.F. and I.M.; writing—review and editing, J.R., R.L., S.T., S.C., V.C. and G.M.; supervision, J.R., R.L., S.T., S.C., V.C. and G.M.; project administration, J.R., R.L., S.T., S.C., V.C. and G.M.; funding acquisition, R.L. All authors have read and agreed to the published version of the manuscript.

**Funding:** This work was supported by the i9Masks Verão com Ciência project (FCT), by the project NORTE-01-0145-FEDER-028178 funded by NORTE 2020 Portugal Regional Operational Program under PORTUGAL 2020 Partnership Agreement through the European Regional Development Fund and the Fundação para a Ciência e Tecnologia (FCT) and by the project PTDC/EEI-EEE/2846/2021, funded by national funds (OE), within the scope of the Scientific Research and Technological Development Projects (IC&DT) program in all scientific domains (PTDC), through the Foundation for Science and Technology, I.P. (FCT, I.P). The research was also supported by FCT with projects reference UIDB/04077/2020, UIDB/00532/2020, UIDB/00319/2020, UIDB/00690/2020, SusTEC (LA/P/0007/2020) and UIDB/04436/2020, by FEDER funds through the COMPETE 2020-Programa Operacional Competitividade e Internacionalização (POCI) with the reference project POCI-01-0145-FEDER-006941.

**Acknowledgments:** V.C. thanks for her grant from FCT with reference UI/BD/151028/2021. S.C. thanks FCT for her contract funding provided through 2020.00215. CEECIND.

**Conflicts of Interest:** The authors declare no conflict of interest.

## References


1. Wang, W.; Xu, Y.; Gao, R.; Lu, R.; Han, K.; Wu, G.; Tan, W. Detection of SARS-CoV-2 in Different Types of Clinical Specimens. *JAMA* **2020**, *323*, 1843–1844. [CrossRef] [PubMed]
2. WHO. 2020. Available online: https://www.who.int/docs/default-source/coronaviruse/situation-reports/20200211-sitrep-22-ncov.pdf (accessed on 17 December 2020).
3. Manigandan, S.; Wu, M.T.; Ponnusamy, V.K.; Raghavendra, V.B.; Pugazhendhi, A.; Brindhadevi, K. A systematic review on recent trends in transmission, diagnosis, prevention and imaging features of COVID-19. *Process Biochem.* **2020**, *98*, 233–240. [CrossRef] [PubMed]
4. Lu, R.; Zhao, X.; Li, J.; Niu, P.; Yang, B.; Wu, H.; Wang, W.; Song, H.; Huang, B.; Zhu, N.; et al. Genomic characterisation and epidemiology of 2019 novel coronavirus: Implications for virus origins and receptor binding. *Lancet* **2020**, *395*, 565–574. [CrossRef]
5. Guan, W.; Ni, Z.; Hu, Y.; Liang, W.; Ou, C.; He, J.; Liu, L.; Shan, H.; Lei, C.; Hui, D.S.C.; et al. Clinical Characteristics of Coronavirus Disease 2019 in China. *N. Engl. J. Med.* **2020**, *382*, 1708–1720. [CrossRef]
6. Santantonio, T.A.; Messina, G. Update on Coronavirus Disease 2019 (COVID-19). *Open Neurol. J.* **2020**, *14*, 4–5. [CrossRef]
7. Cheng, M.P.; Papenburg, J.; Desjardins, M.; Kanjilal, S.; Quach, C.; Libman, M.; Dittrich, S.; Yansouni, C.P. Diagnostic Testing for Severe Acute Respiratory Syndrome–Related Coronavirus 2. *Ann. Intern. Med.* **2020**, *172*, 726–734. [CrossRef]
8. Kashir, J.; Yaqinuddin, A. Loop mediated isothermal amplification (LAMP) assays as a rapid diagnostic for COVID-19. *Med. Hypotheses* **2020**, *141*, 109786. [CrossRef]
9. Zhang, L.; Liang, X.; Li, Y.; Zheng, H.; Qu, W.; Wang, B.; Luo, H. Diagnostic assays for COVID-19: A narrative review. *J. Bio-X Res.* **2020**, *3*, 123–134. [CrossRef]
10. Li, Z.; Yi, Y.; Luo, X.; Xiong, N.; Liu, Y.; Li, S.; Sun, R.; Wang, Y.; Hu, B.; Chen, W.; et al. Development and clinical application of a rapid IgM-IgG combined antibody test for SARS-CoV-2 infection diagnosis. *J. Med. Virol.* **2020**, *92*, 1518–1524. [CrossRef]
11. Diao, B.; Wen, K.; Chen, J.; Liu, Y.; Yuan, Z.; Han, C.; Chen, J.; Pan, Y.; Chen, L.; Dan, Y.; et al. Diagnosis of Acute Respiratory Syndrome Coronavirus 2 Infection by Detection of Nucleocapsid Protein. *MedRxiv* **2020**. [CrossRef]
12. Nguyen, N.N.T.; McCarthy, C.; Lantigua, D.; Camci-Unal, G. Development of Diagnostic Tests for Detection of SARS-CoV-2. *Diagnostics* **2020**, *10*, 905. [CrossRef]
13. Xia, S.; Chen, X. Single-copy sensitive, field-deployable, and simultaneous dual-gene detection of SARS-CoV-2 RNA via modified RT–RPA. *Cell Discov.* **2020**, *6*, 37. [CrossRef]
14. Lambert-Niclot, S.; Cuffel, A.; Le Pape, S.; Vauloup-Fellous, C.; Morand-Joubert, L.; Roque-Afonso, A.-M.; Le Goff, J.; Delaugerre, C. Evaluation of a Rapid Diagnostic Assay for Detection of SARS-CoV-2 Antigen in Nasopharyngeal Swabs. *J. Clin. Microbiol.* **2020**, *58*, e00977-20. [CrossRef]
15. Nour, M.; Cömert, Z.; Polat, K. A Novel Medical Diagnosis model for COVID-19 infection detection based on Deep Features and Bayesian Optimization. *Appl. Soft Comput.* **2020**, *97*, 106580. [CrossRef]
16. Zhu, X.; Wang, X.; Han, L.; Chen, T.; Wang, L.; Li, H.; Li, S.; He, L.; Fu, X.; Chen, S.; et al. Multiplex reverse transcription loop-mediated isothermal amplification combined with nanoparticle-based lateral flow biosensor for the diagnosis of COVID-19. *Biosens. Bioelectron.* **2020**, *166*, 112437. [CrossRef]
17. Sengupta, J.; Hussain, C.M. Graphene-based field-effect transistor biosensors for the rapid detection and analysis of viruses: A perspective in view of COVID-19. *Carbon Trends* **2021**, *2*, 100011. [CrossRef]
18. Qiu, G.; Gai, Z.; Tao, Y.; Schmitt, J.; Kullak-Ublick, G.A.; Wang, J. Dual-Functional Plasmonic Photothermal Biosensors for Highly Accurate Severe Acute Respiratory Syndrome Coronavirus 2 Detection. *ACS Nano* **2020**, *14*, 5268–5277. [CrossRef]
19. Cui, F.; Zhou, H.S. Diagnostic methods and potential portable biosensors for coronavirus disease 2019. *Biosens. Bioelectron.* **2020**, *165*, 112349. [CrossRef]
20. Caruso, D.; Zerunian, M.; Polici, M.; Pucciarelli, F.; Polidori, T.; Rucci, C.; Guido, G.; Bracci, B.; De Dominicis, C.; Laghi, A. Chest CT Features of COVID-19 in Rome, Italy. *Radiology* **2020**, *296*, E79–E85. [CrossRef]

21. Wang, S.; Hossain, M.Z.; Shinozuka, K.; Shimizu, N.; Kitada, S.; Suzuki, T.; Ichige, R.; Kuwana, A.; Kobayashi, H. Graphene field-effect transistor biosensor for detection of biotin with ultrahigh sensitivity and specificity. *Biosens. Bioelectron.* **2020**, *165*, 112363. [CrossRef]

22. Dong, D.; Tang, Z.; Wang, S.; Hui, H.; Gong, L.; Lu, Y.; Xue, Z.; Liao, H.; Chen, F.; Yang, F.; et al. The Role of Imaging in the Detection and Management of COVID-19: A Review. *IEEE Rev. Biomed. Eng.* **2021**, *14*, 16–29. [CrossRef] [PubMed]

23. Liu, J.; Babka, A.M.; Kearney, B.J.; Radoshitzky, S.R.; Kuhn, J.H.; Zeng, X. Molecular detection of SARS-CoV-2 in formalin-fixed, paraffin-embedded specimens. *JCI Insight* **2020**, *5*, e139042. [CrossRef] [PubMed]

24. Seo, G.; Lee, G.; Kim, M.J.; Baek, S.-H.; Choi, M.; Ku, K.B.; Lee, C.-S.; Jun, S.; Park, D.; Kim, H.G.; et al. Rapid Detection of COVID-19 Causative Virus (SARS-CoV-2) in Human Nasopharyngeal Swab Specimens Using Field-Effect Transistor-Based Biosensor. *ACS Nano* **2020**, *14*, 5135–5142. [CrossRef] [PubMed]

25. Yan, C.; Cui, J.; Huang, L.; Du, B.; Chen, L.; Xue, G.; Li, S.; Zhang, W.; Zhao, L.; Sun, Y.; et al. Rapid and visual detection of 2019 novel coronavirus (SARS-CoV-2) by a reverse transcription loop-mediated isothermal amplification assay. *Clin. Microbiol. Infect.* **2020**, *26*, 773–779. [CrossRef] [PubMed]

26. Jiao, J.; Duan, C.; Xue, L.; Liu, Y.; Sun, W.; Xiang, Y. DNA nanoscaffold-based SARS-CoV-2 detection for COVID-19 diagnosis. *Biosens. Bioelectron.* **2020**, *167*, 112479. [CrossRef] [PubMed]

27. Miripour, Z.S.; Sarrami-Forooshani, R.; Sanati, H.; Makarem, J.; Taheri, M.S.; Shojaeian, F.; Eskafi, A.H.; Abbasvandi, F.; Namdar, N.; Ghafari, H.; et al. Real-time diagnosis of reactive oxygen species (ROS) in fresh sputum by electrochemical tracing; correlation between COVID-19 and viral-induced ROS in lung/respiratory epithelium during this pandemic. *Biosens. Bioelectron.* **2020**, *165*, 112435. [CrossRef]

28. Ishige, T.; Murata, S.; Taniguchi, T.; Miyabe, A.; Kitamura, K.; Kawasaki, K.; Nishimura, M.; Igari, H.; Matsushita, K. Highly sensitive detection of SARS-CoV-2 RNA by multiplex rRT-PCR for molecular diagnosis of COVID-19 by clinical laboratories. *Clin. Chim. Acta* **2020**, *507*, 139–142. [CrossRef]

29. Bordi, L.; Piralla, A.; Lalle, E.; Giardina, F.; Colavita, F.; Tallarita, M.; Sberna, G.; Novazzi, F.; Meschi, S.; Castilletti, C.; et al. Rapid and sensitive detection of SARS-CoV-2 RNA using the Simplexa™ COVID-19 direct assay. *J. Clin. Virol.* **2020**, *128*, 104416. [CrossRef]

30. Huang, Z.; Tian, D.; Liu, Y.; Lin, Z.; Lyon, C.J.; Lai, W.; Fusco, D.; Drouin, A.; Yin, X.; Hu, T.; et al. Ultra-sensitive and high-throughput CRISPR-p owered COVID-19 diagnosis. *Biosens. Bioelectron.* **2020**, *164*, 112316. [CrossRef]

31. Corman, V.M.; Landt, O.; Kaiser, M.; Molenkamp, R.; Meijer, A.; Chu, D.K.; Bleicker, T.; Brünink, S.; Schneider, J.; Schmidt, M.L.; et al. Detection of 2019 novel coronavirus (2019-nCoV) by real-time RT-PCR. *Eurosurveillance* **2020**, *25*, 2000045. [CrossRef]

32. Pfefferle, S.; Reucher, S.; Nörz, D.; Lütgehetmann, M. Evaluation of a quantitative RT-PCR assay for the detection of the emerging coronavirus SARS-CoV-2 using a high throughput system. *Eurosurveillance* **2020**, *25*, 2000152. [CrossRef]

33. Azzi, L.; Carcano, G.; Gianfagna, F.; Grossi, P.; Gasperina, D.D.; Genoni, A.; Fasano, M.; Sessa, F.; Tettamanti, L.; Carinci, F.; et al. Saliva is a reliable tool to detect SARS-CoV-2. *J. Infect.* **2020**, *81*, e45–e50. [CrossRef]

34. Pan, Y.; Li, X.; Yang, G.; Fan, J.; Tang, Y.; Zhao, J.; Long, X.; Guo, S.; Zhao, Z.; Liu, Y.; et al. Serological immunochromatographic approach in diagnosis with SARS-CoV-2 infected COVID-19 patients. *J. Infect.* **2020**, *81*, e28–e32. [CrossRef]

35. Martinaud, C.; Hejl, C.; Igert, A.; Bigaillon, C.; Bonnet, C.; Mérens, A.; Wolf, A.; Foissaud, V.; Leparc-Goffart, I. Evaluation of the Quotient®MosaiQ™ COVID-19 antibody microarray for the detection of IgG and IgM antibodies to SARS-CoV-2 virus in humans. *J. Clin. Virol.* **2020**, *130*, 104571. [CrossRef]

36. Creager, H.M.; Cabrera, B.; Schnaubelt, A.; Cox, J.L.; Cushman-Vokoun, A.M.; Shakir, S.M.; Tardif, K.D.; Huang, M.-L.; Jerome, K.R.; Greninger, A.L.; et al. Clinical evaluation of the BioFire® Respiratory Panel 2.1 and detection of SARS-CoV-2. *J. Clin. Virol.* **2020**, *129*, 104538. [CrossRef]

37. Lee, J.-H.; Choi, M.; Jung, Y.; Lee, S.K.; Lee, C.-S.; Kim, J.; Kim, J.; Kim, N.H.; Kim, B.-T.; Kim, H.G. A novel rapid detection for SARS-CoV-2 spike 1 antigens using human angiotensin converting enzyme 2 (ACE2). *Biosens. Bioelectron.* **2021**, *171*, 112715. [CrossRef]

38. Mavrikou, S.; Moschopoulou, G.; Tsekouras, V.; Kintzios, S. Development of a Portable, Ultra-Rapid and Ultra-Sensitive Cell-Based Biosensor for the Direct Detection of the SARS-CoV-2 S1 Spike Protein Antigen. *Sensors* **2020**, *20*, 3121. [CrossRef]

39. Huang, C.; Wen, T.; Shi, F.-J.; Zeng, X.-Y.; Jiao, Y.-J. Rapid Detection of IgM Antibodies against the SARS-CoV-2 Virus via Colloidal Gold Nanoparticle-Based Lateral-Flow Assay. *ACS Omega* **2020**, *5*, 12550–12556. [CrossRef]

40. Wang, C.; Yang, X.; Zheng, S.; Cheng, X.; Xiao, R.; Li, Q.; Wang, W.; Liu, X.; Wang, S. Development of an ultrasensitive fluorescent immunochromatographic assay based on multilayer quantum dot nanobead for simultaneous detection of SARS-CoV-2 antigen and influenza A virus. *Sens. Actuators B Chem.* **2021**, *345*, 130372. [CrossRef]

41. Karakuş, E.; Erdemir, E.; Demirbilek, N.; Liv, L. Colorimetric and electrochemical detection of SARS-CoV-2 spike antigen with a gold nanoparticle-based biosensor. *Anal. Chim. Acta* **2021**, *1182*, 338939. [CrossRef]

42. Kim, H.-Y.; Lee, J.-H.; Kim, M.J.; Park, S.C.; Choi, M.; Lee, W.; Ku, K.B.; Kim, B.T.; Park, E.C.; Kim, H.G.; et al. Development of a SARS-CoV-2-specific biosensor for antigen detection using scFv-Fc fusion proteins. *Biosens. Bioelectron.* **2021**, *175*, 112868. [CrossRef]

43. Chen, Z.; Wu, Q.; Chen, J.; Ni, X.; Dai, J. A DNA Aptamer Based Method for Detection of SARS-CoV-2 Nucleocapsid Protein. *Virol. Sin.* **2020**, *35*, 351–354. [CrossRef]

44. Deng, H.; Jayawardena, A.; Chan, J.; Tan, S.M.; Alan, T.; Kwan, P. An ultra-portable, self-contained point-of-care nucleic acid amplification test for diagnosis of active COVID-19 infection. *Sci. Rep.* **2021**, *11*, 15176. [CrossRef]
45. Zou, M.; Su, F.; Zhang, R.; Jiang, X.; Xiao, H.; Yan, X.; Yang, C.; Fan, X.; Wu, G. Rapid point-of-care testing for SARS-CoV-2 virus nucleic acid detection by an isothermal and nonenzymatic Signal amplification system coupled with a lateral flow immunoassay strip. *Sens. Actuators B Chem.* **2021**, *342*, 129899. [CrossRef]
46. Ma, H.; Zeng, W.; He, H.; Zhao, D.; Yang, Y.; Jiang, D.; Qi, P.Y.; He, W.; Zhao, C.; Yi, R.; et al. COVID-19 diagnosis and study of serum SARS-CoV-2 specific IgA, IgM and IgG by chemiluminescence immunoanalysis. *MedRxiv* **2020**. [CrossRef]
47. Rashed, M.Z.; Kopechek, J.A.; Priddy, M.C.; Hamorsky, K.T.; Palmer, K.E.; Mittal, N.; Valdez, J.; Flynn, J.; Williams, S.J. Rapid detection of SARS-CoV-2 antibodies using electrochemical impedance-based detector. *Biosens. Bioelectron.* **2021**, *171*, 112709. [CrossRef]
48. Baek, Y.H.; Um, J.; Antigua, K.J.C.; Park, J.-H.; Kim, Y.; Oh, S.; Kim, Y.; Choi, W.-S.; Kim, S.G.; Jeong, J.H.; et al. Development of a reverse transcription-loop-mediated isothermal amplification as a rapid early-detection method for novel SARS-CoV-2. *Emerg. Microbes Infect.* **2020**, *9*, 998–1007. [CrossRef]
49. Awal, M.A.; Masud, M.; Hossain, M.S.; Bulbul, A.A.-M.; Mahmud, S.M.H.; Bairagi, A.K. A Novel Bayesian Optimization-Based Machine Learning Framework for COVID-19 Detection From Inpatient Facility Data. *IEEE Access* **2021**, *9*, 10263–10281. [CrossRef]
50. Mahmud, T.; Rahman, M.A.; Fattah, S.A. CovXNet: A multi-dilation convolutional neural network for automatic COVID-19 and other pneumonia detection from chest X-ray images with transferable multi-receptive feature optimization. *Comput. Biol. Med.* **2020**, *122*, 103869. [CrossRef]
51. Panwar, H.; Gupta, P.K.; Siddiqui, M.K.; Morales-Menendez, R.; Singh, V. Application of deep learning for fast detection of COVID-19 in X-rays using nCOVnet. *Chaos Solitons Fractals* **2020**, *138*, 109944. [CrossRef]
52. Khan, A.I.; Shah, J.L.; Bhat, M.M. CoroNet: A deep neural network for detection and diagnosis of COVID-19 from chest X-ray images. *Comput. Methods Programs Biomed.* **2020**, *196*, 105581. [CrossRef] [PubMed]
53. Tuncer, T.; Dogan, S.; Ozyurt, F. An automated Residual Exemplar Local Binary Pattern and iterative ReliefF based COVID-19 detection method using chest X-ray image. *Chemom. Intell. Lab. Syst.* **2020**, *203*, 104054. [CrossRef] [PubMed]
54. Toğaçar, M.; Ergen, B.; Cömert, Z. COVID-19 detection using deep learning models to exploit Social Mimic Optimization and structured chest X-ray images using fuzzy color and stacking approaches. *Comput. Biol. Med.* **2020**, *121*, 103805. [CrossRef] [PubMed]
55. Yan, T.; Wong, P.K.; Ren, H.; Wang, H.; Wang, J.; Li, Y. Automatic distinction between COVID-19 and common pneumonia using multi-scale convolutional neural network on chest CT scans. *Chaos Solitons Fractals* **2020**, *140*, 110153. [CrossRef]
56. Shaban, W.M.; Rabie, A.H.; Saleh, A.I.; Abo-Elsoud, M.A. A new COVID-19 Patients Detection Strategy (CPDS) based on hybrid feature selection and enhanced KNN classifier. *Knowl.-Based Syst.* **2020**, *205*, 106270. [CrossRef]
57. Diaz-Lara, A.; Golino, D.; Preece, J.E.; Al Rwahnih, M. Development of RT-PCR degenerate primers to overcome the high genetic diversity of grapevine virus T. *J. Virol. Methods* **2020**, *282*, 113883. [CrossRef]
58. Jääskeläinen, A.J.; Sironen, T.; Kaloinen, M.; Kakkola, L.; Julkunen, I.; Hewson, R.; Weidmann, M.W.; Mirazimi, A.; Watson, R.; Vapalahti, O. Comparison of Zaire ebolavirus realtime RT-PCRs targeting the nucleoprotein gene. *J. Virol. Methods* **2020**, *284*, 113941. [CrossRef]
59. Peñarrubia, L.; Ruiz, M.; Porco, R.; Rao, S.N.; Juanola-Falgarona, M.; Manissero, D.; López-Fontanals, M.; Pareja, J. Multiple assays in a real-time RT-PCR SARS-CoV-2 panel can mitigate the risk of loss of sensitivity by new genomic variants during the COVID-19 outbreak. *Int. J. Infect. Dis.* **2020**, *97*, 225–229. [CrossRef]
60. Yüce, M.; Filiztekin, E.; Özkaya, K.G. COVID-19 diagnosis—A review of current methods. *Biosens. Bioelectron.* **2021**, *172*, 112752. [CrossRef]
61. Zhu, H.; Zhang, H.; Ni, S.; Korabečná, M.; Yobas, L.; Neuzil, P. The vision of point-of-care PCR tests for the COVID-19 pandemic and beyond. *TrAC Trends Anal. Chem.* **2020**, *130*, 115984. [CrossRef]
62. Wang, Y.; Song, W.; Zhao, Z.; Chen, P.; Liu, J.; Li, C. The impacts of viral inactivating methods on quantitative RT-PCR for COVID-19. *Virus Res.* **2020**, *285*, 197988. [CrossRef]
63. Thompson, D.; Lei, Y. Mini review: Recent progress in RT-LAMP enabled COVID-19 detection. *Sens. Actuators Rep.* **2020**, *2*, 100017. [CrossRef] [PubMed]
64. Zhang, Y.; Odiwuor, N.; Xiong, J.; Sun, L.; Nyaruaba, R.O.; Wei, H.; Tanner, N.A. Rapid Molecular Detection of SARS-CoV-2 (COVID-19) Virus RNA Using Colorimetric LAMP. *MedRxiv* **2020**. [CrossRef]
65. Yang, W.; Dang, X.; Wang, Q.; Xu, M.; Zhao, Q.; Zhou, Y.; Zhao, H.; Wang, L.; Xu, Y.; Wang, J.; et al. Rapid Detection of SARS-CoV-2 Using Reverse transcription RT-LAMP method. *MedRxiv* **2020**. [CrossRef]
66. Lucia, C.; Federico, P.-B.; Alejandra, G.C. An ultrasensitive, rapid, and portable coronavirus SARS-CoV-2 sequence detection method based on CRISPR-Cas12. *BioRxiv* **2020**. [CrossRef]
67. Broughton, J.P.; Deng, X.; Yu, G.; Fasching, C.L.; Servellita, V.; Singh, J.; Miao, X.; Streithorst, J.A.; Granados, A.; Sotomayor-Gonzalez, A.; et al. CRISPR–Cas12-based detection of SARS-CoV-2. *Nat. Biotechnol.* **2020**, *38*, 870–874. [CrossRef]
68. Mashimo, T. Rapid and accurate detection of novel coronavirus SARS-CoV-2 using CRISPR-Cas3. *MedRxiv* **2020**. [CrossRef]
69. Joung, J.; Ladha, A.; Saito, M.; Segel, M.; Bruneau, R.; Huang, M.W.; Kim, N.-G.; Yu, X.; Li, J.; Walker, B.D.; et al. Point-of-care testing for COVID-19 using SHERLOCK diagnostics. *MedRxiv* **2020**. [CrossRef]
70. Song, K.-M.; Lee, S.; Ban, C. Aptamers and Their Biological Applications. *Sensors* **2012**, *12*, 612–631. [CrossRef]

71. Song, Y.; Song, J.; Wei, X.; Huang, M.; Sun, M.; Zhu, L.; Lin, B.; Shen, H.; Zhu, Z.; Yang, C. Discovery of Aptamers Targeting the Receptor-Binding Domain of the SARS-CoV-2 Spike Glycoprotein. *Anal. Chem.* **2020**, *92*, 9895–9900. [CrossRef]
72. Torabi, R.; Ranjbar, R.; Halaji, M.; Heiat, M. Aptamers, the bivalent agents as probes and therapies for coronavirus infections: A systematic review. *Mol. Cell. Probes* **2020**, *53*, 101636. [CrossRef]
73. Van Elslande, J.; Houben, E.; Depypere, M.; Brackenier, A.; Desmet, S.; André, E.; Van Ranst, M.; Lagrou, K.; Vermeersch, P. Diagnostic performance of seven rapid IgG/IgM antibody tests and the Euroimmun IgA/IgG ELISA in COVID-19 patients. *Clin. Microbiol. Infect.* **2020**, *26*, 1082–1087. [CrossRef]
74. Xiao, Q.; Xu, C. Research progress on chemiluminescence immunoassay combined with novel technologies. *TrAC Trends Anal. Chem.* **2020**, *124*, 115780. [CrossRef]
75. Li, F.; You, M.; Li, S.; Hu, J.; Liu, C.; Gong, Y.; Yang, H.; Xu, F. Paper-based point-of-care immunoassays: Recent advances and emerging trends. *Biotechnol. Adv.* **2020**, *39*, 107442. [CrossRef]
76. Peeling, R.W.; Olliaro, P.L.; Boeras, D.I.; Fongwen, N. Scaling up COVID-19 rapid antigen tests: Promises and challenges. *Lancet Infect. Dis.* **2021**, *21*, e290–e295. [CrossRef]
77. Smithgall, M.C.; Dowlatshahi, M.; Spitalnik, S.L.; Hod, E.A.; Rai, A.J. Types of Assays for SARS-CoV-2 Testing: A Review. *Lab. Med.* **2020**, *51*, e59–e65. [CrossRef]
78. Giri, B.; Pandey, S.; Shrestha, R.; Pokharel, K.; Ligler, F.S.; Neupane, B.B. Review of analytical performance of COVID-19 detection methods. *Anal. Bioanal. Chem.* **2021**, *413*, 35–48. [CrossRef]
79. Vengesai, A.; Midzi, H.; Kasambala, M.; Mutandadzi, H.; Mduluza-Jokonya, T.L.; Rusakaniko, S.; Mutapi, F.; Naicker, T.; Mduluza, T. A systematic and meta-analysis review on the diagnostic accuracy of antibodies in the serological diagnosis of COVID-19. *Syst. Rev.* **2021**, *10*, 155. [CrossRef]
80. Nicol, T.; Lefeuvre, C.; Serri, O.; Pivert, A.; Joubaud, F.; Dubée, V.; Kouatchet, A.; Ducancelle, A.; Lunel-Fabiani, F.; Le Guillou-Guillemette, H. Assessment of SARS-CoV-2 serological tests for the diagnosis of COVID-19 through the evaluation of three immunoassays: Two automated immunoassays (Euroimmun and Abbott) and one rapid lateral flow immunoassay (NG Biotech). *J. Clin. Virol.* **2020**, *129*, 104511. [CrossRef]
81. Abid, S.A.; Muneer, A.A.; Al-Kadmy, I.M.S.; Sattar, A.A.; Beshbishy, A.M.; Batiha, G.E.-S.; Hetta, H.F. Biosensors as a future diagnostic approach for COVID-19. *Life Sci.* **2021**, *273*, 119117. [CrossRef]
82. Djaileb, A.; Charron, B.; Jodaylami, M.H.; Coutu, J.; Stevenson, K.; Forest, S.; Live, L.S.; Pelletier, J.N.; Masson, J.-F. A rapid and quantitative serum test for SARS-CoV-2 antibodies with portable surface plasmon resonance sensing. *ChemRxiv* **2020**. [CrossRef]
83. Yasri, S.; Wiwanitkit, V. Sustainable materials and COVID-19 detection biosensor: A brief review. *Sens. Int.* **2022**, *3*, 100171. [CrossRef] [PubMed]
84. Das, N.N.; Kumar, N.; Kaur, M.; Kumar, V.; Singh, D. Automated Deep Transfer Learning-Based Approach for Detection of COVID-19 Infection in Chest X-rays. *Irbm* **2022**, *43*, 114–119. [CrossRef] [PubMed]
85. Chen, J.; See, K.C. Artificial Intelligence for COVID-19: Rapid Review. *J. Med. Internet Res.* **2020**, *22*, e21476. [CrossRef]
86. Nayak, S.R.; Nayak, D.R.; Sinha, U.; Arora, V.; Pachori, R.B. Application of deep learning techniques for detection of COVID-19 cases using chest X-ray images: A comprehensive study. *Biomed. Signal Process. Control* **2021**, *64*, 102365. [CrossRef]
87. Ozturk, T.; Talo, M.; Yildirim, E.A.; Baloglu, U.B.; Yildirim, O.; Acharya, U.R. Automated detection of COVID-19 cases using deep neural networks with X-ray images. *Comput. Biol. Med.* **2020**, *121*, 103792. [CrossRef]
88. Hassantabar, S.; Ahmadi, M.; Sharifi, A. Diagnosis and detection of infected tissue of COVID-19 patients based on lung X-ray image using convolutional neural network approaches. *Chaos Solitons Fractals* **2020**, *140*, 110170. [CrossRef]
89. Brunese, L.; Mercaldo, F.; Reginelli, A.; Santone, A. Explainable Deep Learning for Pulmonary Disease and Coronavirus COVID-19 Detection from X-rays. *Comput. Methods Programs Biomed.* **2020**, *196*, 105608. [CrossRef]
90. Fang, Y.; Zhang, H.; Xie, J.; Lin, M.; Ying, L.; Pang, P.; Ji, W. Sensitivity of Chest CT for COVID-19: Comparison to RT-PCR. *Radiology* **2020**, *296*, E115–E117. [CrossRef]
91. Xie, X.; Zhong, Z.; Zhao, W.; Zheng, C.; Wang, F. Chest CT for Typical 2019-nCoV Pneumonia: Relationship to Negative RT-PCR Testing. *Radiology* **2020**, *296*, E41–E45. [CrossRef]
92. Hope, M.D.; Raptis, C.A.; Henry, T.S. Chest Computed Tomography for Detection of Coronavirus Disease 2019 (COVID-19): Don't Rush the Science. *Ann. Intern. Med.* **2020**, *173*, 147–148. [CrossRef]
93. Song, F.; Shi, N.; Shan, F.; Zhang, Z.; Shen, J.; Lu, H.; Ling, Y.; Jiang, Y.; Shi, Y. Emerging 2019 Novel Coronavirus (2019-nCoV) Pneumonia. *Radiology* **2020**, *295*, 210–217. [CrossRef]
94. Basiri, A.; Heidari, A.; Nadi, M.F.; Fallahy, M.T.P.; Nezamabadi, S.S.; Sedighi, M.; Saghazadeh, A.; Rezaei, N. Microfluidic devices for detection of RNA viruses. *Rev. Med. Virol.* **2021**, *31*, 1–11. [CrossRef]
95. Kaziz, S.; Saad, Y.; Bouzid, M.; Selmi, M.; Belmabrouk, H. Enhancement of COVID-19 detection time by means of electrothermal force. *Microfluid. Nanofluidics* **2021**, *25*, 86. [CrossRef]
96. Faustino, V.; Rodrigues, R.O.; Pinho, D.; Costa, E.; Santos-Silva, A.; Miranda, V.; Amaral, J.S.; Lima, R. A Microfluidic Deformability Assessment of Pathological Red Blood Cells Flowing in a Hyperbolic Converging Microchannel. *Micromachines* **2019**, *10*, 645. [CrossRef]
97. Catarino, S.O.; Rodrigues, R.O.; Pinho, D.; Miranda, J.M.; Minas, G.; Lima, R. Blood Cells Separation and Sorting Techniques of Passive Microfluidic Devices: From Fabrication to Applications. *Micromachines* **2019**, *10*, 593. [CrossRef]

98. Faustino, V.; Catarino, S.O.; Lima, R.; Minas, G. Biomedical microfluidic devices by using low-cost fabrication techniques: A review. *J. Biomech.* **2016**, *49*, 2280–2292. [CrossRef]

99. Pinho, D.; Faustino, V.; Catarino, S.O.; Pereira, A.I.; Minas, G.; Pinho, F.T.; Lima, R. Label-free multi-step microfluidic device for mechanical characterization of blood cells: Diabetes type II. *Micro Nano Eng.* **2022**, *16*, 100149. [CrossRef]

100. Pinho, D.; Carvalho, V.; Gonçalves, I.M.; Teixeira, S.; Lima, R. Visualization and Measurements of Blood Cells Flowing in Microfluidic Systems and Blood Rheology: A Personalized Medicine Perspective. *J. Pers. Med.* **2020**, *10*, 249. [CrossRef]

101. Carvalho, V.; Bañobre-López, M.; Minas, G.; Teixeira, S.F.C.F.; Lima, R.; Rodrigues, R.O. The integration of spheroids and organoids into organ-on-a-chip platforms for tumour research: A review. *Bioprinting* **2022**, *27*, e00224. [CrossRef]

102. Carvalho, V.; Rodrigues, R.O.; Lima, R.A.; Teixeira, S. Computational Simulations in Advanced Microfluidic Devices: A Review. *Micromachines* **2021**, *12*, 1149. [CrossRef]

103. Gonçalves, I.M.; Carvalho, V.; Rodrigues, R.O.; Pinho, D.; Teixeira, S.F.C.F.; Moita, A.; Hori, T.; Kaji, H.; Lima, R.; Minas, G. Organ-on-a-Chip Platforms for Drug Screening and Delivery in Tumor Cells: A Systematic Review. *Cancers* **2022**, *14*, 935. [CrossRef]

104. Carvalho, V.; Gonçalves, I.; Lage, T.; Rodrigues, R.O.; Minas, G.; Teixeira, S.F.C.F.; Moita, A.S.; Hori, T.; Kaji, H.; Lima, R.A. 3D Printing Techniques and Their Applications to Organ-on-a-Chip Platforms: A Systematic Review. *Sensors* **2021**, *21*, 3304. [CrossRef]

105. Burdino, E.; Cerutti, F.; Panero, F.; Allice, T.; Gregori, G.; Milia, M.G.; Cavalot, G.; Altavilla, A.; Aprà, F.; Ghisetti, V. SARS-CoV-2 microfluidic antigen point-of-care testing in Emergency Room patients during COVID-19 pandemic. *J. Virol. Methods* **2022**, *299*, 114337. [CrossRef]

106. Jhou, Y.-R.; Wang, C.-H.; Tsai, H.-P.; Shan, Y.-S.; Lee, G.-B. An integrated microfluidic platform featuring real-time reverse transcription loop-mediated isothermal amplification for detection of COVID-19. *Sens. Actuators B Chem.* **2022**, *358*, 131447. [CrossRef]

107. Wu, Y.; Hu, B.; Ma, X.; Wang, Y.; Li, W.; Wang, S. Enhancement of Binding Kinetics on Affinity Substrates Using Asymmetric Electroosmotic Flow on a Sinusoidal Bipolar Electrode. *Micromachines* **2022**, *13*, 207. [CrossRef]

108. Torrente-Rodríguez, R.M.; Lukas, H.; Tu, J.; Min, J.; Yang, Y.; Xu, C.; Rossiter, H.B.; Gao, W. SARS-CoV-2 RapidPlex: A Graphene-Based Multiplexed Telemedicine Platform for Rapid and Low-Cost COVID-19 Diagnosis and Monitoring. *Matter* **2020**, *3*, 1981–1998. [CrossRef]

109. Basu, A.; Pamreddy, A.; Singh, P.; Sharma, K. An Adverse Outcomes Approach to Study the Effects of SARS-CoV-2 in 3D Organoid Models. *J. Mol. Biol.* **2022**, *434*, 167213. [CrossRef]

110. Si, L.; Bai, H.; Rodas, M.; Cao, W.; Oh, C.Y.; Jiang, A.; Moller, R.; Hoagland, D.; Oishi, K.; Horiuchi, S.; et al. A human-airway-on-a-chip for the rapid identification of candidate antiviral therapeutics and prophylactics. *Nat. Biomed. Eng.* **2021**, *5*, 815–829. [CrossRef]

111. Salahudeen, A.A.; Choi, S.S.; Rustagi, A.; Zhu, J.; van Unen, V.; de la O, S.M.; Flynn, R.A.; Margalef-Català, M.; Santos, A.J.M.; Ju, J.; et al. Progenitor identification and SARS-CoV-2 infection in human distal lung organoids. *Nature* **2020**, *588*, 670–675. [CrossRef]

*micromachines*

*Article*

# The Effect of Dynamic, In Vivo-like Oxaliplatin on HCT116 Spheroids in a Cancer-on-Chip Model Is Representative of the Response in Xenografts

Job Komen [1,*], Sanne M. van Neerven [2], Elsbeth G. B. M. Bossink [2], Nina E. de Groot [2], Lisanne E. Nijman [2], Albert van den Berg [1], Louis Vermeulen [2] and Andries D. van der Meer [3]

1    BIOS Lab on a Chip Group, MESA+ Institute for Nanotechnology, University of Twente, 7500 AE Enschede, The Netherlands; a.vandenberg@utwente.nl
2    Laboratory for Experimental Oncology and Radiobiology, Center for Experimental and Molecular Medicine, Cancer Center Amsterdam and Amsterdam Gastroenterology, Endocrinology and Metabolism, Amsterdam University Medical Centers, 1105 AZ Amsterdam, The Netherlands; s.m.vanneerven@amsterdamumc.nl (S.M.v.N.); e.g.b.m.bossink@utwente.nl (E.G.B.M.B.); n.e.degroot@amsterdamumc.nl (N.E.d.G.); l.e.nijman@amsterdamumc.nl (L.E.N.); l.vermeulen@amsterdamumc.nl (L.V.)
3    Applied Stem Cell Technologies, TechMed Centre, University of Twente, 7500 AE Enschede, The Netherlands; andries.vandermeer@utwente.nl
*    Correspondence: j.komen@utwente.nl

**Citation:** Komen, J.; van Neerven, S.M.; Bossink, E.G.B.M.; de Groot, N.E.; Nijman, L.E.; van den Berg, A.; Vermeulen, L.; van der Meer, A.D. The Effect of Dynamic, In Vivo-like Oxaliplatin on HCT116 Spheroids in a Cancer-on-Chip Model Is Representative of the Response in Xenografts. *Micromachines* **2022**, 13, 739. https://doi.org/10.3390/mi13050739

Academic Editors: Violeta Carvalho, Senhorinha Teixeira and João Eduardo P. Castro Ribeiro

Received: 1 April 2022
Accepted: 4 May 2022
Published: 6 May 2022

**Publisher's Note:** MDPI stays neutral with regard to jurisdictional claims in published maps and institutional affiliations.

**Abstract:** The cancer xenograft model in which human cancer cells are implanted in a mouse is one of the most used preclinical models to test the efficacy of novel cancer drugs. However, the model is imperfect; animal models are ethically burdened, and the imperfect efficacy predictions contribute to high clinical attrition of novel drugs. If microfluidic cancer-on-chip models could recapitulate key elements of the xenograft model, then these models could substitute the xenograft model and subsequently surpass the xenograft model by reducing variation, increasing sensitivity and scale, and adding human factors. Here, we exposed HCT116 colorectal cancer spheroids to dynamic, in vivo-like, concentrations of oxaliplatin, including a 5 day drug-free period, on-chip. Growth inhibition on-chip was comparable to existing xenograft studies. Furthermore, immunohistochemistry showed a similar response in proliferation and apoptosis markers. While small volume changes in xenografts are hard to detect, in the chip-system, we could observe a temporary growth delay. Lastly, histopathology and a pharmacodynamic model showed that the cancer spheroid-on-chip was representative of the proliferating outer part of a HCT116 xenograft, thereby capturing the major driver of the drug response of the xenograft. Hence, the cancer-on-chip model recapitulated the response of HCT116 xenografts to oxaliplatin and provided additional drug efficacy information.

**Keywords:** cancer-on-chip; xenograft; microfluidic; colorectal cancer; pharmacodynamics; pharmacokinetics; drug efficacy; oxaliplatin

## 1. Introduction

Cancer is the leading cause of death in the Western world [1]. Continuous improvements in treatment have extended survival times; nevertheless, high attrition rates in clinical trials of novel drugs hamper treatment improvement [2,3]. One reason for high attrition rates is the limited translational quality of in vitro and animal models [4].

One of the most used animal models in cancer drug discovery is the mouse xenograft model [5]. In this model, human cancer cells are implanted in an immune-deficient mouse. The xenograft adds certain physiologic aspects compared to in vitro assays, such as pharmacokinetics, homeostasis, and three-dimensional, vascularized tumor growth. However, the predictive power of the model toward human cancer response has limitations due to the reduction in biological complexity, species differences, and experimental variation [6].

Despite these shortcomings, the xenograft model remains a key assay for in vivo efficacy testing of novel drugs before clinical trials.

Basic human cell culture models are no direct alternative for xenograft models in pre-clinical efficacy testing, as they are incapable of capturing all relevant aspects of an in vivo tumor. Recently, microfluidic 'cancer-on-chip' technology has led to major improvements in modeling tumor physiology in vitro. Cancer-on-chip models can recapitulate essential in vivo characteristics of cancer, such as continuous perfusion to maintain homeostasis, angiogenesis, controllable oxygen and nutrient gradients, interaction with stromal cells, and mechanical stimulation [7–12]. Recently, the possibility to offer dynamic, in vivo-like drug concentrations has been added to this repertoire [12–14]. Despite all these promising aspects, cancer-on-chip technology has not yet been sufficiently validated to significantly replace existing preclinical, in vivo assays. Validation against xenograft models will be an essential step for cancer-on-chip models to be used as valid alternatives in preclinical efficacy testing. Moreover, such validation will offer an indispensable foundation for further development of even more advanced cancer-on-chip models with added human complexity [15].

Comparison of the drug response in cancer-on-chip models to xenografts is budding, and multiple studies have performed such side-by-side experiments [16–18]. These first studies were based on cancer-on-chip models that included vascularization, stromal cell coculture [18,19], patient-derived cells [16], and dynamic drug concentrations [17]. The results of the side-by-side comparisons were highly encouraging, with cancer-on-chip models and xenografts based on the same cells responding similarly upon treatment with the same drugs. However, these early studies focused on live–dead stains [16], volume-based growth readouts [17], in situ immunofluorescence [17], or gene expression [18] for comparison between cancer-on-chip and xenograft models. An important readout in xenograft studies is provided by histopathology and immunohistochemistry, which provides data on cancer cell morphology and tissue structure. Furthermore, immunostaining of tissue sections reduces the probability of signal distortion compared to in situ immunofluorescence, where dampening of the signal can occur due to the thickness of the cancer tissues [20]. Therefore, side-by-side histological analysis of both cancer-on-chip and xenograft models will be an indispensable step in the validation of cancer-on-chip models.

Here, we performed a side-by-side comparison between the response to oxaliplatin of HCT116 colorectal cancer spheroids in a cancer-on-chip model versus HCT116 xenografts in existing studies. We evaluated the effect of oxaliplatin on volume growth and proliferation and apoptosis markers. Moreover, we explored how histological information can be used to scale the overall growth data of spheroids in the cancer-on-chip model to xenografts to provide an improved comparison with data from the in vivo xenograft.

## 2. Materials and Methods

### 2.1. Device Design and Fabrication

The microfluidic chip consisted of two detachable parts (Figure 1a,b). The top part contained a 21 × 2 × 1 mm (length × width × height) straight channel for continuous perfusion of nutrients and drug. The bottom part contained a single U-shaped well for holding a single spheroid. The well had a diameter of 2 mm and a depth of 0.8 mm to facilitate placement and growth of spheroids from 0.5 mm to 1 mm diameter (Figure 1a,b).

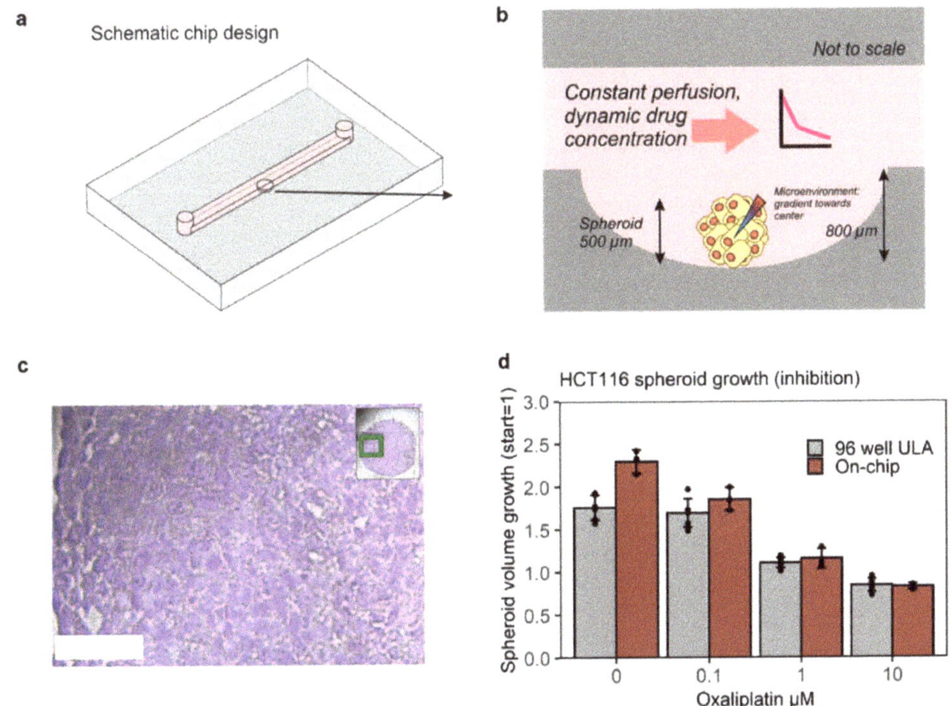



**Figure 1. Design and validation of the chip system for mimicking oxaliplatin efficacy in a HCT116 xenograft model.** (**a**,**b**) Chip design. The two-part chip consists of a 21 × 2 × 1 mm (l × w × h) top channel for continuous perfusion and dynamic drug concentrations and a U-shaped well for spheroid placement and growth. (**c**) H&E staining shows low differentiation and little extracellular matrix in the spheroid on-chip, as in HCT-116 xenografts. Scale bar = 50 µm. (**d**) Growth inhibition after 48 h exposure to constant concentrations of oxaliplatin is not confounded by the chip, although control growth is slightly higher, as shown by comparison to a 96-well ultralow-attachment (ULA) plate; $n = 3$ spheroids for on-chip, $n = 6$ spheroids for 96 well; error bars represent the standard deviation (SD).

Chip parts were made of polydimethylsiloxane (PDMS). Chips were fabricated using PDMS in a 10:1 weight ratio of base to curing agent (Sylgard 184, Dow Corning, USA). The PDMS was poured on a micro-milled mold (Datron Neo, Germany) and cured at 60 °C overnight. Inlet and outlet holes were punched using Harris UniCore punchers of 1 mm diameter. Medical-grade adhesive, double-sided tape AR Care 8939 (Adhesive Research, Limerick, Ireland), 0.11 mm thickness, was applied to the bottom part before coating the well. For ensuring non-adhesion of the spheroids to the well, a coating of Pluronic F127 was used [21]. Pluronics attach to the hydrophobic PDMS with a central hydrophobic block, and the hydrophilic tails form a hydrophilic brush which prevents binding of proteins and cells [21,22]. Pluronics have been shown to prevent cell adhesion for up to 4 weeks, and they remain intact in microfluidic channels after flow [23,24]. Pluronic F127 has been used to prevent adhesion of spheroids in chips for up to 10 days [21]. Pluronic F127 20 mg/mL in deionized (DI) water was applied to the wells and incubated for 2 h at 37 °C [21]. To ensure homogeneous coating, Pluronic F127 was prevented from drying out by a droplet volume of 6 µL and the placement of a PDMS cap over the well. Chips were placed in a box inside the incubator with Kimwipes soaked in phosphate-buffered saline (PBS, ThermoFisher, Waltham, MA, USA) to prevent drying out. After incubation, the Pluronic was removed, and the well was flushed with 2 µL of medium before placing the spheroids.

### 2.2. Cell and Spheroid Culture

HCT-116 colorectal cancer cells (Sigma-Aldrich, Burlington, MA, USA), with a passage number under 30, were cultured in McCoy 5A medium (ThermoFisher, USA) with 5% fetal bovine serum (FBS, ThermoFisher) and 1% penicillin–streptomycin (ThermoFisher, USA). Four days before the start of the chip experiment, 2000 HCT116 cells in 200 μL of medium per well were placed in a Nunclon Sphera ultralow-absorption round-bottom plate (ThermoFisher, USA). Cells were spun at 390 G for 5 min and placed in a humidified incubator at 37 °C.

### 2.3. Chip Experiments

Spheroids were collected from the 96 well plate with a wide-bore P200 pipette tip set at 50 μL (VWR, Radnor, PA, USA). In several seconds, the spheroid sunk to the bottom of the tip, and the tip was subsequently brought in contact with the medium in the well, resulting in the transfer of the spheroid. Subsequently, the top part of the chip was placed on the adhesive tape, which was already attached to the bottom part. Chips were filled with medium at a flow of 20 μL/min. Chips were continuously perfused at a flow rate of 2 μL/min using a syringe pump (Harvard PhD2000, Fargo, ND, USA).

### 2.4. Histopathology

Spheroids were taken off-chip for histopathology by removing the top part of the chip, by inserting tweezers between the top part and the adhesive tape, without touching the channel with the tweezers. A PDMS ring with an inner diameter of 8 mm and a thickness of 3 mm was placed around the well and gently filled with 150 μL of PBS. The spheroid was harvested with the p200 wide bore pipette tip and transferred to 4% paraformaldehyde (PFA, Sigma, USA) in PBS. Spheroids were placed in 70% alcohol overnight, dehydrated, transferred to a biopsy foam pad, embedded in paraplast, and cut into 0.4 μm sections. Tissue sections were rehydrated and stained with hematoxylin and eosin (H&E) to evaluate the tissue structure or a diaminobenzidine (DAB) staining for apoptosis (cleaved Caspase-3) or proliferation (Ki67) with a hematoxylin counterstain. Antibodies used for the DAB staining were primary antibodies cleaved Caspase-3 (ASP-175) (1:1000, Cell Signaling #9661, USA) or Ki67 (1:1000, Sigma #SAB5500134, USA) and secondary anti-rabbit antibody linked to horseradish peroxidase (HRP). Slides were dehydrated and mounted with Pertex (VWR, Amsterdam, The Netherlands).

### 2.5. Imaging and Analyses

Brightfield images of the spheroids on-chip were taken at t = 0, t = 2 days, and t = 7 days at $10\times$ magnification (Leica DM IRM, Wetzlar, Germany). Spheroid size was determined with open-source software Fiji/ImageJ using the algorithm developed by Ivanov et al. (Supplementary Figure S1) [25]. Spheroid growth inhibition was based on the formula as described in the xenograft literature [26], and was defined as

$$1 - \frac{\text{average}\left(Volume\ end_{treatment} - Volume\ start_{treatment}\right)}{\text{average}\left(Volume\ end_{control} - Volume\ start_{control}\right)}.$$

Spheroid cell counts were carried out by harvesting the spheroids from the chips as described in the histopathology section; however, instead of fixation, spheroids were placed in a 96-well plate in 100 μL of trypsin for 20 min at 37 °C. After resuspension, cells were counted with a Luna automated cell counter (Logos Biosystems, Anyang-si, South Korea).

Immunohistochemistry analyses were conducted with open-source software Qupath on $20\times$ brightfield images [27]. For Ki67 detection, colors were first deconvoluted for each image with the "estimate stain vectors" function. Nucleus detection was based on the parameters used by Robertson et al. [28]. The cutoff for Ki67 positivity was 0.3. CC3 positivity was based on the percentage of positive pixels within the spheroid at a cutoff value of 0.5.

*2.6. Xenograft Data*

To compare growth inhibition on-chip to published xenograft studies, a literature search was conducted. Inclusion criteria were as follows: indexation in PubMed, analysis period, and treatment start at a tumor starting volume of 50 mm$^3$ or larger, subcutaneous placement of tumors, and more than one treatment cycle during the analysis period of 2–5 weeks [26,29–34]. An overview of tumor growth (inhibition) data is provided in Supplementary Table S1 and Supplementary Figure S2.

To compare the histology and immunohistochemistry between the untreated on-chip spheroids and HCT116 xenografts, xenograft specimens obtained in an existing xenograft experiment were used [35]. Briefly, $1 \times 10^6$ human CRC cells were injected subcutaneously in NOD/SCID IL2R gamma$^{-/-}$ (NSG) mice. When tumors reached a volume of ~300 mm$^3$, mice were sacrificed and xenografts were stored in formalin-fixed, paraffin-embedded (FFPE) blocks.

## 3. Results and Discussion

### 3.1. Chip Design and Validation for Mimicking Xenograft Drug Response On-Chip

To mimic drug efficacy in a xenograft model on-chip, important biological factors and readouts should be comparable [15]. Injection of the colorectal cancer cell line HCT116 below the skin of the mouse led to a fast-growing tumor with poor differentiation (Supplementary Figure S3). When the tumor achieved a measurable volume (50–300 mm$^3$), oxaliplatin was administered, typically one to several times a week for multiple cycles [26,29–34]. Oxaliplatin concentration in the mouse was characterized by a peak and an exponential decline, with limited free, non-protein-bound drug, measurable 48 h after administration (Figure 2a) [36,37]. During the experiment, the diameter of the tumor was measured with calipers, and the volume was derived. Growth inhibition was based on the volume at the end of the experiment of the control group versus the treated group. Isolated tumors were further evaluated with histopathology, e.g., on proliferation and apoptosis markers [26,29–34].

To incorporate these key biological parameters and readouts, especially histopathology, a detachable two-part chip was designed (Figure 1a,b). The top part consisted of a straight channel to supply continuous perfusion of nutrients and dynamic drug concentrations [14]. The bottom part contained a 2 mm diameter, 0.8 mm deep U-shaped well to facilitate placement and unimpeded growth of spheroids (Figure 1b). The well was able to hold the spheroids under flow without the need for fixation with the extracellular matrix.

The chip was reversibly bonded by medical-grade adhesive tape to enable off-chip standard histopathology. Spheroids were taken out of the chip, formalin-fixed, paraffin-embedded (FFPE), cut in sections, and stained with hematoxylin–eosin (HE), or hematoxylin and markers for proliferation (Ki67) and apoptosis (cleaved Caspase-3, CC3). HE staining showed that spheroids contained poorly differentiated cancer cells, with limited extracellular matrix, as observed in viable regions of the HCT116 xenografts (Figure 1c, Supplementary Figure S3).

To mimic oxaliplatin concentrations over time found in the blood of mice, dynamic control of solute concentrations was needed. Dynamic control of oxaliplatin with a molecular weight of 397 Dalton (Da) and estimated diffusion coefficient of $8.2 \times 10^{-6}$ cm$^2 \cdot$s$^{-1}$ was validated with fluorescein, a small molecule with a molecular weight of 332 Da and diffusion coefficient estimates of $4$–$5.7 \times 10^{-6}$ cm$^2 \cdot$s$^{-1}$ [38,39]. Concentration steps similar to oxaliplatin were programmed, and the fluorescence signal was quantified and compared to expected fluorescence changes. Measured fluorescence follows expected fluorescence, taking into account expected lags in concentration changes due to Poiseuille flow (Supplementary Figure S6). Dynamic concentration control of oxaliplatin on-chip has also been described elsewhere in more detail [14].

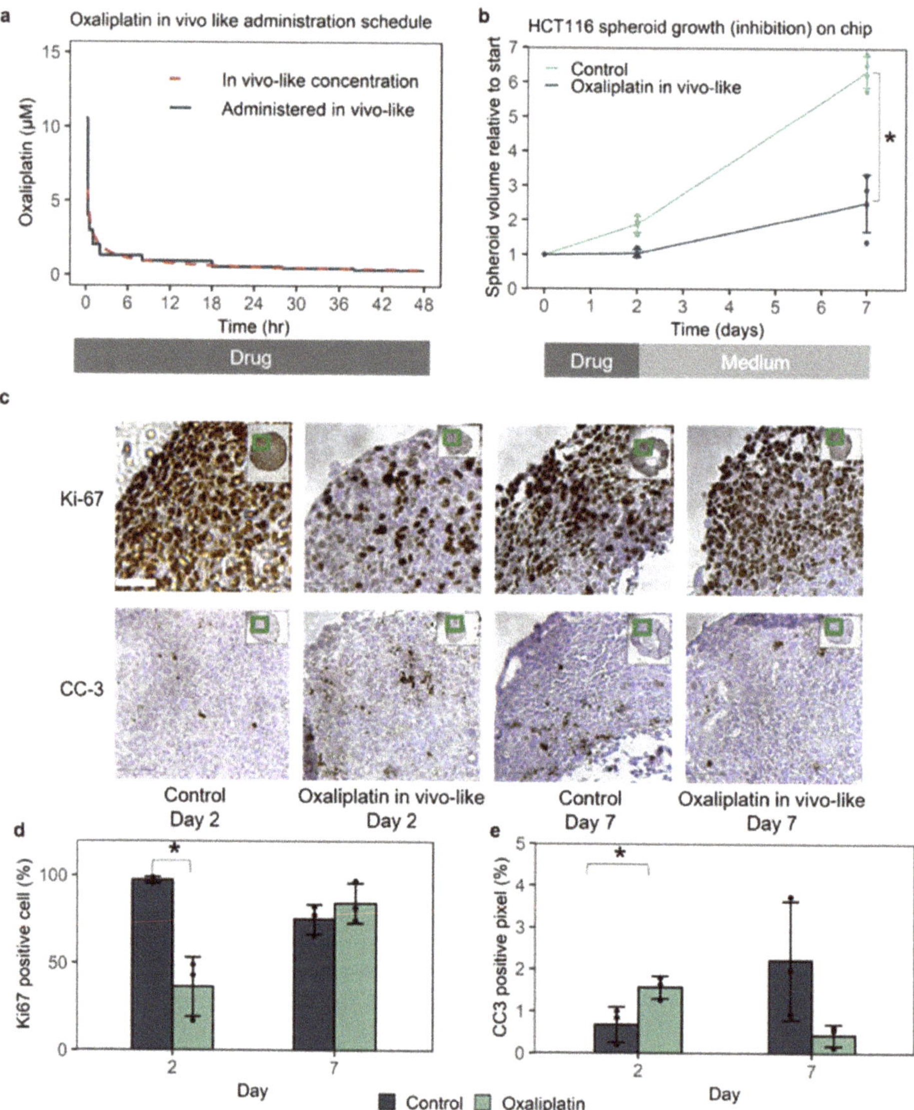



**Figure 2. In vivo-like oxaliplatin leads to 70% growth inhibition on-chip based on spheroid volume, with a temporary reduction of proliferation and increase in apoptosis.** (**a**) In vivo-like oxaliplatin concentration over time found in mice [36], corrected for protein binding [37], translated into an in vivo-like drug administration schedule. (**b**) Dynamic, in vivo-like oxaliplatin concentration led to temporary, partial growth inhibition of the spheroid. $N = 4$ spheroids per condition. Error bars represent the SD. * $p < 0.001$ according to a two-sided, unequal variance $t$-test. (**c**) In vivo-like oxaliplatin reversibly decreased proliferation (Ki67) for 2 days, and increased apoptosis (CC-3), albeit from a low level. Scale bar = 50 µm. (**d,e**) Quantification of staining with Qupath [27,28] between control and treated chips after 2 and 7 days. Depicted data points represent sections from two separate spheroids. * $p < 0.05$ according to two-sided, unequal variance $t$-test with $N = 3$ sections per condition. Error bars represent the SD.

The well was coated with Pluronic F127 to prevent adhesion and non-spheroidal growth of the spheroid [21]. The U shape of the well allowed for central placement of the spheroid-by-spheroid transfer via fluid–fluid contact. Central placement prevented automated imaging challenges and distortion of diffusion of nutrients to the spheroid.

An algorithm in Fiji/ImageJ based on earlier work was used to quantify the spheroid surface and derive volume (growth) (Supplementary Figure S1) [25]. Due to the non-adhesion of spheroids, the outer diameter accurately reflected spheroid volume. Trypsinization of spheroids of different sizes and subsequent automated cell counting confirmed the correlation between volume derived from the outer diameter and cell number growth (Supplementary Figure S4).

The chip system should not confound control growth and drug effect [40]. Ideally, cancer-on-chip technology can bridge the gap between standard in vitro assays and in vivo models. It is especially important to verify that cancer cell growth is not disrupted on-chip, and that the chip does not interfere with drug efficacy. Therefore, spheroids were exposed to constant oxaliplatin concentrations on-chip and in an ultralow-attachment plate with U-shaped wells ('Sphera'). Average starting diameters across treatment conditions on-chip and in the well plate ranged from 0.45 to 0.48 mm (Supplementary Figure S5). Although control growth was slightly higher on-chip (2.3 fold in 48 h) than growth in Sphera plates (1.8 fold), likely due to constant perfusion, growth under treatment was very similar (Figure 1d). A two-way analysis of variance (ANOVA) for unbalanced data [41] led to $p$-values of $2.2 \times 10^{-16}$, $1.5 \times 10^{-4}$, and $3.2 \times 10^{-4}$ for the factors of oxaliplatin dose, chip versus 96-well plate, and interaction, respectively. The significance of interaction is driven by the control growth conditions, which also drives the overall significance of the difference between chip and well plates, as the $p$-values of chip versus 96-well plate and interaction were 0.15 and 0.29, respectively, when the control condition was left out of the ANOVA. Hence, the chip system did not negatively affect cancer spheroid growth or efficacy of oxaliplatin.

The growth (inhibition) results for validating the chip system also clearly showed a difference in growth of HCT116 cells in spheroids versus monolayers (2D). Untreated doubling time increases from 20 h to 48 h for HCT116 spheroids with a typical diameter of 0.5 mm [42].

*3.2. In Vivo-like Oxaliplatin Led to 70% Growth Inhibition On-Chip, with a Temporary Halt of Growth*

To mimic growth inhibition in xenograft studies, drug exposure in mice needed to be recapitulated. Dosages in in vivo studies varied from 5–20 mg/kg per week, given in 1–3 dosages per week [26,29–34]. Here, we used oxaliplatin concentrations measured in blood plasma in mice over time after a single dose of 8 mg/kg [36]. As the only drug which is not bound to plasma proteins is active [43], the total oxaliplatin concentration was corrected for the fraction of oxaliplatin bound to plasma proteins. As the free oxaliplatin concentration over time in mice was found to be only 33% of total drug concentration, total plasma levels of oxaliplatin found in mice were decreased by 67%, which resulted in the unbound oxaliplatin concentration over time [36,37]. The in vivo (unbound) oxaliplatin concentration over time was translated to a drug administration schedule, and both are shown in Figure 2a.

Growth inhibition is the primary readout of xenograft experiments. Exposure of HCT116 spheroids to in vivo-like oxaliplatin on-chip resulted in growth inhibition of 70%, as measured over a 7 day period (Figure 2b). Growth inhibition of 70% found on-chip matches well with average growth inhibition of 60% (range 30–80%) found in xenografts (Supplementary Figure S2, Table S1) [26,29–34].

Immunohistochemistry can give further insight into the molecular mechanisms of growth inhibition of the drug [32,44]. Immunohistochemical markers were quantified for the entire spheroid section using Qupath software, which has been used for scoring Ki67 in clinical breast cancer samples (Figure 2c,d, Supplementary Figures S7 and S8) [27,28]. Immunohistochemistry indicated that drug treatment led to both a reduction in proliferation marker Ki67 and an increase in apoptosis marker CC3 after 2 days compared to control spheroids at 2 days (Figure 2c–e). Oxaliplatin treatment of HCT116 xenografts also led to a decrease in Ki67 and increase in CC3 [32,33,44].

The in vivo-like drug exposure led to a growth stop for the first 2 days, after which growth recovered. The recovery of growth was accompanied by an increase in the proliferation marker Ki67 and a decrease in CC3 after 7 days, as compared to treated spheroids after 2 days (Figure 2c–e). Comparing the Ki67 and CC3 status of the treated spheroid with the untreated spheroid after 7 days was less straightforward, as the untreated spheroid had, due to its size, a decreased proliferation rate and a higher baseline apoptosis, especially surrounding the necrotic core.

To our knowledge, the temporary growth reduction and subsequent growth recovery have not been described for HCT116 xenografts treated with oxaliplatin in in vivo studies, possibly due to the limited precision of measuring tumor volume through the skin with calipers, on a millimeter scale, and the number of animals needed for histological analyses at multiple timepoints. For HCT116 xenografts treated with gemcitabine, daily histological sections did show a similar full growth stop and gradual recovery over six days, as measured with BrdUrd [45]. Knowledge of recovery times of cancer cells in vivo could support future design of clinical dosage schedules, which, for example, is once per 2–3 weeks for oxaliplatin, although side-effects also have to be taken into account [46,47].

### 3.3. The Cancer-On-Chip Model Recapitulates Drug Response as It Is Representative of Proliferating Cells in the HCT116 Xenograft

In the previous section, it was shown that growth inhibition on-chip and in xenografts was comparable. Nevertheless, even though spheroids share similarities with xenografts, such as spheroidal shape and solute gradients, spheroids also differ from xenografts in several ways, e.g., spheroids are avascular and typically have a diameter $10\times$ smaller than xenografts. Furthermore, although large spheroids grow more slowly than 2D cultured cells, spheroids still typically grow faster than xenografts of the same cell line.

The structural and growth differences between xenografts and spheroids raises the question of whether the spheroid is representative of the whole xenograft tumor or only for specific areas.

The untreated spheroids on-chip were 95% Ki67-positive after 2 days, which declined to ~75% after 7 days (Figure 2c,d, Supplementary Figure S7). Histopathological analysis of untreated HCT116 xenografts showed that only a fraction (46%) of the xenograft was proliferating (Figure 3a, Supplementary Figure S10). The proliferating cells clustered in distinct regions toward the periphery of the tumor, likely due to differences in perfusion and interactions with the host stroma [48–50]. These proliferating regions contained ~80% Ki67-positive cells (Figure 3b), whereas the nonproliferating regions contained <5% Ki67-positive cells (Supplementary Figure S12). With H&E staining, the non-proliferative regions were also characterized by eosinophilic staining, nuclear condensation, and loss of nuclei, suggesting necrosis (Supplementary Figure S3) [51].

The presence of proliferating and nonproliferating regions in the HCT116 xenograft not only provides an explanation for the different control growth rates, but can also explain the comparable growth inhibition found in both the cancer-on-chip model and the xenografts. The effect of oxaliplatin on the nonviable, nonproliferating, regions will be limited, as they are not growing, poorly perfused, and already in a (pre)necrotic state (Supplementary Figure S3) [49]. In the viable, proliferating, regions ~80% of cells were Ki67-positive, for which the spheroid-on-chip should be representative. Hence, the effect of oxaliplatin in xenografts is likely dominated by the effect on proliferating cells, which are present on-chip.

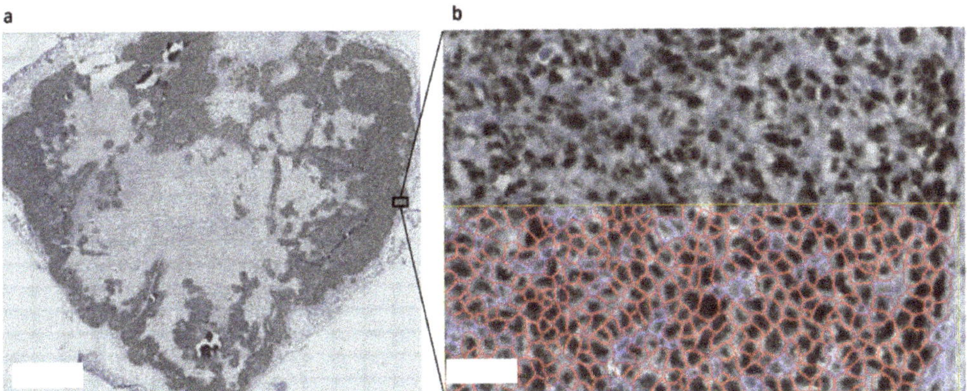

**Figure 3. An HCT116 xenograft with a proliferating shell and a nonproliferating core.** (a) HCT116 xenografts stained for proliferation marker Ki67 (brown) contained peripheral regions with high proliferation and large areas of nonproliferating tissue toward the core. Scale bar = 2 mm. (b) Proliferating areas were ~80% Ki67-positive. Ki67 quantification was performed by Qupath software; positive cells are shown in red [27]. Scale bar = 50 μm.

*3.4. A Pharmacodynamic Model Further Validates the Representativeness of On-Chip Growth Inhibition*

The representativeness of the spheroid for the proliferating cells in the xenograft provided a qualitative argument for why the drug effect is recapitulated on-chip. A pharmacodynamic ('the effect of drug on tissue') model, in which on-chip growth is mapped to the representative parts of the xenograft to predict treated and untreated growth, could further validate the ability of the cancer-on-chip model to mimic the effect of oxaliplatin on HCT116 xenografts, as well as identify translational discrepancies.

As HCT116 xenografts consist of proliferating and non-proliferating cells, the pharmacodynamic model employed here consisted of two populations, a proliferating and nonproliferating subpopulation [52,53]. The subpopulations were geometrically divided (Figure 4a), as the vast majority of proliferating cells were located on the periphery of the xenografts (Figures 3a and 4a, Supplementary Figure S10). Constant radius growth was assumed for the modeled xenograft, as constant radius growth has been described for other colorectal cancer xenografts [50,54] and fits with growth curves for HCT116 xenografts [26,29–34].

Radius growth per day ($\alpha$) was based on the thickness of the proliferating shell found in xenografts and the on-chip growth of spheroids in the first 2 days (Supplementary Figures S10 and S11 and accompanying data). For untreated modeled xenografts, this resulted in a radius growth of 0.25 mm/day and a tumor volume growth in 14 days from 100 mm$^3$ to 1100 mm$^3$, while, for modeled xenografts treated with oxaliplatin, the resulting radius growth was 0.11 mm/day and the final tumor volume was 376 mm$^3$ (Figure 4b).

Modeled xenograft growth with and without oxaliplatin treatment fell within the range of growth found in xenograft studies (Figure 4b); untreated tumor volume growth in existing HCT116 xenograft studies over 14 days was 5–13-fold, and xenografts treated with oxaliplatin grew 1.5–7.5-fold over 14 days (Figure 4b, Supplementary Table S1) [26,29–34]. Hence, the outcome of the pharmacodynamic model for untreated and treated xenograft growth was on the high side, but within the range found in existing xenograft studies.

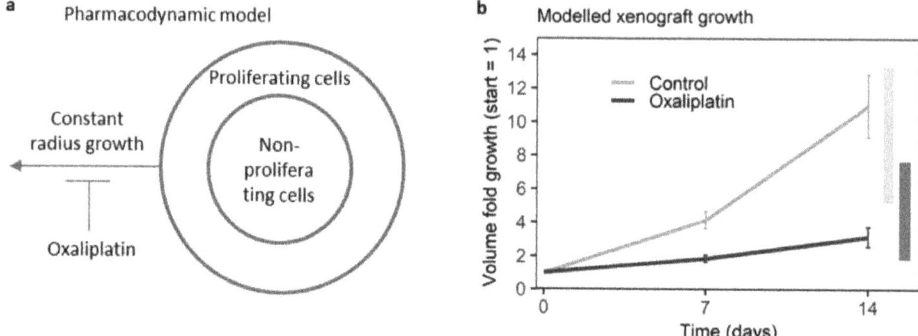



**Figure 4. The pharmacodynamic model of HCT116 xenograft growth.** (a) The modeled xenograft consisted of a shell of proliferating cells. Constant radius growth was derived from on-chip spheroid growth and the thickness of the proliferating shell in untreated HCT xenografts (see text). (b) Modeled growth with and without treatment. Error bars represents the standard error of the mean (SEM), based on individual growth found in $n$ = 4 chips per condition. Shaded bars on the right represent the range found in xenograft studies [26,29–34].

The formula describing xenograft growth employed here was as follows [50]:

$$V(t) = 4/3\,\pi\,(\alpha t + \beta)^3,$$

where $V$ is the volume of the tumor in mm$^3$ at day $t$, $\alpha$ represents the radius growth per day in mm/day, and $\beta$ is the starting radius in mm. Starting volume was set at 100 mm$^3$ ($\beta$ = 2.9 mm), an often-used starting volume in xenograft studies (Supplementary Table S1).

Potential reasons for the relatively high growth predictions could be a slightly higher proliferation rate in spheroids in the first 2–3 days, a decline in radius growth per day at higher tumor volumes in xenografts, higher external pressure in xenografts [55] leading to potential compression of the necrotic core during growth, and measurement errors due to interference of skin and connective tissue with caliper measurement [56], which could have a larger influence on smaller starting tumors.

However, according to the immunohistochemistry showing that on-chip spheroids were representative of the proliferating cells, and the pharmacodynamic model resulting in modeled xenograft growth (inhibition) on the high side but within the range of actual xenograft studies, remaining potential translational discrepancies provide room for future optimization but do not seem to interfere with the ability of the cancer-on-chip model to represent the effect of oxaliplatin on HCT116 xenografts.

With further validation, cancer-on-chip models could, thus, complement cancer xenograft models. Different cell lines, dosage schedules, and drug combinations can be tested for further validation of novel drug efficacy without increasing animal use. Furthermore cancer-on-chip models have the potential to decrease the variability of in-vivo assays as there is less unwanted biological variation from the animal host. Moreover, continuous monitoring of cancer growth-on-chip could elucidate drug exposure and response relations. To mimic human cancer more closely, in addition to expanding cell lines to have greater coverage of clinical variety, human aspects can be added in a stepwise fashion, such as human pharmacokinetics, stroma, and immune components, as such cancer-on-chip models have the potential to progressively complement and substitute the cancer xenograft model.

## 4. Conclusions

A cancer-on-chip model was designed and validated to incorporate key elements of the cancer xenograft assay. On-chip growth inhibition of HCT116 colorectal cancer spheroids at in vivo-like oxaliplatin concentrations was comparable to in vivo xenograft growth inhibition, while providing novel insights into the temporal response. Furthermore, immunohistochemistry performed off-chip indicated a decrease in proliferation marker Ki67 and an increase in apoptosis marker CC3 right after treatment.

As the size and structure of xenografts and spheroids differ, we further evaluated the representativeness of the spheroid on-chip. Immunohistochemical staining suggested that the on-chip HCT116 spheroid was representative of the proliferating cells in the outer shell of the xenograft, which likely determine the growth and growth inhibition in the xenograft. Furthermore, a pharmacodynamic model with a proliferating and nonproliferating cell population, in which constant radius growth was based on growth (inhibition) found on-chip, resulted in comparable growth (inhibition) to HCT116 xenografts treated with oxaliplatin, thus providing additional support for the translational relevance of the cancer-on-chip model.

**Supplementary Materials:** The following supporting information can be downloaded at https://www.mdpi.com/article/10.3390/mi13050739/s1. Figure S1: Spheroid size determination by the Fiji algorithm. Table S1: Xenograft studies used for comparing growth inhibition and growth. Figure S2: Growth inhibition from existing HCT116 xenograft studies plotted against dose per week. Figure S3: HCT116 xenograft tissue stained with hematoxylin-eosin (HE). Figure S4: Correlation between image based tumor volume derivation of the spheroids and automated cell count after trypsinization. Figure S5: Equality of spheroid start diameter per treatment condition, on-chip and in the 96 well plate ("sphera"). Figure S6: Validation of dynamic drug control on chip. Figure S7: Ki67 (proliferation) quantification of HCT116 spheroid sections in Qupath. Figure S8: Detection of apoptosis marker cleaved caspase 3 (CC3). Figure S9: Limited cleaved caspase 3 (CC3) staining at the tumor edge of untreated xenografts. Figure S10: Macroscopic view of HCT116 xenografts stained with HE (left) or Ki67 (right). Figure S11: Quantitative analysis of Ki67 positivity. Figure S12: Less than 5% Ki67 positive cells in non-proliferating regions of the untreated HCT116 xenograft.

**Author Contributions:** Conceptualization, J.K., S.M.v.N., A.v.d.B., A.D.v.d.M. and L.V.; methodology, J.K., S.M.v.N., N.E.d.G., L.E.N., E.G.B.M.B., A.D.v.d.M. and L.V.; software, J.K.; validation, J.K. and A.D.v.d.M.; formal analysis, J.K.; investigation, J.K., S.M.v.N., N.E.d.G., L.E.N. and E.G.B.M.B.; resources, A.v.d.B., A.D.v.d.M. and L.V.; data curation, J.K.; writing—original draft preparation, J.K.; writing—review and editing, S.M.v.N., A.v.d.B., A.D.v.d.M. and L.V.; visualization, J.K., S.M.v.N. and E.G.B.M.B.; supervision, A.v.d.B., A.D.v.d.M. and L.V.; project administration, J.K. and S.M.v.N.; funding acquisition A.v.d.B., A.D.v.d.M. and L.V. All authors have read and agreed to the published version of the manuscript.

**Funding:** This research was funded by the Netherlands Organ-on-Chip Initiative (NOCI), an NWO Gravitation Project funded by the Ministry of Education, Culture, and Science of the Government of the Netherlands, under Grant 024.003.001.

**Data Availability Statement:** Not applicable.

**Conflicts of Interest:** The authors declare no conflict of interest. The funders had no role in the design of the study; in the collection, analyses, or interpretation of data; in the writing of the manuscript, or in the decision to publish the results.

# References

1. Dagenais, G.R.; Leong, D.P.; Rangarajan, S.; Lanas, F.; Lopez-Jaramillo, P.; Gupta, R.; Diaz, R.; Avezum, A.; Oliveira, G.B.F.; Wielgosz, A.; et al. Variations in common diseases, hospital admissions, and deaths in middle-aged adults in 21 countries from five continents (PURE): A prospective cohort study. *Lancet* **2020**, *395*, 785–794. [CrossRef]
2. Abbema, D.V.; Vissers, P.; Vos-Geelen, J.; Lemmens, V.; Janssen-Heijnen, M.; Tjan-Heijnen, V. Trends in Overall Survival and Treatment Patterns in Two Large Population-Based Cohorts of Patients with Breast and Colorectal Cancer. *Cancers* **2019**, *11*, 1239. [CrossRef] [PubMed]
3. Wong, C.H.; Siah, K.W.; Lo, A.W. Estimation of clinical trial success rates and related parameters. *Biostatistics* **2019**, *20*, 273–286. [CrossRef] [PubMed]
4. McIntyre, R.E.; Buczacki, S.J.; Arends, M.J.; Adams, D.J. Mouse models of colorectal cancer as preclinical models. *BioEssays News Rev. Mol. Cell. Dev. Biol.* **2015**, *37*, 909–920. [CrossRef]

5.  Gengenbacher, N.; Singhal, M.; Augustin, H.G. Preclinical mouse solid tumour models: Status quo, challenges and perspectives. *Nat. Rev. Cancer* **2017**, *17*, 751–765. [CrossRef]

6.  Hutchinson, L.; Kirk, R. High drug attrition rates—Where are we going wrong? *Nat. Rev. Clin. Oncol.* **2011**, *8*, 189–190. [CrossRef]

7.  Hung, P.J.; Lee, P.J.; Sabounchi, P.; Lin, R.; Lee, L.P. Continuous perfusion microfluidic cell culture array for high-throughput cell-based assays. *Biotechnol. Bioeng.* **2005**, *89*, 1–8. [CrossRef]

8.  Jeong, S.Y.; Lee, J.H.; Shin, Y.; Chung, S.; Kuh, H.J. Co-Culture of Tumor Spheroids and Fibroblasts in a Collagen Matrix-Incorporated Microfluidic Chip Mimics Reciprocal Activation in Solid Tumor Microenvironment. *PLoS ONE* **2016**, *11*, e0159013. [CrossRef]

9.  Ayuso, J.M.; Virumbrales-Munoz, M.; Lacueva, A.; Lanuza, P.M.; Checa-Chavarria, E.; Botella, P.; Fernandez, E.; Doblare, M.; Allison, S.J.; Phillips, R.M.; et al. Development and characterization of a microfluidic model of the tumour microenvironment. *Sci. Rep.* **2016**, *6*, 36086. [CrossRef]

10.  Hassell, B.A.; Goyal, G.; Lee, E.; Sontheimer-Phelps, A.; Levy, O.; Chen, C.S.; Ingber, D.E. Human Organ Chip Models Recapitulate Orthotopic Lung Cancer Growth, Therapeutic Responses, and Tumor Dormancy In Vitro. *Cell Rep.* **2017**, *21*, 508–516. [CrossRef]

11.  Nashimoto, Y.; Okada, R.; Hanada, S.; Arima, Y.; Nishiyama, K.; Miura, T.; Yokokawa, R. Vascularized cancer on a chip: The effect of perfusion on growth and drug delivery of tumor spheroid. *Biomaterials* **2020**, *229*, 119547. [CrossRef]

12.  Sontheimer-Phelps, A.; Hassell, B.A.; Ingber, D.E. Modelling cancer in microfluidic human organs-on-chips. *Nat. Rev. Cancer* **2019**, *19*, 65–81. [CrossRef]

13.  Lohasz, C.; Frey, O.; Bonanini, F.; Renggli, K.; Hierlemann, A. Tubing-Free Microfluidic Microtissue Culture System Featuring Gradual, in vivo-Like Substance Exposure Profiles. *Front. Bioeng. Biotechnol.* **2019**, *7*, 72. [CrossRef]

14.  Komen, J.; Westerbeek, E.Y.; Kolkman, R.W.; Roesthuis, J.; Lievens, C.; van den Berg, A.; van der Meer, A.D. Controlled pharmacokinetic anti-cancer drug concentration profiles lead to growth inhibition of colorectal cancer cells in a microfluidic device. *Lab A Chip* **2020**, *20*, 3167–3178. [CrossRef]

15.  Komen, J.; van Neerven, S.M.; van den Berg, A.; Vermeulen, L.; van der Meer, A.D. Mimicking and surpassing the xenograft model with cancer-on-chip technology. *EBioMedicine* **2021**, *66*, 103303. [CrossRef]

16.  Ivanova, E.; Kuraguchi, M.; Xu, M.; Portell, A.J.; Taus, L.; Diala, I.; Lalani, A.S.; Choi, J.; Chambers, E.S.; Li, S.; et al. Use of Ex Vivo Patient-Derived Tumor Organotypic Spheroids to Identify Combination Therapies for HER2 Mutant Non-Small Cell Lung Cancer. *Clin. Cancer Res. Off. J. Am. Assoc. Cancer Res.* **2020**, *26*, 2393–2403. [CrossRef]

17.  Petreus, T.; Cadogan, E.; Hughes, G.; Smith, A.; Reddy, V.P.; Lau, A.; O'Connor, M.J.; Critchlow, S.; Ashford, M.; O'Connor, L.O. Tumour-on-chip microfluidic platform for assessment of drug pharmacokinetics and treatment response. *Commun. Biol.* **2021**, *4*, 1001. [CrossRef]

18.  Hachey, S.J.; Movsesyan, S.; Nguyen, Q.H.; Burton-Sojo, G.; Tankazyan, A.; Wu, J.; Hoang, T.; Zhao, D.; Wang, S.; Hatch, M.M.; et al. An in vitro vascularized micro-tumor model of human colorectal cancer recapitulates in vivo responses to standard-of-care therapy. *Lab A Chip* **2021**, *21*, 1333–1351. [CrossRef]

19.  Gioeli, D.; Snow, C.J.; Simmers, M.B.; Hoang, S.A.; Figler, R.A.; Allende, J.A.; Roller, D.G.; Parsons, J.T.; Wulfkuhle, J.D.; Petricoin, E.F.; et al. Development of a multicellular pancreatic tumor microenvironment system using patient-derived tumor cells. *Lab A Chip* **2019**, *19*, 1193–1204. [CrossRef]

20.  Nürnberg, E.; Vitacolonna, M.; Klicks, J.; von Molitor, E.; Cesetti, T.; Keller, F.; Bruch, R.; Ertongur-Fauth, T.; Riedel, K.; Scholz, P.; et al. Routine Optical Clearing of 3D-Cell Cultures: Simplicity Forward. *Front. Mol. Biosci.* **2020**, *7*, 20. [CrossRef]

21.  Liu, W.; Liu, D.; Hu, R.; Huang, Z.; Sun, M.; Han, K. An integrated microfluidic 3D tumor system for parallel and high-throughput chemotherapy evaluation. *Analyst* **2020**, *145*, 6447–6455. [CrossRef]

22.  Hecker, M.; Ting, M.S.H.; Malmström, J. Simple Coatings to Render Polystyrene Protein Resistant. *Coatings* **2018**, *8*, 55. [CrossRef]

23.  Liu, V.A.; Jastromb, W.E.; Bhatia, S.N. Engineering protein and cell adhesivity using PEO-terminated triblock polymers. *J. Biomed. Mater. Res.* **2002**, *60*, 126–134. [CrossRef]

24.  Wang, J.C.; Liu, W.; Tu, Q.; Ma, C.; Zhao, L.; Wang, Y.; Ouyang, J.; Pang, L.; Wang, J. High throughput and multiplex localization of proteins and cells for in situ micropatterning using pneumatic microfluidics. *Analyst* **2015**, *140*, 827–836. [CrossRef]

25.  Ivanov, D.P.; Parker, T.L.; Walker, D.A.; Alexander, C.; Ashford, M.B.; Gellert, P.R.; Garnett, M.C. Multiplexing spheroid volume, resazurin and acid phosphatase viability assays for high-throughput screening of tumour spheroids and stem cell neurospheres. *PLoS ONE* **2014**, *9*, e103817. [CrossRef]

26.  Zhang, R.; Song, X.Q.; Liu, R.P.; Ma, Z.Y.; Xu, J.Y. Fuplatin: An Efficient and Low-Toxic Dual-Prodrug. *J. Med. Chem.* **2019**, *62*, 4543–4554. [CrossRef]

27.  Bankhead, P.; Loughrey, M.B.; Fernández, J.A.; Dombrowski, Y.; McArt, D.G.; Dunne, P.D.; McQuaid, S.; Gray, R.T.; Murray, L.J.; Coleman, H.G.; et al. QuPath: Open source software for digital pathology image analysis. *Sci. Rep.* **2017**, *7*, 16878. [CrossRef]

28.  Robertson, S.; Acs, B.; Lippert, M.; Hartman, J. Prognostic potential of automated Ki67 evaluation in breast cancer: Different hot spot definitions versus true global score. *Breast Cancer Res. Treat.* **2020**, *183*, 161–175. [CrossRef]

29.  Nagaraju, G.P.; Alese, O.B.; Landry, J.; Diaz, R.; El-Rayes, B.F. HSP90 inhibition downregulates thymidylate synthase and sensitizes colorectal cancer cell lines to the effect of 5FU-based chemotherapy. *Oncotarget* **2014**, *5*, 9980–9991. [CrossRef]

30.  Threatt, S.D.; Synold, T.W.; Wu, J.; Barton, J.K. In vivo anticancer activity of a rhodium metalloinsertor in the HCT116 xenograft tumor model. *Proc. Natl. Acad. Sci. USA* **2020**, *117*, 17535–17542. [CrossRef]

31. Liang, J.; Cheng, Q.; Huang, J.; Ma, M.; Zhang, D.; Lei, X.; Xiao, Z.; Zhang, D.; Shi, C.; Luo, L. Monitoring tumour microenvironment changes during anti-angiogenesis therapy using functional MRI. *Angiogenesis* **2019**, *22*, 457–470. [CrossRef] [PubMed]

32. de Bruijn, M.T.; Raats, D.A.; Hoogwater, F.J.; van Houdt, W.J.; Cameron, K.; Medema, J.P.; Borel Rinkes, I.H.; Kranenburg, O. Oncogenic KRAS sensitises colorectal tumour cells to chemotherapy by p53-dependent induction of Noxa. *Br. J. Cancer* **2010**, *102*, 1254–1264. [CrossRef] [PubMed]

33. Xu, K.; Chen, G.; Qiu, Y.; Yuan, Z.; Li, H.; Yuan, X.; Sun, J.; Xu, J.; Liang, X.; Yin, P. miR-503-5p confers drug resistance by targeting PUMA in colorectal carcinoma. *Oncotarget* **2017**, *8*, 21719–21732. [CrossRef]

34. Shelton, J.W.; Waxweiler, T.V.; Landry, J.; Gao, H.; Xu, Y.; Wang, L.; El-Rayes, B.; Shu, H.K. In vitro and in vivo enhancement of chemoradiation using the oral PARP inhibitor ABT-888 in colorectal cancer cells. *Int. J. Radiat. Oncol. Biol. Phys.* **2013**, *86*, 469–476. [CrossRef] [PubMed]

35. Linnekamp, J.F.; Hooff, S.R.V.; Prasetyanti, P.R.; Kandimalla, R.; Buikhuisen, J.Y.; Fessler, E.; Ramesh, P.; Lee, K.; Bochove, G.G.W.; de Jong, J.H.; et al. Consensus molecular subtypes of colorectal cancer are recapitulated in in vitro and in vivo models. *Cell Death Differ.* **2018**, *25*, 616–633. [CrossRef] [PubMed]

36. Li, S.; Chen, Y.; Zhang, S.; More, S.S.; Huang, X.; Giacomini, K.M. Role of organic cation transporter 1, OCT1 in the pharmacokinetics and toxicity of cis-diammine(pyridine)chloroplatinum(II) and oxaliplatin in mice. *Pharm. Res.* **2011**, *28*, 610–625. [CrossRef]

37. Boughattas, N.A.; Hecquet, B.; Fournier, C.; Bruguerolle, B.; Trabelsi, H.; Bouzouita, K.; Omrane, B.; Lévi, F. Comparative pharmacokinetics of oxaliplatin (L-OHP) and carboplatin (CBDCA) in mice with reference to circadian dosing time. *Biopharm. Drug Dispos.* **1994**, *15*, 761–773. [CrossRef]

38. Casalini, T.; Salvalaglio, M.; Perale, G.; Masi, M.; Cavallotti, C. Diffusion and aggregation of sodium fluorescein in aqueous solutions. *J. Phys. Chemistry. B* **2011**, *115*, 12896–12904. [CrossRef]

39. Modok, S.; Scott, R.; Alderden, R.A.; Hall, M.D.; Mellor, H.R.; Bohic, S.; Roose, T.; Hambley, T.W.; Callaghan, R. Transport kinetics of four- and six-coordinate platinum compounds in the multicell layer tumour model. *Br. J. Cancer* **2007**, *97*, 194–200. [CrossRef]

40. Paguirigan, A.L.; Beebe, D.J. From the cellular perspective: Exploring differences in the cellular baseline in macroscale and microfluidic cultures. *Integr. Biol. Quant. Biosci. Nano Macro* **2009**, *1*, 182–195. [CrossRef]

41. Hector, A.; Von Felten, S.; Schmid, B. Analysis of variance with unbalanced data: An update for ecology & evolution. *J. Anim. Ecol.* **2010**, *79*, 308–316. [CrossRef]

42. NCI-60 Screening Methodology. Available online: https://dtp.cancer.gov/discovery_development/nci-60/methodology.htm (accessed on 10 March 2020).

43. Heuberger, J.; Schmidt, S.; Derendorf, H. When is protein binding important? *J. Pharm. Sci.* **2013**, *102*, 3458–3467. [CrossRef]

44. Sun, W.; Li, J.; Zhou, L.; Han, J.; Liu, R.; Zhang, H.; Ning, T.; Gao, Z.; Liu, B.; Chen, X.; et al. The c-Myc/miR-27b-3p/ATG10 regulatory axis regulates chemoresistance in colorectal cancer. *Theranostics* **2020**, *10*, 1981–1996. [CrossRef]

45. Huxham, L.A.; Kyle, A.H.; Baker, J.H.; Nykilchuk, L.K.; Minchinton, A.I. Microregional effects of gemcitabine in HCT-116 xenografts. *Cancer Res.* **2004**, *64*, 6537–6541. [CrossRef]

46. Bokemeyer, C.; Bondarenko, I.; Makhson, A.; Hartmann, J.T.; Aparicio, J.; Braud, F.d.; Donea, S.; Ludwig, H.; Schuch, G.; Stroh, C.; et al. Fluorouracil, Leucovorin, and Oxaliplatin with and Without Cetuximab in the First-Line Treatment of Metastatic Colorectal Cancer. *J. Clin. Oncol.* **2009**, *27*, 663–671. [CrossRef]

47. Koopman, M.; Antonini, N.F.; Douma, J.; Wals, J.; Honkoop, A.H.; Erdkamp, F.L.; de Jong, R.S.; Rodenburg, C.J.; Vreugdenhil, G.; Loosveld, O.J.; et al. Sequential versus combination chemotherapy with capecitabine, irinotecan, and oxaliplatin in advanced colorectal cancer (CAIRO): A phase III randomised controlled trial. *Lancet* **2007**, *370*, 135–142. [CrossRef]

48. Forster, J.C.; Harriss-Phillips, W.M.; Douglass, M.J.; Bezak, E. A review of the development of tumor vasculature and its effects on the tumor microenvironment. *Hypoxia (Auckl. N.Z.)* **2017**, *5*, 21–32. [CrossRef]

49. Kyle, A.H.; Baker, J.H.; Gandolfo, M.J.; Reinsberg, S.A.; Minchinton, A.I. Tissue penetration and activity of camptothecins in solid tumor xenografts. *Mol. Cancer Ther.* **2014**, *13*, 2727–2737. [CrossRef]

50. Lenos, K.J.; Miedema, D.M.; Lodestijn, S.C.; Nijman, L.E.; van den Bosch, T.; Romero Ros, X.; Lourenco, F.C.; Lecca, M.C.; van der Heijden, M.; van Neerven, S.M.; et al. Stem cell functionality is microenvironmentally defined during tumour expansion and therapy response in colon cancer. *Nat. Cell Biol.* **2018**, *20*, 1193–1202. [CrossRef]

51. Elmore, S.A.; Dixon, D.; Hailey, J.R.; Harada, T.; Herbert, R.A.; Maronpot, R.R.; Nolte, T.; Rehg, J.E.; Rittinghausen, S.; Rosol, T.J.; et al. Recommendations from the INHAND Apoptosis/Necrosis Working Group. *Toxicol. Pathol.* **2016**, *44*, 173–188. [CrossRef]

52. Jusko, W.J. A pharmacodynamic model for cell-cycle-specific chemotherapeutic agents. *J. Pharmacokinet. Biopharm.* **1973**, *1*, 175–200. [CrossRef]

53. Lobo, E.D.; Balthasar, J.P. Pharmacodynamic modeling of chemotherapeutic effects: Application of a transit compartment model to characterize methotrexate effects in vitro. *AAPS PharmSci* **2002**, *4*, 212–222. [CrossRef]

54. van der Heijden, M.; Miedema, D.M.; Waclaw, B.; Veenstra, V.L.; Lecca, M.C.; Nijman, L.E.; van Dijk, E.; van Neerven, S.M.; Lodestijn, S.C.; Lenos, K.J.; et al. Spatiotemporal regulation of clonogenicity in colorectal cancer xenografts. *Proc. Natl. Acad. Sci. USA* **2019**, *116*, 6140–6145. [CrossRef]

55. Stylianopoulos, T.; Martin, J.D.; Chauhan, V.P.; Jain, S.R.; Diop-Frimpong, B.; Bardeesy, N.; Smith, B.L.; Ferrone, C.R.; Hornicek, F.J.; Boucher, Y.; et al. Causes, consequences, and remedies for growth-induced solid stress in murine and human tumors. *Proc. Natl. Acad. Sci. USA* **2012**, *109*, 15101–15108. [CrossRef] [PubMed]

56. Ayers, G.D.; McKinley, E.T.; Zhao, P.; Fritz, J.M.; Metry, R.E.; Deal, B.C.; Adlerz, K.M.; Coffey, R.J.; Manning, H.C. Volume of preclinical xenograft tumors is more accurately assessed by ultrasound imaging than manual caliper measurements. *J. Ultrasound Med. Off. J. Am. Inst. Ultrasound Med.* **2010**, *29*, 891–901. [CrossRef] [PubMed]

*Article*

# Stimulus-Evoked Activity Modulation of In Vitro Engineered Cortical and Hippocampal Networks

Francesca Callegari [1,†], Martina Brofiga [1,2,†], Fabio Poggio [1] and Paolo Massobrio [1,3,*]

1   Department of Informatics, Bioengineering, Robotics and Systems Engineering (DIBRIS), University of Genova, 16145 Genova, Italy; francesca.callegari@edu.unige.it (F.C.); martina.brofiga@dibris.unige.it (M.B.); fabio.poggio@edu.unige.it (F.P.)
2   ScreenNeuroPharm s.r.l., 18038 Sanremo, Italy
3   National Institute for Nuclear Physics (INFN), 16146 Genova, Italy
*   Correspondence: paolo.massobrio@unige.it; Tel.: +39-010-335-2761
†   These authors contributed equally to this work.

**Abstract:** The delivery of electrical stimuli is crucial to shape the electrophysiological activity of neuronal populations and to appreciate the response of the different brain circuits involved. In the present work, we used dissociated cortical and hippocampal networks coupled to Micro-Electrode Arrays (MEAs) to investigate the features of their evoked response when a low-frequency (0.2 Hz) electrical stimulation protocol is delivered. In particular, cortical and hippocampal neurons were topologically organized to recreate interconnected sub-populations with a polydimethylsiloxane (PDMS) mask, which guaranteed the segregation of the cell bodies and the connections among the sub-regions through microchannels. We found that cortical assemblies were more reactive than hippocampal ones. Despite both configurations exhibiting a fast (<35 ms) response, this did not uniformly distribute over the MEA in the hippocampal networks. Moreover, the propagation of the stimuli-evoked activity within the networks showed a late (35–500 ms) response only in the cortical assemblies. The achieved results suggest the importance of the neuronal target when electrical stimulation experiments are performed. Not all neuronal types display the same response, and in light of transferring stimulation protocols to in vivo applications, it becomes fundamental to design realistic in vitro brain-on-a-chip devices to investigate the dynamical properties of complex neuronal circuits.

**Keywords:** cortical neurons; hippocampal neurons; electrical stimulation; Micro-Electrode Arrays; engineered neuronal networks; polydimethylsiloxane; microchannels

**Citation:** Callegari, F.; Brofiga, M.; Poggio, F.; Massobrio, P. Stimulus-Evoked Activity Modulation of In Vitro Engineered Cortical and Hippocampal Networks. *Micromachines* **2022**, *13*, 1212. https://doi.org/10.3390/mi13081212

Academic Editor: Nam-Trung Nguyen

Received: 30 June 2022
Accepted: 28 July 2022
Published: 29 July 2022

**Publisher's Note:** MDPI stays neutral with regard to jurisdictional claims in published maps and institutional affiliations.



## 1. Introduction

The mammalian nervous system is characterized by the existence of an intrinsic spontaneous activity that displays complex patterns, such as oscillations, rhythms at well-defined frequencies, spikes, and bursts lasting from a tenth to hundreds of milliseconds [1–4]. The network organization is deemed to be the cause of the genesis of such dynamics, as demonstrated by several studies both at the in vivo and in vitro level ([5,6] and references therein). Furthermore, it is clear and well-established that the nervous system modulates its intrinsic spontaneous activity when it receives sensory stimuli [7,8]. Typically, the response to stimuli depends on the single-cell specialization and on the connectivity of the network itself: the stimulation of a hub neuron or of a highly-connected local cluster of neurons will produce different effects than the stimulation of sparse weakly connected assemblies. Therefore, it has emerged how two fundamental features can influence the response of a neuronal ensemble, namely, the neuronal type (e.g., cortical, hippocampal, thalamic, etc.) and the topological properties (e.g., degree of modularity or clusterization, existence of long-range connections and/or hubs, etc.) of the involved network.

A possible way to investigate the response of neuronal assemblies is to use dissociated neuronal cultures coupled to Micro-Electrode Arrays (MEAs). In this simplified experimental model, neuronal networks can freely form synaptic connections during their natural development by self-organizing in complex topologies such as rich-club [9] and scale-free [10]. However, its great potential is that it is possible to engineer the substrate where neurons grow to design well-defined circuits by promoting topological properties such as modularity [11], ordered lattices [12], or pushing the directionality of the connections [13]. This engineering process guarantees the modulation of the emerging neuronal dynamics [14] as well as precise control over the interacting neuronal types [15]. Since the early 2000s, these models were used to characterize the neuronal response to different electrical stimulation protocols, which differ in the frequency and (partially) in the amplitude of the delivered stimuli.

High-frequency stimulation protocols were tested on dissociated neuronal networks coupled to MEAs, with the aim to induce forms of long-term plasticity (potentiation and depression). Typically, plasticity protocols exploit tetanic stimulations. From the pioneering work of Jimbo and coworkers, who proved that it was possible to induce long-lasting changes in the network responses [16], variations in these high-frequency stimulation patterns were used for learning protocols [17] or with the coexistence of a low-frequency stimulation pulse to induce long-term network potentiation [18]. This last protocol was also able to induce a reshaping of the underlying functional connectivity by increasing the number of functional connections [19].

At the same time, stimulation protocols based on the delivery of low-frequency pulses also was investigated to appreciate the modes of response of a neuronal network. By providing a low-frequency stimulation (less than 1 Hz), Eytan and coworkers found a transient enhanced response of the network as a function of the stimulated subnetwork (i.e., stimulation sites) [20]. Some years later, Vajda and colleagues demonstrated that low-frequency stimulation shaped the patterns of spontaneous electrophysiological activity by changing features such as the shape of the network bursts [21]. More recently, the interplay between low-frequency stimulation patterns and the underlying functional connectivity was investigated, too [22].

In the present work, we investigated the low-frequency response of two different neuronal populations (namely cortical and hippocampal ones) plated over MEAs with a modular connectivity. The development of three interconnected assemblies was obtained thanks to a polymeric mask (realized in polydimethylsiloxane, PDMS) reversibly coupled to the surface of the MEAs. We investigated and compared the stimuli-evoked response of the cortical and hippocampal populations as a function of the location of the stimulation sites (i.e., stimulated assembly). This analysis was carried out by taking into account the different synaptic excitatory responses of the network, namely, the early phase—mediated by the $\alpha$-amino-3-hydroxy-5-methyl-4-isoxazolepropionic acid (AMPA) receptors, lasting no more than 35 ms; and the late phase—N-methyl-D-aspartate (NMDA) mediated, lasting hundreds of milliseconds.

Our results show that the cortical assemblies are generally more reactive than the hippocampal ones to the delivery of low-frequency stimulation protocols. Moreover, although both the cortical and hippocampal cultures exhibited a fast (<35 ms) response, this did not propagate uniformly in the hippocampal networks. In addition, we found that only cortical assemblies displayed a significant late (35–500 ms) response. The achieved results suggest the need to develop brain-on-a-chip models as test beds for the characterization of neuronal dynamics and for studying the efficacy of stimulation protocols to interact with the neuronal structures. Using the same principles and observations, devices such as the one we employed in the present work could help getting more reliable results, also in chemical-based modulations (neuromodulation). To put it in perspective, to achieve a more comprehensive knowledge of the nervous system, next-generation devices should include multi-functional components for the monitoring not only of the electrophysiological activity but also of the relevant parameters relative to the delivery of the selected chemicals, and for the fine control over the experimental environment by means of perfusion motorized systems [23].

## 2. Materials and Methods

### 2.1. Polymeric Device

A three-compartment polydimethylsiloxane (PDMS) mask was used for the experiments. It consists of one larger chamber (6.5 mm wide, 4.3 mm long, with 16 μL capacity) and two smaller ones (3.4 mm in diameter and with 8 μL capacity) interconnected with arrays of 28 regularly spaced (50 μm) microchannels (Figure 1A) [24]. The microchannel dimensions (10 μm wide, 250 μm long, and 5 μm high) prevent the migration of cell bodies among the compartments while allowing the crossing of neurites only [25].

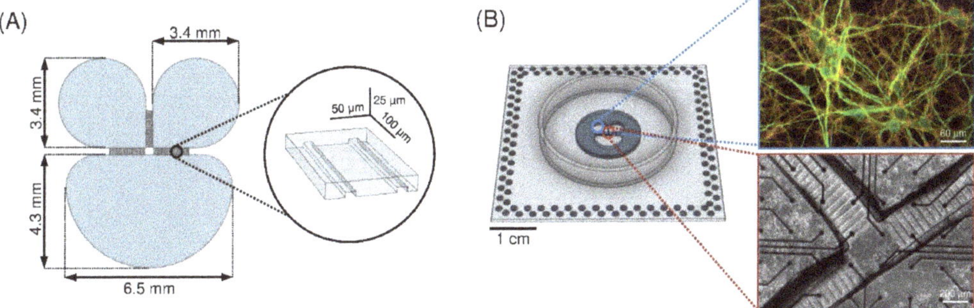

**Figure 1.** (**A**) Sketch of the silicon master (negative mold of the PDMS mask) used to create the polymeric device. In the inset, the dimensions of the microchannels of the PDMS mask are reported. (**B**) PDMS mask reversibly bonded to the MEA surface. (***Right, top***) Close-up of the cortical network (DIV 18) inside the compartment where microtubule-associated protein 2 (MAP2, green) and the vesicular glutamate transporter (Vglut1, red) are labeled. (***Right, bottom***) Close-up of the channel region of a cortical network organized in three interconnected sub-populations at DIV 18.

The mask was fabricated by conventional soft lithography molding techniques (Scriba Nanotechnologies S.r.l., Bologna, Italy). The SU-8/silicon master was obtained by consecutive steps of UV photolithography performed on different photoresistant layers of different thicknesses. The pattern of microchannels was created upon a 5-μm-thick SU 2005 layer. The cell chambers structures were aligned and superimposed on a 300-μm-thick SU 2075 layer. The prebaking and postbaking time were optimized to preserve the structure of the microchannels. After development, a thin layer (400 nm) of polyacrylic acid was spin-coated upon the master to enhance the peel-off of the PDMS replica.

The replicas were obtained by pouring a PDMS solution (curing agent 1:10 *w/w* in prepolymer) on the mold followed by heat curing in dry oven at 80 °C for 20 min. The masks were then aligned and reversibly bounded to planar MEAs (Figure 1B) with 120 electrodes (30 μm in diameter and 200 μm in spacing, arranged in a 12 × 12 array). In order to obtain a reversible sealing between the PDMS mask and the MEA substrate, a plasma oxygen treatment (60 W for 10 s) was applied to the surfaces to be coupled. Then, the treated surfaces were coupled and the achieved device (i.e., MEA with PDMS mask) was placed at 60 °C for 1 hour. To ensure a good hydrophilization of the microchannels, after sterilization the devices underwent oxygen-plasma treatment (120 W for 60 s).

### 2.2. Cell Culture Preparation

Primary cultures of Sprague-Dawley rat embryonic (E18-E19) cortical and hippocampal neurons were obtained by enzymatic digestion of hippocampi and cortices with a Trypsin (0.125%) and DNAse (0.03%) solution (in Hanks' solution). The process was quenched by adding fetal bovine serum (FBS, 10%) supplemented medium (Neurobasal). Single cells were obtained with a mechanical trituration of the tissues with a fine-tipped Pasteur pipette. The cell suspension was diluted in serum-free medium supplemented with 2% B-27 Supplement, 1% stable L-Glutamine, and 1% PenStrep. Glia proliferation was not prevented as glia cells are known to be necessary for the development of healthy neuronal populations [26,27]. Cells were plated on 120-channel MEAs, which were previously coupled to masks (Section 2.1), sterilized in dry oven for 3 h, plasma treated, and coated with poly-L-ornithine (100 μg/mL).

Cells of the same neuronal type were plated on MEAs at a defined density (1500 cell/mm$^2$ for cortical and 1300 cell/mm$^2$ for hippocampal cells) to avoid effects on the network development due to different plating densities [2]. The samples were incubated at 37 °C, with 5% $CO_2$ and 95% humidity. After 5 days, and twice a week afterward, half volume of the medium was replaced with BrainPhys medium, supplemented with 2% NeuroCult SM1, 1% Glutamax, and 1% PenStrep solution. Neurons were let to organize into a morphologically and functionally mature network for 3 weeks before experiments were performed (Figure 1B, *right*). To maintain the same experimental conditions during the recording phase, we controlled both the temperature (37 °C) and the flow of humidified gas (5% $CO_2$, 20% $O_2$, 75% $N_2$) supplied to the cell culture environment.

### 2.3. Ethical Approval

The experimental protocol for in vitro cultures was approved by the European Animal Care Legislation (2010/63/EU) and the Italian Ministry of Health, in accordance with D.L. 116/1992 and the guidelines of the University of Genova (Prot. 75F11.N.6JI, 08/08/18).

### 2.4. Experimental Protocol

The electrophysiological activity of the neuronal networks was recorded at DIV 18–19 using the MEA2100 system (Multi Channel Systems, MCS, Reutlingen, Germany) with a sampling frequency of 10 kHz. Recordings were performed outside the incubator. To minimize the thermal and mechanical stress on the cells, excessive evaporation, and unphysiological changes to the pH of the cultures, the MEA lodging was heated (37 °C) and a constant slow flow of humidified gas (5% $CO_2$, 20% $O_2$, 75% $N_2$) was supplied to the cell culture environment. MC_Rack software (MCS) was used for data recording and for delivering the electrical stimulation on individual electrodes.

Each experimental session lasted 25 min. First, 10 min of spontaneous activity was recorded to evaluate the basal dynamics (i.e., presence of spikes and bursts, number of active electrodes, etc.) of the network. Then, one electrode for each compartment was chosen to be stimulated for 5 min. The selection of the stimulated electrodes was based on their initial spontaneous activity: only the more active electrodes were chosen as possible stimulating sites, indicating that they were functionally connected to the rest of the network. The stimulating pulse was biphasic (positive phase first); it lasted for 400 μs with a 50% duty cycle and was sent at a frequency of 0.2 Hz. The choice of such a low-frequency stimulation was driven by conventional electrophysiological stimulation protocols based on test stimuli that induce non-plastic responses [16,18]. Overall, the amplitude of each pulse was 1.5 V peak-to-peak relative to the reference electrode.

To make sure that the order of the stimulation did not influence the results, the first compartment to be stimulated was changed for each experimental session.

### 2.5. Dataset

The dataset used in this work consists of spontaneous and stimulus-evoked recordings from 8 different cultures (4 cortical (Cx) and 4 hippocampal (Hp) networks) coming from two preparations. All cultures were recorded at 18–19 day in vitro (DIV).

*2.6. Data Analysis*

Electrophysiological activity was characterized offline by using customized analysis tools developed in MATLAB 2021a (The Mathworks, Natick, MA, USA). The first step was the identification of the spikes exploiting the Precise Timing Spike Detection (PTSD) method [28]. We set three parameters: (i) the peak lifetime period to 2 ms; (ii) the refractory period to 1 ms; and (iii) the adaptive threshold for each electrode to 8 times the standard deviation of thermal and biological noise. Data were not spike sorted, since during a bursting event, a global increase of the activity produces a fast sequence of spikes with different and overlapping shapes, which makes the sorting difficult and unreliable [2].

To evaluate the spontaneous activity of the network, we analyzed the basal spiking activity in terms of Mean Firing Rate (MFR). An electrode was discarded from the analysis if the MFR was lower than 0.1 spike/s. In addition, we characterized the bursting activity. For this purpose, we identified bursts by applying the string method devised in [29], by setting the minimum number of spikes in a burst to 5, and the maximum inter spike interval into bursts at 100 ms. Then, we extracted the following parameters: the Mean Bursting Rate (MBR), i.e., the mean number of bursts per minute; and the Burst Duration (BD), i.e., the temporal duration of the bursts.

To obtain quantitative information about the stimulus-evoked activity, we computed the Post-Stimulus Time Histogram (PSTH), which represents the impulsive response of each electrode to the electrical stimulation. The PSTH was computed considering time windows of 600 ms that follow each stimulus. Then, we divided that interval into 4 ms bins and counted the number of spikes that occurred in each time interval. Eventually, we normalized the obtained histogram by dividing the number of spikes detected for each bin by the number of stimuli. If an electrode had a PSTH with area lower than "1", that channel was removed from the statistics because it was considered inactive (i.e., unable to evoke minimal response to stimulation). Furthermore, in order to use the PSTH area as a measure of the response of each electrode and to be able to compare the data among the different networks and configurations, we normalized the PSTH areas over the maximum obtained value.

The response evoked by an electrical stimulation is typically characterized by two components: an *early* and rapid one, which occurs in a time range between 0 and 35 ms after the stimulus; and a *late* and slower one, which occurs around the 35–500 ms. Previous works demonstrated their presence due to the AMPA- and NMDA-mediated transmission by performing selective blocking with antagonist molecules such as amino-5-phosphonovaleric acid (APV) and cyanquixaline (CNQX), respectively [30]. In order to identify and characterize these responses of the different neuronal populations, we developed a new method to distinguish the *early* component from the *late* one. We classified the PSTH for each electrode with the $k$-mean algorithm and the silhouette analysis (to identify the optimal number of classes). After the PSTH classification, we computed the average response for each class $j$ (obtained by averaging the responses of each active electrode belonging to the same class), and we smoothed it by applying a moving average ($PSTH_{smooth}$, temporal bin of 50 ms). Then, we detected the local maxima of $PSTH_{smooth}$: if a peak was detected within the first 52 ms (*early*, $x_1$), a second one with at least 10% of the prominence of the first peak was sought between 52 and 600 ms (*late*, $x_2$). We searched for the local minimum ($x_{min}$) within the range of the two peaks, and we evaluated the separation between them following the approach devised in [31]:

$$s_j = \frac{PSTH_{smooth,j}(x_{min})}{\sqrt{PSTH_{smooth,j}(x_1)\,PSTH_{smooth,j}(x_2)}} \tag{1}$$

If $s_j$ exceeded an empirical threshold set at 0.3, $x_{min}$ was considered the optimal time threshold for dividing the *early* and *late* responses for the $j$-class. Otherwise, the class was characterized only by the *early* response, and its latency value for the *late* response was set to a value higher than the observation window (i.e., 650 ms).

### 2.7. Statistical Analysis

Data are expressed as the mean $\pm$ standard deviation of the mean. Statistical analysis was performed using a nonparametric Kruskal–Wallis test since data did not follow a normal distribution (evaluated by the Kolmogorov–Smirnov normality test). Differences were considered statistically significant when $p < 0.05$. The box plots representation indicates the 25–75 percentile (box), the standard deviation (whiskers), the mean (square), and the median (line) values.

## 3. Results

The presented results aim to demonstrate that different cell types in vitro respond differently when they receive external stimulation. In the following sections, we explored the intrinsic activity patterns of the two considered cell types for the present study, namely, cortical (Cx) and hippocampal (Hp) cells (Section 3.1). Then, we analyzed the response produced by the delivery of a low-frequency stimuli on the two considered configurations and evaluated their differences both in terms of "amplitude" and of propagation of the evoked activity (Section 3.2).

### 3.1. Interconnected Cortical and Hippocampal Assemblies Display Spontaneous Activity

Firstly, we evaluated the spontaneous electrophysiological activity of the networks to characterize the intrinsic parameters of each network and their differences. Qualitatively, both cortical and hippocampal (Figure 2A, shades of red and green, respectively; the shades of color used in the raster plot representation identify their position in the MEA configuration) dynamics displayed the coexistence of a random spiking activity and highly packed bursts. This ensures that the neuronal populations under consideration had established an adequate, functionally connected network able to maintain sustained spiking and bursting patterns of activity. Without this condition, any stimulation protocol would be ineffective. As the site of simulation was not significant for these analyses, the activity of all three compartments was condensed and considered as one.

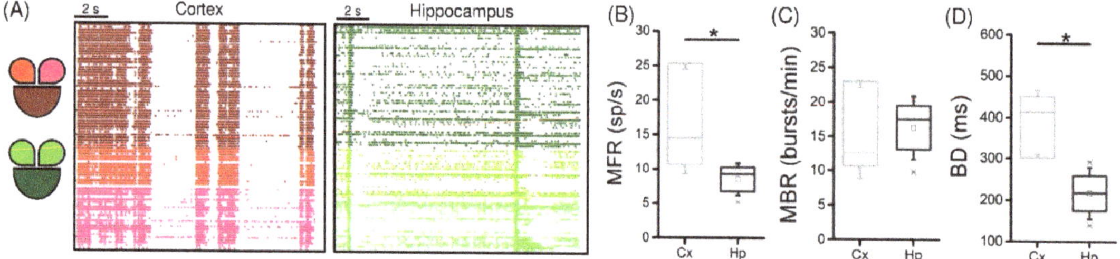


**Figure 2.** Spontaneous network activity of interconnected cortical (Cx) and hippocampal (Hp) assemblies: (**A**) 16 s of spontaneous activity of cortical (shades of **red**) and hippocampal (shades of **green**) interconnected populations; (**B**) MFR; (**C**) MBR; (**D**) BD of the cortical (**light grey**) and the hippocampal (**dark grey**) populations (* refers to $0.01 < p < 0.05$; Kruskal–Wallis nonparametric test).

Generally, the cortical (light grey) populations displayed higher spiking activity (Figure 2B, $p = 0.03$) and scattered data points with respect to the one of hippocampal (dark grey) networks, which on the other hand had very little variability.

Moving on to the characterization of the bursting dynamics of the two populations, we observed that the increase in the firing did not induce a difference in the bursting rate (Figure 2C). However, the different cell types of the two networks brought a change in the shape of the bursts, as they were significantly longer in the Cx case (Figure 2D, $p = 0.03$).

### 3.2. Low-Frequency Electrical Stimulation Produces Different Response Patterns in Cortical and Hippocampal Populations

After the initial assessment of the spontaneous activity of both populations, the stimulus-evoked activity was evaluated by means of the PSTH (cf. Section 2.6). Figure 3 shows normalized PTSHs for a representative cortical (top row) and hippocampal (bottom row) network stimulated in three different electrodes, each belonging to a different compartment (the site of simulation is indicated as a lighting near the related compartment). In particular, the cortical interconnected sub-populations exhibited a clear and marked early response followed by a long late response, independently from the choice of the location of the stimulated channel, as demonstrated by the double peak in the PSTH (Figure 3A–C). On the contrary, the hippocampal sub-networks apparently presented only an early response, which in most cases decayed within the first 200 ms (Figure 3D–F). Moreover, the considered experiments showed that the Cx cultures seemed to be more responsive to external electrical stimulation than the Hp configuration in terms of amplitude of the PSTH areas.

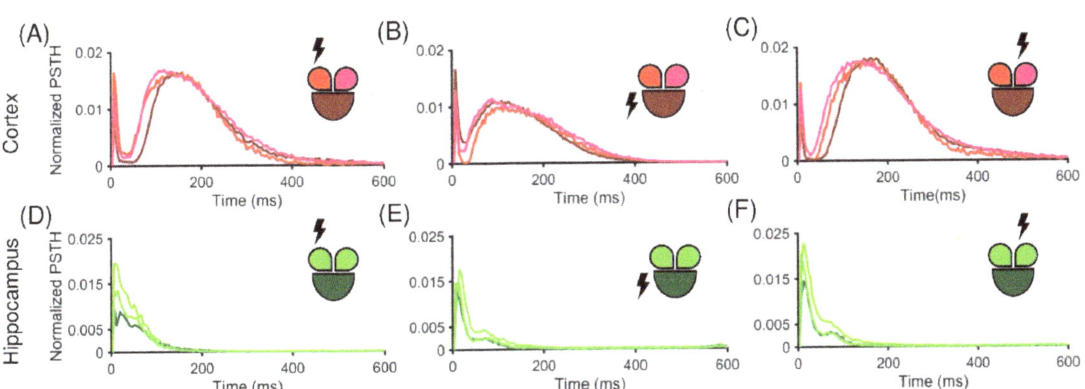

**Figure 3.** Normalized PSTH traces of a representative cortical (**A–C**) and hippocampal (**D–F**) network. The insets show the color-coded configurations. The stimulated compartment is indicated with a lightning sign in the inset. PSTH generated for (**A**) the small left, (**B**) the big central, and (**C**) the small right stimulated compartment in the cortical configuration. (**D–F**) The hippocampal counterpart in the same conditions.

To prove this observation, we computed the area under the PSTH for the two populations. Figure 4A,B depict the color- and height-coded area computed for each electrode in two representative experiments, stimulated in the electrode depicted with a red dot. Qualitatively, it was evident from this example that the Hp population had a lower response in terms of amplitude (Figure 4B) than the one observed in the Cx counterpart (Figure 4A). From the plots, we observed that the representative Hp network displayed a peak of the PSTH area of 40 spikes, while in the cortical population we reached a peak of 120 spikes. Moreover, only few electrodes adjacent to the stimulated site had a high response in the Hp assembly, whereas the cortical response was rather uniform throughout the network. This consideration was confirmed by the cumulative normalized PSTH areas for Cx and Hp (Figure 4C), which were computed by considering the response of the two populations in any configuration. It highlighted a significant difference ($p = 9.9 \times 10^{-9}$) between the two populations.

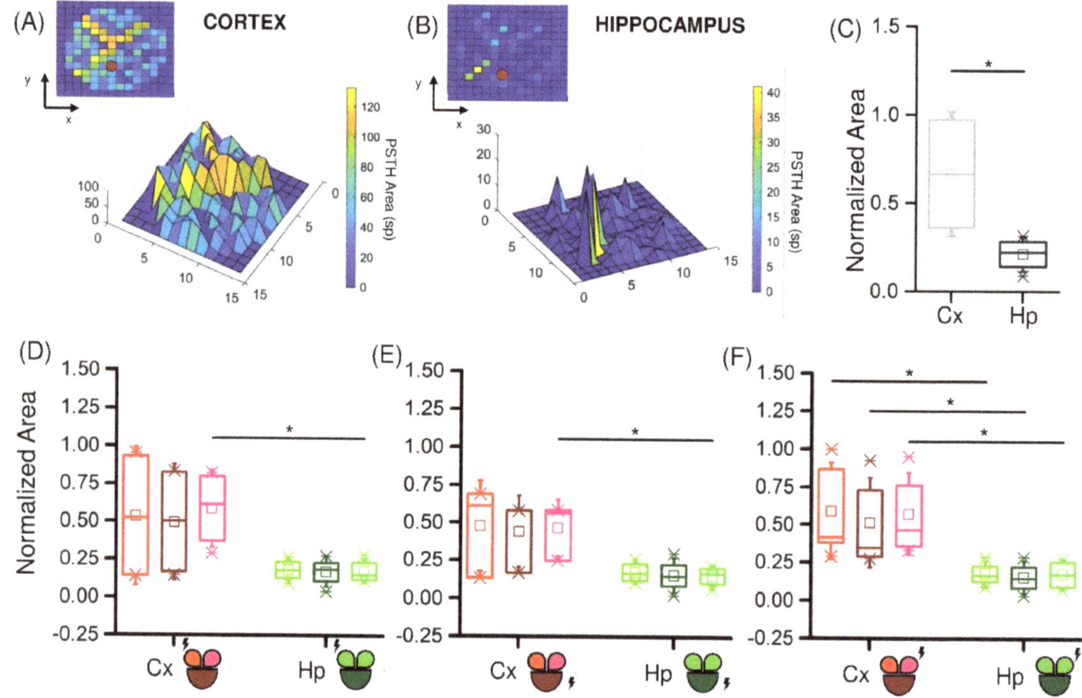

**Figure 4.** Cortical and hippocampal assemblies display different responsiveness when stimulated by low-frequency stimuli. Color- and height-coded representation of the PSTH area of a representative experiment with (**A**) Cx and (**B**) Hp neurons. (**C**) PSTH normalized area distributions of the Cx (**light grey**) and Hp (**dark grey**) assemblies. (**D**–**F**) Relevance of the site of stimulation (highlighted with a lightning sign) in cortical (**red**) and hippocampal (**green**) networks. (* refers to $0.01 < p < 0.05$; Kruskal–Wallis nonparametric test).

Next, we investigated whether the site of stimulation had any effect on how much the network responded. Generally, the position of the stimulus did not produce significant variations in the amplitude of response of the three compartments. However, it had some effect when comparing the two studied configurations. In fact, when the stimuli were delivered in the small compartment, the Cx and Hp responses were statistically different, either only in the non-stimulated (Figure 4D,E) or in all (Figure 4F) the compartments. The larger stimulated compartment instead produced a difference between Cx and Hp only in one of the two smaller compartments.

Another feature that characterizes the response mode of the neurons is how the signal propagates throughout the population. The modular topology of the networks induced by the presence of the PDMS mask and the use of two neuronal types could produce changes in the transmission of the evoked activity, consequently modifying the velocity of involvement of the electrodes. To quantify these possible differences, we extracted from the PSTH the *early* and *late* latency components, as described in Section 2.6. The first interesting result was that most of the Hp networks did not exhibit a late response. On the other hand, the Cx cultures showed both a fast uniformly distributed response (Figure 5A, *left*) and a slow response, whose latency values increased with the distance from the stimulation electrode (Figure 5A, *right*). It is interesting to notice that the uniformity of propagation was not as evident in the Hp configuration (Figure 5B), where the direction of propagation of the signal was not clearly recognizable.

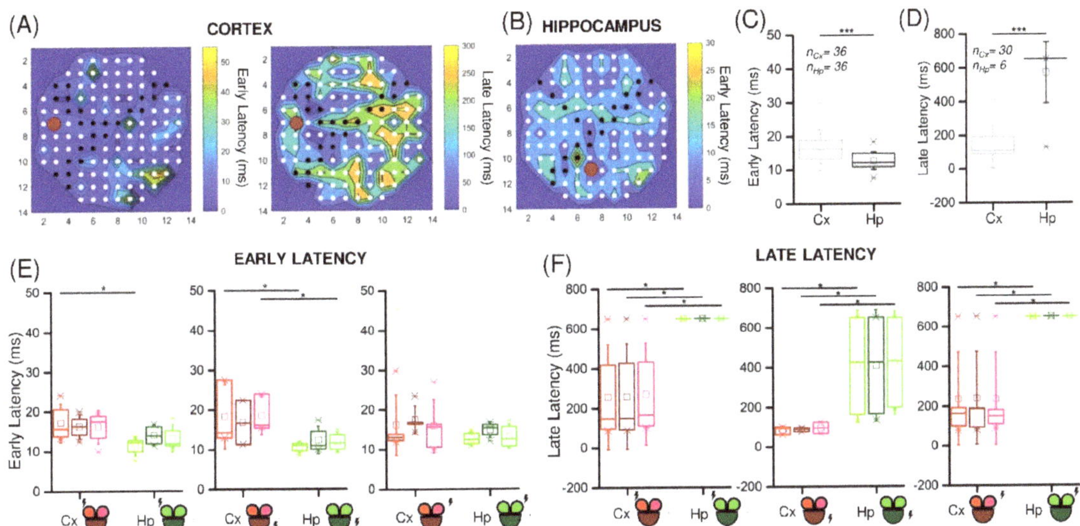

**Figure 5.** Cortical and hippocampal assemblies display different latency values when stimulated by low-frequency stimuli. Color-coded representation of the early and late (if present) latency values of representative (**A**) Cx and (**B**) Hp experiments. The red dot indicates the stimulated electrode. (**C,D**) Early and late latency cumulative values of Cx (**light grey**) and Hp (**dark grey**) assemblies. $n_{Cx}$ and $n_{Hp}$ indicate the number of evoked responses in the Cx and Hp configurations, respectively. (**E,F**) Relevance of the site of stimulation (highlighted with a lightning sign) in cortical (**red**) and hippocampal (**green**) networks in early and late latency values, respectively. (* refers to $0.01 < p < 0.05$, *** to $p < 0.001$; Kruskal–Wallis nonparametric test).

This observation was confirmed by comparing the cumulative latency values for the two populations (Figure 5C,D). In particular, for each stimulation, both the early and the late latency mean values were computed for each compartment. We observed that the early response (Figure 5C) was evoked in every compartment for each stimulus (number of evoked responses: $n_{Cx} = n_{Hp} = 36$) in both the Cx and the Hp populations, which however manifested some differences. In fact, the Hp networks were significantly faster in this phase ($p = 8.7 \times 10^{-5}$). Regarding the late component (Figure 5D), it was present in almost all cases in the Cx ($n_{Cx} = 30$), while it was almost absent in the Hp networks ($n_{Hp} = 6$). Consequently, a statistical difference in the latency values between the two populations was found ($p = 2.3 \times 10^{-9}$). Considering the previous results regarding the stimuli position in the MEA layout, we again analyzed the response time, depending on the position of the stimulated electrode (Figure 5E,F). We did not find statistical differences within the same population, although the stimulated compartment generally had slightly lower latency values. However, we observed significative differences between the Cx and Hp populations, especially in the late response. If in the early latency values the cortex resulted to be statistically slower only in three cases (Figure 5F), the comparison of late latency values gave rise to significant differences in all the cases.

## 4. Discussion

In vivo, small/medium-sized neuronal networks are continuously bombarded by inputs coming from different sources, ranging from external stimuli (e.g., chemical, electromagnetic, nociceptive, cognitive, etc.) to synaptic inputs projected by other neuronal assemblies. This huge number of stimuli continuously triggers and shapes the spontaneous electrophysiological activity of the nervous system. The processing of external inputs depends not only on the features of the presented patterns of stimulation but also on several internal neuronal and network variables defining the dynamical state of the circuit [32]. Although many studies (both theoretical and experimental) have focused on understanding the input–output functions that regulate the dependency between the delivered stimulation and evoked response [33–37], the whole picture is not completely clear yet.

In the present work, we exploited the controllability of an engineered in vitro system made up of interconnected neuronal networks coupled to Micro-Electrode Arrays (MEAs) to explore the role of two of the aforementioned internal neuronal variables on connectivity and neuronal heterogeneity. We applied low-frequency stimulation protocols to cortical and hippocampal neurons, topologically organized by means of a PDMS mask reversibly bounded to the active area of the MEA, to guarantee the segregation of the cell bodies. The role of these physical constraints (with or without microfluidics properties) for the modulation of the spontaneous activity of dissociated neuronal networks has been extensively studied [11,38–40]. In the present work, a mask inspired by the work of [24] was employed to define a neuronal network made up of three interconnected subpopulations to quantify the effect of a low-frequency electrical stimulation. The strategy to use polymeric masks to segregate neuronal populations and modulate their connectivity is a well-consolidated technique that allows imposing a directionality to connectivity (e.g., [13,41]) or to interface different neuronal types (e.g., [42,43]). In the present work, our polymeric mask had only the role to confine three sub-populations without forcing the directionality of their connections. The three-compartment mask devised in Figure 1, was used for both cortical and hippocampal cultures. At this stage of maturation (18–19 DIV), the network is already well-structured and a dense connectivity among the compartments is already present [24], causing a sustained spontaneous spiking and bursting activity (Figure 2). We found that cortical assemblies are more reactive than hippocampal ones to the delivery of low-frequency stimulation protocols (Figure 4C). We can speculate that such behavior could be explained by two factors: an intrinsic higher membrane excitability of cortical neurons or a different network organization, which facilitates synaptic integration in the case of cortical ensembles. In addition, we found that both cortical and hippocampal neurons exhibited a fast (<35 ms) (AMPA mediated) response that, however, did not uniformly propagate over the MEA in the hippocampal networks (Figure 5). Nonetheless, only cortical assemblies displayed a significant late (35–500 ms) response, typically mediated by the presence of excitatory NMDA synaptic receptors (Figure 5). In conclusion, the achieved results suggest the relevance of the neuronal target when an electrical stimulation experiment is designed and performed: the fundamental evidence emerging from the present work is that not all neuronal types display the same modes of response. This becomes fundamental when electrical stimulation protocols are transferred to in vivo experiments. Nowadays, the beneficial effects of electrical stimulation to cure or to reduce the symptoms of some brain impairments are well recognized: injuries [43], strokes [44], and tremors [45] are only some examples that have been treated by means of ad hoc electrical stimulation protocols in the last years. However, if the effect of an external stimulation on the electrophysiological activity and on the cognitive/motor functions is evident, the reasons beneath are still not completely understood [46]. Realistic in vitro models based on the concept of brain-on-a-chip could prove to be a valid support to solve this important question [47]. These hybrid devices (of which the system used in the present work is a simple example) allow performing experiments in a controlled environment, whose complexity should take into account the relevant intrinsic variables, such as the presence of different interacting neuronal populations [15,42], or the existence of topological features in the three-dimensional space [48].

**Author Contributions:** Conceptualization, F.C., M.B., F.P. and P.M.; methodology, M.B., F.C. and P.M.; software, M.B. and F.P.; validation, F.C., M.B. and F.P.; formal analysis, M.B.; writing—original draft preparation, F.C., M.B. and P.M. All authors have read and agreed to the published version of the manuscript.

**Funding:** This research received no external funding.

**Institutional Review Board Statement:** The experimental protocol was approved by the European Animal Care Legislation (2010/63/EU), by the Italian Ministry of Health in accordance with the D.L. 116/1992 and by the guidelines of the University of Genova (Prot. 75F11.N.6JI, 08/08/18).

**Data Availability Statement:** The peak trains of the entire dataset of this paper as well as the customized MATLAB functions used to analyze the data have been deposited in Zenodo. The DOI of the deposited data and code reported in this paper is: https://doi.org/10.5281/zenodo.6882374.

**Conflicts of Interest:** The authors declare no conflict of interest.

## References

1. Arieli, A.; Sterkin, A.; Grinvald, A.; Aertsen, A. Dynamics of Ongoing Activity: Explanation of the Large Variability in Evoked Cortical Responses. *Science* **1996**, *273*, 1868–1871. [CrossRef] [PubMed]
2. Wagenaar, D.A.; Pine, J.; Potter, S.M. An extremely rich repertoire of bursting patterns during the development of cortical cultures. *BMC Neurosci.* **2006**, *7*, 11. [CrossRef] [PubMed]
3. O'Donovan, M.J. The origin of spontaneous activity in developing networks of the vertebrate nervous system. *Curr. Opin. Neurobiol.* **1999**, *9*, 94–104. [CrossRef]
4. Buzsáki, G.; Draguhn, A. Neuronal oscillations in cortical networks. *Science* **2004**, *304*, 1926–1929. [CrossRef] [PubMed]
5. Sporns, O. Structure and function of complex brain networks. *Dialogues Clin. Neurosci.* **2013**, *15*, 247. [CrossRef] [PubMed]
6. Poli, D.; Pastore, V.P.; Massobrio, P. Functional connectivity in in vitro neuronal assemblies. *Front. Neural Circuits* **2015**, *9*, 57. [CrossRef] [PubMed]
7. Wagenaar, A.D.; Madhavan, R.; Pine, J.; Potter, S.M. Controlling Bursting in Cortical Cultures with Closed-Loop Multi-Electrode Stimulation. *J. Neurosci.* **2005**, *25*, 680. [CrossRef]
8. Massobrio, P.; Baijon, P.L.; Maccione, A.; Chiappalone, M.; Martinoia, S. Activity modulation elicited by electrical stimulation in networks of dissociated cortical neurons. In Proceedings of the 2007 29th Annual International Conference of the IEEE Engineering in Medicine and Biology Society, Lyon, France, 22–26 August 2007; pp. 3008–3011. [CrossRef]
9. Schroeter, M.S.; Charlesworth, P.; Kitzbichler, M.G.; Paulsen, O.; Bullmore, E.T. Emergence of rich-club topology and coordinated dynamics in development of hippocampal functional networks in vitro. *J. Neurosci.* **2015**, *35*, 5459–5470. [CrossRef]
10. Pastore, V.P.; Massobrio, P.; Godjoski, A.; Martinoia, S. Identification of excitatory-inhibitory links and network topology in large-scale neuronal assemblies from multi-electrode recordings. *PLoS Comput. Biol.* **2018**, *14*, e1006381. [CrossRef]
11. Shein-Idelson, M.; Ben-Jacob, E.; Hanein, Y. Engineered neuronal circuits: A new platform for studying the role of modular topology. *Front. Neuroeng.* **2011**, *4*, 10. [CrossRef]
12. Marconi, E.; Nieus, T.; Maccione, A.; Valente, P.; Simi, A.; Messa, M.; Dante, S.; Baldelli, P.; Berdondini, L.; Benfenati, F. Emergent Functional Properties of Neuronal Networks with Controlled Topology. *PLoS ONE* **2012**, *7*, e34648. [CrossRef]
13. Pan, L.; Alagapan, S.; Franca, E.; Leondopulos, S.S.; DeMarse, T.B.; Brewer, G.J.; Wheeler, B.C. An in vitro method to manipulate the direction and functional strength between neural populations. *Front. Neural Circuits* **2015**, *9*, 32. [CrossRef]
14. Yamamoto, H.; Moriya, S.; Ide, K.; Hayakawa, T.; Akima, H.; Sato, S.; Kubota, S.; Tanii, T.; Niwano, M.; Teller, S.; et al. Impact of modular organization on dynamical richness in cortical networks. *Sci. Adv.* **2018**, *4*, eaau4914. [CrossRef]
15. Dauth, S.; Maoz, B.M.; Sheehy, S.P.; Hemphill, M.A.; Murty, T.; Macedonia, M.K.; Greer, A.M.; Budnik, B.; Parker, K.K. Neurons derived from different brain regions are inherently different in vitro: A novel multiregional brain-on-a-chip. *J. Neurophysiol.* **2017**, *117*, 1320–1341. [CrossRef]
16. Jimbo, Y.; Tateno, T.; Robinson, H.P.C. Simultaneous Induction of Pathway-Specific Potentiation and Depression in Networks of Cortical Neurons. *Biophys. J.* **1999**, *76*, 670–678. [CrossRef]
17. Ruaro, M.E.; Bonifazi, P.; Torre, V. Toward the neurocomputer: Image processing and pattern recognition with neuronal cultures. *IEEE Trans. Biomed. Eng.* **2005**, *52*, 371–383. [CrossRef]
18. Chiappalone, M.; Massobrio, P.; Martinoia, S. Network plasticity in cortical assemblies. *Eur. J. Neurosci.* **2008**, *28*, 221–237. [CrossRef]
19. Poli, D.; Massobrio, P. High-frequency electrical stimulation promotes reshaping of the functional connections and synaptic plasticity in in vitro cortical networks. *Phys. Biol.* **2018**, *15*, 06LT01. [CrossRef]
20. Eytan, D.; Marom, S. Dynamics and effective topology underlying synchronization in networks of cortical neurons. *J. Neurosci.* **2006**, *26*, 8465–8476. [CrossRef]
21. Vajda, I.; Van Pelt, J.; Wolters, P.; Chiappalone, M.; Martinoia, S.; Van Someren, E.; Van Ooyen, A. Low-frequency stimulation induces stable transitions in stereotypical activity in cortical networks. *Biophys. J.* **2008**, *94*, 5028–5039. [CrossRef]
22. le Feber, J.; Stegenga, J.; Rutten, W.L.C. The Effect of Slow Electrical Stimuli to Achieve Learning in Cultured Networks of Rat Cortical Neurons. *PLoS ONE* **2010**, *5*, e8871. [CrossRef]
23. Brofiga, M.; Massobrio, P. Brain-on-a-Chip: Dream or Reality? *Front. Neurosci.* **2022**, *16*, 837623. [CrossRef]
24. Brofiga, M.; Pisano, M.; Tedesco, M.; Boccaccio, A.; Massobrio, P. Functional Inhibitory Connections Modulate the Electrophysiological Activity Patterns of Cortical-Hippocampal Ensembles. *Cereb. Cortex* **2022**, *32*, 1866–1881. [CrossRef]
25. Taylor, A.M.; Rhee, S.W.; Tu, C.H.; Cribbs, D.H.; Cotman, C.W.; Jeon, N.L. Microfluidic Multicompartment Device for Neuroscience Research. *Langmuir* **2003**, *19*, 1551–1556. [CrossRef]

26. Pfrieger, F.W.; Barres, B.A. Synaptic efficacy enhanced by glial cells in vitro. *Science* **1997**, *277*, 1684–1687. [CrossRef]
27. Araque, A.; Parpura, V.; Sanzgiri, R.P.; Haydon, P.G. Tripartite synapses: Glia, the unacknowledged partner. *Trends Neurosci.* **1999**, *22*, 208–215. [CrossRef]
28. Maccione, A.; Gandolfo, M.; Massobrio, P.; Novellino, A.; Martinoia, S.; Chiappalone, M. A novel algorithm for precise identification of spikes in extracellularly recorded neuronal signals. *J. Neurosci. Methods* **2009**, *177*, 241–249. [CrossRef]
29. Chiappalone, M.; Novellino, A.; Vajda, I.; Vato, A.; Martinoia, S.; van Pelt, J. Burst detection algorithms for the analysis of spatio-temporal patterns in cortical networks of neurons. *Neurocomputing* **2005**, *65–66*, 653–662. [CrossRef]
30. Shahaf, G.; Eytan, D.; Gal, A.; Kermany, E.; Lyakhov, V.; Zrenner, C.; Marom, S. Order-Based Representation in Random Networks of Cortical Neurons. *PLoS Comput. Biol.* **2008**, *4*, e1000228. [CrossRef]
31. Pasquale, V.; Martinoia, S.; Chiappalone, M. Stimulation triggers endogenous activity patterns in cultured cortical networks. *Sci. Rep.* **2017**, *7*, 9080. [CrossRef] [PubMed]
32. Buonomano, D.V.; Maass, W. State-dependent computations: Spatiotemporal processing in cortical networks. *Nat. Rev. Neurosci.* **2009**, *10*, 113–125. [CrossRef] [PubMed]
33. Giugliano, M.; la Camera, G.; Fusi, S.; Senn, W. The response of cortical neurons to in vivo-like input current: Theory and experiment: II. Time-varying and spatially distributed inputs. *Biol. Cybern.* **2008**, *99*, 303–318. [CrossRef] [PubMed]
34. Barkai, O.; Butterman, R.; Katz, B.; Lev, S.; Binshtok, A.M. The input-output relation of primary nociceptive neurons is determined by the morphology of the peripheral nociceptive terminals. *J. Neurosci.* **2020**, *40*, 9346–9363. [CrossRef] [PubMed]
35. Nieus, T.; D'Andrea, V.; Amin, H.; Di Marco, S.; Safaai, H.; Maccione, A.; Berdondini, L.; Panzeri, S. State-dependent representation of stimulus-evoked activity in high-density recordings of neural cultures. *Sci. Rep.* **2018**, *8*, 5578. [CrossRef]
36. La Camera, G.; Giugliano, M.; Senn, W.; Fusi, S. The response of cortical neurons to in vivo-like input current: Theory and experiment. *Biol. Cybern.* **2008**, *99*, 279–301. [CrossRef]
37. Peyrin, J.M.; Deleglise, B.; Saias, L.; Vignes, M.; Gougis, P.; Magnifico, S.; Betuing, S.; Pietri, M.; Caboche, J.; Vanhoutte, P.; et al. Axon diodes for the reconstruction of oriented neuronal networks in microfluidic chambers. *Lab Chip* **2011**, *11*, 3663–3673. [CrossRef]
38. Park, J.; Kim, S.; Park, S.I.; Choe, Y.; Li, J.; Han, A. A microchip for quantitative analysis of CNS axon growth under localized biomolecular treatments. *J. Neurosci. Methods* **2014**, *221*, 166–174. [CrossRef]
39. Poli, D.; Pastore, V.P.; Martinoia, S.; Massobrio, P. From functional to structural connectivity using partial correlation in neuronal assemblies. *J. Neural Eng.* **2016**, *13*, 026023. [CrossRef]
40. DeMarse, T.B.; Pan, L.; Alagapan, S.; Brewer, G.J.; Wheeler, B.C. Feed-Forward Propagation of Temporal and Rate Information between Cortical Populations during Coherent Activation in Engineered In Vitro Networks. *Front. Neural Circuits* **2016**, *10*, 32. [CrossRef]
41. Brofiga, M.; Pisano, M.; Callegari, F.; Massobrio, P. Exploring the contribution of thalamic and hippocampal input on cortical dynamics in a brain-on-a-chip model. *IEEE Trans. Med. Robot. Bionics* **2021**, *3*, 315–327. [CrossRef]
42. Chang, C.; Furukawa, T.; Asahina, T.; Shimba, K.; Kotani, K.; Jimbo, Y. Coupling of in vitro Neocortical-Hippocampal Coculture Bursts Induces Different Spike Rhythms in Individual Networks. *Front. Neurosci.* **2022**, *16*, 873664. [CrossRef]
43. Guggenmos, D.J.; Azin, M.; Barbay, S.; Mahnken, J.D.; Dunham, C.; Mohseni, P.; Nudo, R.J. Restoration of function after brain damage using a neural prosthesis. *Proc. Natl. Acad. Sci. USA* **2013**, *110*, 21177–21182. [CrossRef]
44. Volz, L.J.; Rehme, A.K.; Michely, J.; Nettekoven, C.; Eickhoff, S.B.; Fink, G.R.; Grefkes, C. Shaping Early Reorganization of Neural Networks Promotes Motor Function after Stroke. *Cereb. Cortex* **2016**, *26*, 2882–2894. [CrossRef]
45. Mendonça, M.D.; Meira, B.; Fernandes, M.; Barbosa, R.; Bugalho, P. Deep brain stimulation for lesion-related tremors: A systematic review and meta-analysis. *Parkinsonism Relat. Disord.* **2018**, *47*, 8–14. [CrossRef]
46. Siebner, H.R.; Funke, K.; Aberra, A.S.; Antal, A.; Bestmann, S.; Chen, R.; Classen, J.; Davare, M.; Di Lazzaro, V.; Fox, P.T.; et al. Transcranial magnetic stimulation of the brain: What is stimulated?—A consensus and critical position paper. *Clin. Neurophysiol.* **2022**, *140*, 59–97. [CrossRef]
47. Brofiga, M.; Pisano, M.; Raiteri, R.; Massobrio, P. On the road to the brain-on-a-chip: A review on strategies, methods, and applications. *J. Neural Eng.* **2021**, *18*, 41005. [CrossRef]
48. Brofiga, M.; Pisano, M.; Tedesco, M.; Raiteri, R.; Massobrio, P. Three-dimensionality shapes the dynamics of cortical interconnected to hippocampal networks. *J. Neural Eng.* **2020**, *17*, 56044. [CrossRef]

*micromachines*

MDPI

*Article*

# A Poly-(ethylene glycol)-diacrylate 3D-Printed Micro-Bioreactor for Direct Cell Biological Implant-Testing on the Developing Chicken Chorioallantois Membrane

Eric Lutsch, Andreas Struber, Georg Auer, Thomas Fessmann and Günter Lepperdinger *

System-Precision-on-Chip—SPOC Laboratories, Department of Biosciences and Medical Biology, University Salzburg, Hellbrunnerstrasse 34, A-5020 Salzburg, Austria; eric.lutsch@plus.ac.at (E.L.); andreas.struber@plus.ac.at (A.S.); georg.auer@gmail.com (G.A.); thomas.fessmann@stud.sbg.ac.at (T.F.)
* Correspondence: guenter.Lepperdinger@plus.ac.at

**Abstract:** Advancements in biomaterial manufacturing technologies calls for improved standards of fabrication and testing. Currently 3D-printable resins are being formulated which exhibit the potential to rapidly prototype biocompatible devices. For validation purposes, 3D-printed materials were subjected to a hierarchical validation onto the chorioallantoic membrane of the developing chicken, better known as the HET CAM assay. Working along these lines, prints made from poly-(ethylene glycol)-diacrylate (PEGDA), which had undergone appropriate post-print processing, outperformed other commercial resins. This material passed all tests without displaying adverse effects, as experienced with other resin types. Based on this finding, the micro bioreactors (MBR) design, first made of PDMS and that also passed with cell tests on the HET-CAM, was finally printed in PEGDA, and applied in vivo. Following this workflow shows the applicability of 3D-printable resins for biomedical device manufacturing, consents to adherence to the present standards of the 3R criteria in material research and development, and provides flexibility and fast iteration of design and test cycles for MBR adaptation and optimization.

**Keywords:** in vivo micro bioreactor; additive manufacturing; poly-(ethylene glycol)-diacrylate; biocompatibility

**Citation:** Lutsch, E.; Struber, A.; Auer, G.; Fessmann, T.; Lepperdinger, G. A Poly-(ethylene glycol)-diacrylate 3D-Printed Micro-Bioreactor for Direct Cell Biological Implant-Testing on the Developing Chicken Chorioallantois Membrane. *Micromachines* **2022**, *13*, 1230. https://doi.org/10.3390/mi13081230

Academic Editors: Violeta Carvalho, Senhorinha Teixeira and João Eduardo P. Castro Ribeiro

Received: 29 June 2022
Accepted: 28 July 2022
Published: 31 July 2022

**Publisher's Note:** MDPI stays neutral with regard to jurisdictional claims in published maps and institutional affiliations.





## 1. Introduction

Technologies used in medical biology are constantly evolving through incorporation of new techniques, e.g., 3D-printing [1]. Additive manufacturing (AM) provides a novel means for device production not only in engineering but also for biomedical applications. This manufacturing technology eases fabrication, which, based on traditional methods such as machining or molding, would otherwise be very challenging. In line with such desideration and prior to applications as complex as for clinical use, processed material and composites need to be carefully selected. This may eventually also include animal testing. By now, contemporary AM procedures have been developed, which allow almost everybody to print any type of object. To guide translation of this enabling technology, valid standards for material processing and eventually testing must be established.

Three-dimensional-printing approaches that support and enable a plethora of biomedical applications have previously been described [2,3]. To date, this technology is increasingly employed for the prototyping of devices that are used for characterization of living cells in culture. Eventually, the printing of pieces of implantable devices by means of AM is also greatly desired.

For successful implementation of 3D-printed objects, an appropriate evaluation of 3D-printed polymers after photo-polymerization is both critical and pivotal for understanding the physicochemical properties of a biomaterial, especially in the context of effectuating cellular physiology. Bioassays need to be applied for sensitive evaluation of chemicals that potentially leach out from 3D-prints. The manufactured materials are submersed in cell culture media for prolonged periods and the conditioned media including leachables are applied on 2D cell cultures [4]. Hence, this type of assay may also inform about cytocompatibility in the case when cells are in close contact or even adhere to biomaterials. Despite this important first reference point, it will however not yield full information regarding the material's biocompatibility. Recent studies investigating the biocompatibility of 3D-printable resins most commonly assess cytocompatibility regarding cytotoxic assays [4,5]. As the field lacks specifications for labels and declarations regarding cyto-/biocompatibility, printable material is currently synthesized, selected, and tested for applications in a biological context [6–8].

As a first conducive step towards reaching these goals, we here addressed questions not only for rapid and inexpensive manufacturing of devices suitable for research purposes in cell and tissue culture, but also in combination with organ-on-a-chip technology, a technology that has seen rapid advancements in both technology and application recently [9–11]. Although it is a pertinent aspect in deepening our understanding concerning distinct functionalities of cellular ensembles, in particular that of human cells, we here promote the concept of chip-in-an-organ, or implantable bioreactor technology, as just being a further step towards advancing knowledge of living systems [12]. For this purpose, we employed a reliable 3R-compliant test, the hen's egg test chorioallantoic membrane (HET-CAM) assay. In due course of this attempt, we also wanted to increase the standards of this in vivo test by designing a suitable micro bioreactor (MBR) and cost-effective peripheral instrumentation, such as a micropump system for automatic medium change within the reactor, and a camera system inside the incubator for live imaging of the chicken embryo. Features of the HET-CAM bioassay are its simple accessibility, high sensitivity, resourcefulness, low cost, and above that, the fact that growing blood vessels are directly contacted. Moreover, quantitative results can be obtained applying analysis of macroscopic images of CAM development as well as evaluation of vessel character based on microscopic techniques [13,14]. Therefore, the HET-CAM assay is an effective monitoring tool for assessing in vivo biocompatibility [15] and observing complex biological interactions [13]. As such, it provides an easy approach for testing chemicals [16], responses to biomaterials [17,18], living cells [19], and xenografts [20]. To date a wide variety of different types of MBR systems exist with different levels of complexity and versatile applications [21]. Hence, the objectives of this work were to conceptualize a scheme for easing the design and the prototyping by means of AM of fluidic MBRs for their use in an in vivo environment.

## 2. Materials and Methods

### 2.1. Reactor Prototyping and Additive Manufacturing

The 3D-printing was performed with the aid of SLA printers (Phrozen Sonic Mini and Phrozen Sonic Mini 4K according to the manufacturer's instructions; Phrozen Technology; Taiwan) and thermoplastic extrusion printers (Prusa i3MK3S, Prusa Research a.s., Prague, Czech Republic). The 3D objects were either designed with the open-source programs OpenSCAD (OpenSCAD.org) and FreeCAD (freecadweb.org) or with AutoCAD (Autodesk, USA). The 3D digital vector files were processed with the open-source software Chitubox (chitubox.com). The protocol for PDMS HET-CAM reactor prototyping can be found in Supplementary Material and Methods/Results 1.

### 2.2. Cytotoxicity and Cytocompatibility

To determine the cytotoxicity and cytocompatibility of the 3D-printed materials, two different analytical tools were used. The first measure taken was the performance of the colorimetric MTT assay which determines cytotoxicity. Therefore, complete medium was conditioned in the 3D printed vessels for 24 h and subsequently added to a 96-well plate containing adherent hFOB cells. After another 24 h, the analysis was performed; for detailed protocol see Supplementary Material and Methods/Results 2—Medium conditioning.

Secondly, passaging experiments were conducted in direct contact to the materials. In this setting, hFOB cells were first cultivated in standard tissue culture dishes (TC), followed by detachment and transfer into 3D-printed wells for further cultivation, followed by detachment from 3D-printed wells and transfer back into tissue culture dishes. The full protocol can be found in Supplementary Material and Methods/Results 2—Cellular attachment and proliferation. Briefly, cells were seeded at a density of $1 \times 106$ cells and cultivated in standard 10-cm tissue culture (TC) dishes until reaching confluency of ~80%. After passaging the cells for the first time, they were transferred into printed resin wells PEGDA (3 min and 6 min UV exposure), Rapid Clear (3 min UV), as well as Phrozen Black (3 min and 6 min UV) at a density of $3 \times 105$ cells, and subsequently incubated therein for up to 48 h. The cell culture supernatant was collected and transferred into 6-well plates after 24 h and 48 h of incubation. Attached cells residing in the wells were detached with Trypsin/EDTA (Second passage) and transferred into 6-well plates. Pictures were acquired directly after transfer (2–3 h), and at different time points during culture.

### 2.3. Hen's Egg Test—Chorioallantoic Membrane Assay (HET-CAM)—Biocompatibility

A hen's egg test chorioallantoic membrane assay was done to investigate the biocompatibility of the 3D prints. Consequently, the 3D prints were placed on the CAM surface of a chicken embryo for 72 h. The full protocol can be found in Supplementary Material and Methods/Results 2—Hen's egg test—chorioallantoic membrane assay (HET-CAM).

### 2.4. Micro Bioreactor—Periphery and Instrumentation

The MBR was manufactured with the aid of an SLA printer and PEGDA resin. The reactor consists of an inlet and an outlet which are connected to a central chamber via channels. The MBR is coupled to a pump system, which enables application of, e.g., factors or cell suspension respectively.

## 3. Results

### 3.1. PDMS Micro Bioreactor for HET-CAM

The technology of 3D-printing offers the advantage of easily generating objects with complex shapes and a network of channels. For proof of concept and feasibility, a material with validated biocompatibility was applied, polydimethylsiloxane (PDMS). This material is widely used for research purposes and is also used for many clinical applications [22]. For PDMS reactor production, molds were made by means of fused deposition molding (FDM) (Figure 1). To create a leak-free connection between the reactor and the CAM, different materials and methods were tested. N-butyl-cyanoacrylate super glue provided firm attachment of PDMS to the CAM without causing apparent interferences (Figure S1A). hFOB 1.19 cells expressing green fluorescent protein could be injected into the reactor and adhering cell clusters residing on the CAM could be observed (Figure 1G). For a detailed description see Supplementary Material and Methods/Results 1—PDMS HET-CAM Reactor Prototyping.

### 3.2. Cytocompatibility of 3D Prints

PDMS casting of complex geometries is greatly restricted, as nested or winding networks of channels or undercut shapes with skewed or slanted walls cannot be manufactured. Hence, rapid prototyping methods applying common AM technology were contemplated and resins amenable for 3D-printing had to be selected. In addition to commercial resins, Monocure Rapid 3D Resins (Clear or Black; RC or RB), Phrozen Black resin (PB), and eSUN water washable resin (eS), poly-(ethylene glycol)-diacrylate (PEGDA; P) has also been evaluated. These polyethylene glycol (PEG) derivatives have already been successfully used in a variety of tissue engineering and drug delivery-based applications. PEGDA, according to its higher polymeric chain length and thus molecular weight (MW), displays increasing grades of elasticity. All selected resins were stated to be odorless, emitting no volatile organic chemicals, displaying shrinkage below 0.5%, rapidly curing, and exhibiting high tensile strength.

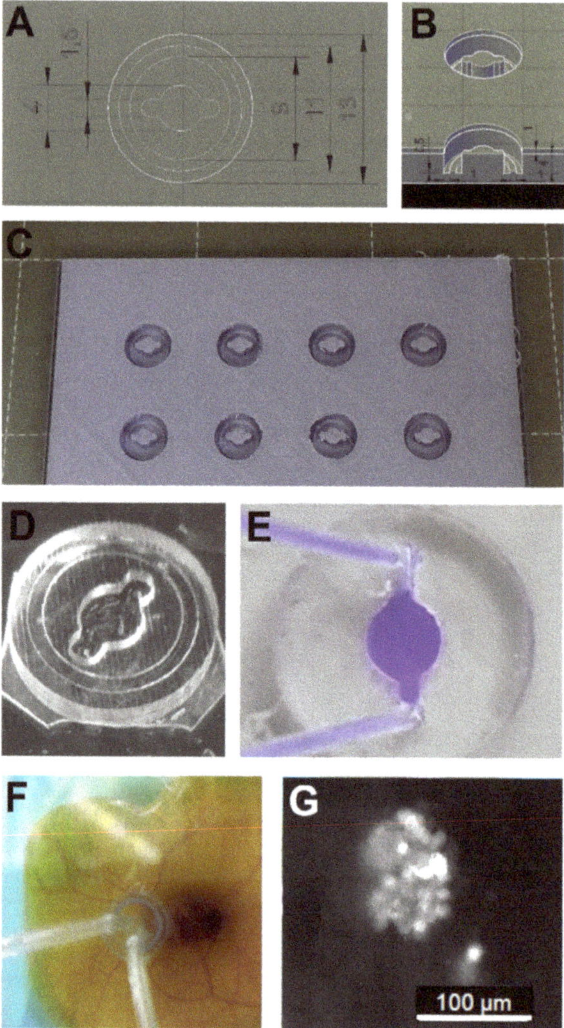

**Figure 1. Polydimethylsiloxane (PDMS) micro bioreactor (MBR) prototyping:** (**A,B**) The design was dome-shaped exhibiting a central flow chamber with two extensions for inlet and outlet connectivity. (**C**) An FDM mold was used for PDMS casting. (**D,E**) PDMS casts could be bonded leak-free with tissue adhesive to a plastic surface and connected to a pumping system via tubing. (**F**) PDMS MBR mounted onto an exo ovo cultivated HET-CAM could be operated without leakage for up to 2 days. (**G**) Human fetal osteoblasts expressing green fluorescent protein (hFOB:GFP) were injected into a PDMS chamber that had been planted onto a HET-CAM. After adherence to the CAM, cell clusters could be observed.

In the context of cytocompatibility testing, the cellular metabolic activity of human fetal osteoblast cell line 1.19 (hFOB) and the human osteosarcoma-derived cell line SaOS-2 were assessed and compared. hFOB cells appeared to be the more sensitive biosensor with an overall higher metabolic activity compared to SaOS-2 (Figure 2A). Working along these lines, post-print processing procedures for the individual printed materials could be accomplished (Figure 2B/Table S1). Since hFOB cells reacted particularly sensitively, dilutions of conditioned media were also evaluated (Figures 2C and S2A).

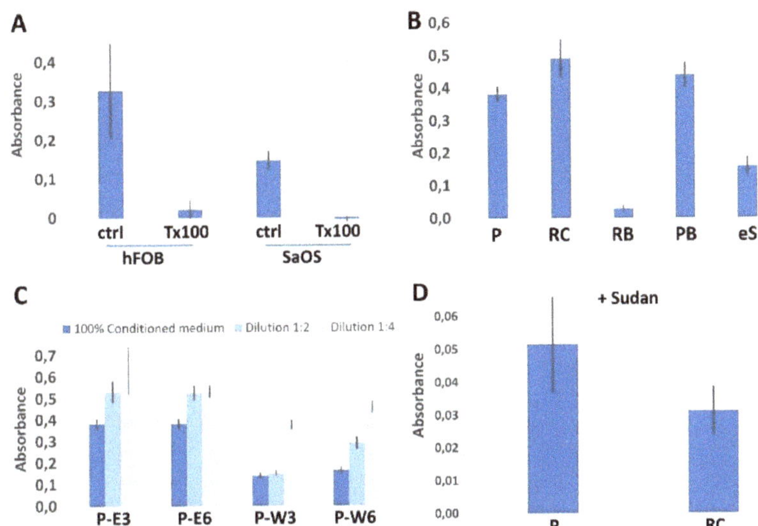

**Figure 2. Cytocompatibility:** (**A**) Metabolic activities were determined for hFOB 1.19 and SaOS-2 cells. Cells were treated for 24 h with growth medium, which had been conditioned with printed materials for 24 h. Controls were complete medium (ctrl), or medium containing 0.1% Triton-X-100 (T × 100). Conversion of MTT was assessed by measuring absorbance values at 492 nm. (**B**) Due to the broad dynamic range, cytocompatibility assessment was performed in hFOB cells applying media conditioned in wells PEGDA (P), Rapid Clear (RC), Rapid Black (RB), Phrozen Black (PB), and eSUN (eS). (**C**) For refining post-processing steps after 3D-printing, PEGDA wells were rinsed either with ethanol (E) or commercial resin wash (W) in combination with UV curing for indicated times in minutes. Media were diluted before treatment as indicated. (**D**) Addition of Sudan I to enhance manufacturing of the translucent resins PEGDA and Rapid Clear resulted in low vitality. Error bars indicate standard deviation; n = 6.

RC resin showed positive results after sonication in ethanol in combination with UV curing for 3 min. Comparable results were achieved when sonicating in resin wash and UV curing for 6 min, yet polymerized RB resin appeared toxic regardless of extensive washing procedures. eS resin had been selected because it could be printed at high resolution and because it is water washable. However, all attempts failed to render the printed polymer suitable for application in cell culture. In stark contrast, cells incubated in media conditioned with PB exhibited enhanced metabolic activities compared to cells cultured in complete medium. Simply washing with ethanol and curing with UV light for 6 min resulted in high vitality. Treating hFOB cells with undiluted conditioned media derived from PB yielded 120% vitality compared to normal complete medium. We therefore further optimized the post-printing procedure for PB yielding best results by first sonicating in deionized water (MilliQ) for 1 h and a subsequent sonication in ethanol for 1 h (Figure 2B). Applying these test conditions for measuring the content of toxic compounds in post washing solutions, optimized washing procedures regarding times and solvents for the individual materials could be acquired (Figures 2C and S2B and Table S1).

We were able to show that rinsing PEGDA with ethanol instead of commercially available resin wash solution resulted in better cellular vitality (Figure 2C). Finally, positive results were obtained for PEGDA and RC, as well as for PB, which prompted us to continue further evaluating these three materials. In contrast to PB, PEGDA and RC are translucent and thus suboptimal for producing high precision printing results. To render these resins suitable for high resolution printing, the commonly used UV absorber Sudan I was mixed with the resins before printing. We were however unsuccessful in establishing a post-printing protocol, which yielded the material compatible for the use in cell culture.

To further investigate cellular vitality, adherence, and proliferation, the different resins were used to manufacture cell culture vessels with a diameter of 35 mm and a height of 6 mm, in which hFOB 1.19 cells were cultivated for passaging experiments (Figure 3A).

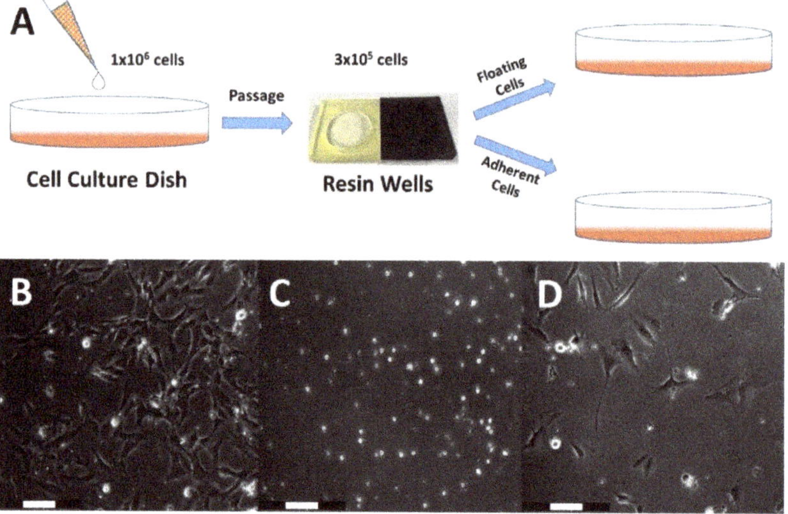


**Figure 3. HET-CAM material testing:** (**A**) Schematic representation of the workflow regarding ex ovo chicken embryo cultivation and material testing highlighting specific days of embryonic development (Day 0–10). (**B**) Full view of an embryo at EDD 9 with mounted PDMS rings for holding specimen in place; (**C**) 3D-printed PEGDA (black arrow) and (**D**) Phrozen Black flat rings were grafted to the CAM surface and cultivated for 72 h. Pictures were taken at embryonic day 10. In case of PEGDA, no apparent signs of adverse reactions or hemorrhage were detectable during the entire observation period, whereas Phrozen Black entailed severe defects and hemorrhage, as well as lysis endpoints.

Comparing all three resins, the benchmark resin PEGDA showed by far the best results throughout the entire series of experiments (Figure 3B). PEGDA exhibited highest cellular vitality considering the number of viable cells in both cell fractions, either floating in the supernatant, or adherent residing in the printed culture vessels. Notably such a result could be exclusively observed for PEGDA. Cells derived from wells made of RC (Figure 3C) showed no apparent survival in both cell fractions. PB displayed only very low compatibility for cell cultivation as only a few cells remained viable (Figure 3D). Furthermore, after trypsinization of the remaining cells that resided in the wells, a significantly higher number of surviving cells was found in PEGDA wells compared to the other tested resins. At day 9 of cultivation after passaging, cells from resin wells made of PEGDA which had been UV-cured for 3 min reached 100% confluency. This result motivated us to continue with an in vivo analysis of PEGDA and PB resin to possibly endorse a potentially high degree of biocompatibility for these materials.

### 3.3. Material Testing on HET-CAM

After having assessed the cytotoxicity of the 3D-printed materials in vitro, putative effects after implantation into a living system, which is made to vastly grow, were determined. In this study, we intended to take advantage of testing the developing embryos fast growing vascular network. This developmental phase is highly sensible as the CAM grows and newly formed vessels sprout by means of endothelial cell proliferation. Due to these facts, the HET-CAM assay was performed ex ovo, in sterilized (70% EtOH; multiple hours of UV light) weighing dishes covered with provided lids or glass (Figure 4A). Pre-treatment of the xenografts included sterilization with 70% ethanol, subsequent washing with PBS, and incubation in complete medium. On embryonic development day 6 (EDD6), sterilized PDMS rings were placed over major vessels of the CAM surface to keep implanted specimen in place. On EDD7, 3D-printed xenografts were fit into the rings. For PEGDA, a test series of two biological replicates consisting of 4 technical replicates (n = 8) was conducted. After 72 h of incubation (Figure 4B), the experiments were terminated, both embryos showed no visible signs of disruption (Figure 4C). Based on all gathered results and observations, we consider PEGDA to be a biocompatible resin for biomedical applications or tissue engineering experiments. Within 72 h of CAM surface treatment with PB resin rings, the embryo had died. The vascular network collapsed completely, and clear signs of hemorrhage were observed (Figure 4D). Compatibility testing with respect to HET-CAM assays has been conducted beyond EDD10 by many research groups to date. At this developmental stage, the embryo evolves first consensual movements, approximately between EDD6 and EDD7, and first sensory neurons are formed from EDD8. Due to these facts, all experimental analyses in this study were terminated at EDD10, as the embryos begin to sense pain [23].

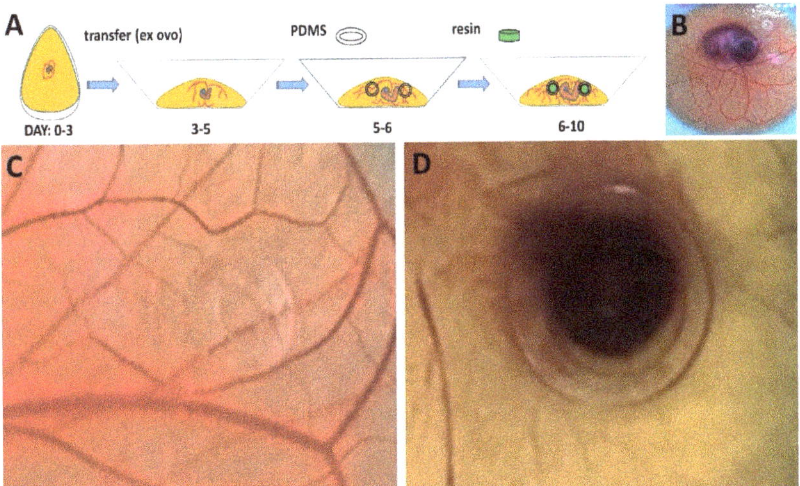



**Figure 4. Cell cultivation in 3D-printed vessels.** (**A**) Schematic representation of the workflow for this experiment. (**B**) PEGDA UV 3 min at day 2 after transfer from the resin well to 6 well plate. Cells from this well reached confluency at day 9 after transfer. (**C**) Rapid Clear UV 3 min at day 2 after transfer. No surviving cells were visible. (**D**) Phrozen Black UV 3 min at day 3 after transfer. The number of viable cells after transfer is much lower compared to PEGDA. Proliferation potential seemed to be inhibited.

### 3.4. PEGDA Micro-Bioreactor for HET CAM and Cell Biological Applications

Next, a MBR for use in an in vivo environment to be produced from PEGDA-250 resin was redesigned (Figure 5). We sought to create a windowed perfusion MBR to ease observation of events taking place within the chamber once applied onto the CAM. A coverslip was laser-cut into the appropriate size and glued in as a glass ceiling of the MBR using a small amount of PEGDA resin, applied to the rims of the glass, and subsequently cured in a UV chamber. The perfusion MBR enables different application settings e.g.,

(i) cell seeding into the reactor chamber and monitoring of the interaction, (ii) flushing the chamber with different biogenic factors, (iii) application of gradients of reactants (low to high concentrations), (iv) testing materials incorporated into the chamber, or (v) vascularization of 3D cell culture (spheroids). The design and dimensions of the PEGDA MBR is depicted in Figure 5A,B. Post-print processing was done according to the presented protocol (Table S1). Thereafter, the glass window was fit on top of the chamber and MBRs were conditioned as described before. The MBR was tested on different substrates, i.e., tissue culture dishes and test dummies made of alginate-gelatin hydrogel, before applying on the CAM. The MBR was conjugated to the CAM of a chicken embryo at EDD6 with tissue adhesive (Surgibond) and stably remained there for at least 24 h.

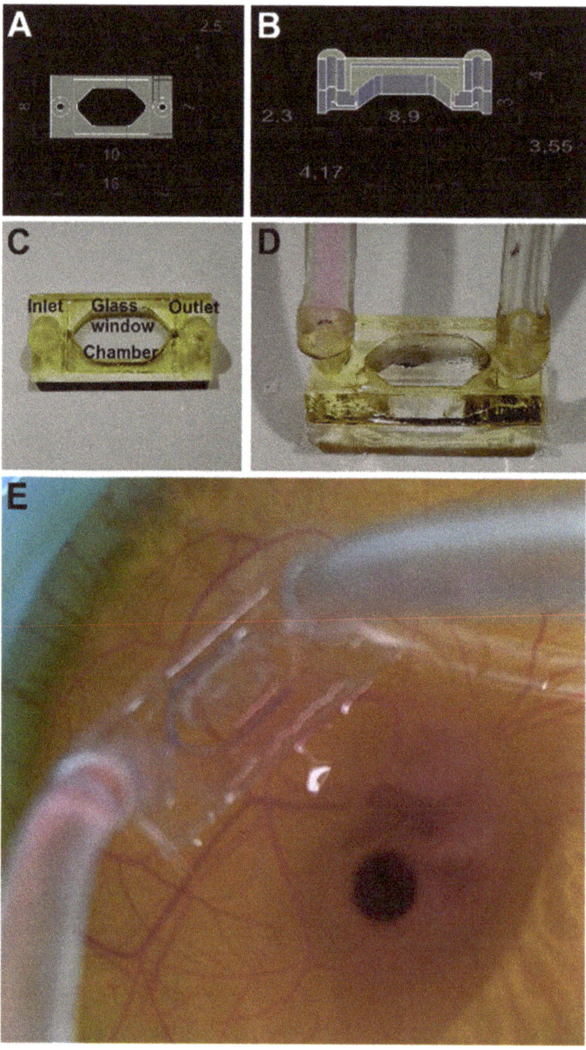


**Figure 5. PEGDA micro bioreactor prototyping and HET-CAM application:** (**A,B**) 3D CAD designs are depicted together with size marks. (**C,D**) The 3D prints exhibit inlet and outlet tubing connectors for media supply, and atop a glass ceiling. (**E**) A PEGDA MBR with already mounted tubing system was grafted on a CAM at embryonic day 7.

## 4. Discussion

Polymer extrusion is probably the simplest 3D-printing technique. It begins with a simple plastic filament that is heat-molten in a printing head and dispensed through a nozzle that reconstructs the desired object in a delineated fashion. Stereolithographic printers have significant advantages in particular regarding printing resolution, efficiency, and working conditions [24]. They most often comprise of a UV light source with a display and a printing platform. Energy-rich UV light activates polymerization of a liquid resin, which in due course turns the resin into a rigid polymer. When first passing light through a liquid crystal display that depicts a high resolution black-and-white-image, polymerization only takes place behind the translucent bright pixels. Hence, 3D objects can be built layer-by-layer through high resolution spatial photo-polymerization. As the rigid print body adheres to the printing platform, objects that exhibit petite features can also be rapidly manufactured [25].

The first MBR 3D-prototyping experiments which were previously conducted by our research team, conceptualized for proof of concept, utilizing FDM printing and PDMS casting were successful, suggesting moving towards SLA printing to enable a more detailed manufacturing of fluidic devices. High resolution printing provides several advantages for biomedical device fabrication, thus is limited by several factors such as the lack of biocompatibility.

We therefore firstly concentrated on introducing a novel workflow to meet reasonable biocompatibility standards, starting with post-print processing of 3D prints, followed by the assessment of cellular vitality in vitro, and eventually concluding the investigation with in vivo tests. Hence, this selective evaluation method enables determining adverse effects (i) exerted by toxic compounds, which leach out from the material, (ii) triggered also by surface properties when provided as a substrate for cell adherence and growth, and (iii) together when brought into a complex environment of a developing biological system. This analytical series exceeds the common standard by introducing further measures regarding cytocompatibility and in vivo compatibility. Only together these analyses provide a reliable first indication whether the material can actually be considered biocompatible. Indeed, bioreactor technology has already been introduced and applied to assess cytocompatibility and various aspects of biocompatibility [26]. Currently, 3D-printing technology is rapidly advancing in the field of biomedicine, in particular being adopted for tissue engineering approaches [27], including technical refinements which enable the production of very small-sized bioreactors [28] or inclusion of microstructures [29].

As most resin formulations and compositions are proprietary, biocompatibility must be independently checked after printing and crosslinking of the monomeric compounds together with a sufficient post-print processing of the polymeric structure. As outlined above, it is imperative to also include in vivo analyses [30]. As a first step, in particular to safeguard the integrity of cells and tissue, we set out to optimize washing and post-print-processing steps by applying various cleansing solvents and incubation times together with further post-curing UV exposures. Post-print processing was validated by means of the ISO 10993-5:2009 standard assay on two different cell lines hFOB 1.19 and SaOS-2. We chose hFOB cells because they have been derived from fetal tissue and it could be shown, by us and others, that they represent a multipotent surrogate in vitro model for stromal mesenchymal stem cells [31,32]. We were able to show that the hFOB1.19 cells more sensitively indicate cytotoxicity than SaOS-2 cancer cells.

Biocompatibility was further assessed on a living organism in vivo with the aid of HET-CAM assays. Biocompatibility is defined as the ability of a material to perform its desired functions with respect to a medical therapy, to induce an appropriate host response in a specific application and to interact with living systems without having any risk of injury, toxicity, or rejection by the immune system and undesirable or inappropriate local or systemic effects [33]. This bioassay is very sensitive as the measures of the irritation score is given in time and hemorrhage, lysis, or coagulation can already be observed after a few minutes of material application. In this context, a wide variety of material formulations, such as collagen, gelatin sponges, various hydrogels, discs made of synthetic materials as well as drug-delivery methods in combination with small molecules, growth factors, or biosimilars, could be evaluated [34]. Furthermore, application of cells, most often derived from tumors as well as transgenic cells derived thereof, has been undertaken to study alteration of vascular growth and pattern [35]. Combining growth factors with cells in synthetic or natural scaffold materials is at the focus of tissue engineering, hence the CAM assay has also been adopted for this type of innovative research [36]. Work along similar lines of in vivo evaluation substantiated the previous findings from cell culture experiments. PEGDA was tolerated by the embryo during the full period of 72 h, with no visible signs of adverse effects. PB resin triggered adverse effects during a 48 h-incubation by leading to clear signs of hemorrhage and lysis. A superficial evaluation of the latter resin would have otherwise yielded a potentially positive result and release for application in biological research.

We expect that device prototyping and production by means of additive manufacturing will greatly promote PEGDA as a material applied in bioanalysis and regenerative medicine. Covalent binding of polyethylene glycol to a substrate is known as PEGylation [37]. PEGylation of nanoparticles applied in cancer treatment resulted in longer plasma half-life. It likely circumvents early clearing through activation of the complement system. Moreover, plasma clearance works through mitigating the opsonization of compounds [38]. It could however be shown that PEG-based hydrogels are degraded under oxidative conditions. This takes place both in vitro and in vivo [39]. It should be further noted that PEG-coatings elicited an anti-PEG based immune response [38]. After all, determining the grade of biocompatibility is an issue of primarily assessing incompatibility. That said, the question remains to be answered whether implanted material is actively segregated from surrounding tissue driven by a chronified foreign body reaction [40]. Taken together, to the best of our knowledge, this is the first report showing that 3D-printed MBR technology can be applied and used on the CAM of a chicken embryo. We are currently using the PEGDA MBR for test with human dermal microvascular endothelial cell line (HMEC-1) to investigate induced neovascularization and incorporation of HMEC-1 into the CAM vascular network to generate a novel vascularized organ-on-chip platform.

## 5. Conclusions

We herein introduced a selective evaluation procedure for the assessment of biomaterial properties that are displayed after photopolymerization and washing of various 3D-printable resins. The interconnected method provides means for grading cytocompatibility before performing elaborate in vivo biocompatibility evaluations. Commercially available resins, for which neither formulation nor composition have been disclosed, could be evaluated in reference to a well-characterized material in the field, PEGDA. The step-by-step procedure facilitates certification of polymers with respect to their applicability in the context of specific analytical methods in basic cell biology and applied biomedicine. It could be shown that prints made of PEGDA are likely exhibiting a high degree of biocompatibility as this material showed no obvious signs of adverse reactions in a living and growing embryological setting. The method could also track down the suitability of additives, which, when mixed into resins, enhanced printing performance but had a pronounced impact on cells even after polymerization and extensive cleansing of the otherwise well-tolerated biomaterial. The MBR system comprises both an inlet for cell seeding and distribution of media or factors as well as an outlet for removing waste by means of controlled micropumps. The reactor has a glass ceiling for better visualization and monitoring within the chamber. Together with compatibility assessment, the prototypic design of novel in vivo MBR technology now enables complex organoid experiments, which can address a growing vascular system or propagate 3D-cell biology approaches taken into a microfluidic environment for the control of extrinsic parameters.

**Supplementary Materials:** The following supporting information can be downloaded at: https://www.mdpi.com/article/10.3390/mi13081230/s1, Figure S1: Experimental setup to test of the in-vivo reactor on HET-CAM. (A) The reactor was prepared by using a pipette tip to distribute the adhesive on the bottom surface of the reactor. (B) The reactor was lowered onto the CAM. (C) After a connection had been established, the reactor was flushed with cell suspensions. After flushing the reactor with the periphery device labeled with (D) (red arrow), the embryo was placed back into the incubator. (E) Camera system for un-interrupted monitoring of the MBR in ex ovo HET-CAM setting. (F,G) Images taken immediately after start of incubation and 24 h later.; Figure S2: Cellular vitality index examined in hFOB cells after 24 h incubation with conditioned media derived from corresponding resin vessels used in this study (PEGDA, Rapid Clear, Rapid Black, Phrozen Black, and eSUN). Controls were complete medium (CTR GM) and a toxic control (CTR TX) containing detergent (complete medium + 0.1% Triton-X). The value of the control con-taining complete medium was set to 100% and the percentage values of conditioned media de-rived from the printed wells were calculated in reference to controls. Results exerting oxidoreduc-tase activities less than 70% are considered potentially toxic. (A) Various post-print processes were tested for each resin to achieve the best possible vitality for the cells. Illustrated is an example for PEGDA. (B) Based on vigorous testing, an optimal condition was determined for each resin. De-picted are the results of MTT analysis of optimized post-print processing methods for each resin used. (C) Optimized post-print processes for translucent resins PEGDA and Rapid Clear with Su-dan I. Despite different post-print processing, it was not possible to render resins mixed with Su-dan I cytocompatible; Figure S3: (A) Bartels mp-Multiboard micropump setup. With the mp-Multiboard 4 micropumps, a flow sensor, a pressure sensor, a thermal conductivity sensor, and two valves can be operated in parallel. In this experiment, only one pump was used. Liquid such as cell culture medium can be drawn and pumped into the reactor chamber, exchanging depleted medium with fresh medium at the same time. (B) The frequency and voltage can be set in the software. The voltage can be set individually for each pump. The frequency applies to all pumps. (C) With the integrated timer mode, pumps can be switched on and off at defined time points. (D) The reactor can also be sup-plied with medium using a self-made syringe pump. With the self-programmed Pure-Data soft-ware, the syringe pump can be controlled very precisely. Another advantage is that only a few components are needed and the pump costs significantly less than 100 €. (E) Pure-Data software: With this self-written program, the parameters can be defined very precisely; Figure S4: Web-based camera system setup inside the incubator, running on a Raspberry Pi. The motor (M) is mounted on the acrylic glass plate with the rotor facing down. One end of the rotat-ing arm (RA) is connected with the rotor and the other end with the Raspberry Pi camera (C). The motor can be programmed to halter the rotating arm at defined positions at defined time points for the camera to take macrographs; Figure S5: Dashboard of the experimental camera setup. When setting up the automation for an experiment, the rotation of the camera can be set to four different positions (0, 90, 180, 270) in correlation to degrees in a circle. A time interval between each macrographic acquisition can be freely selected; Figure S6: Website for HET-CAM observation. Settings are shown on the left side of the screen. HET-CAM: The positions can be moved manually by clicking on the cursor symbols. With the camera symbol, pictures can be taken. The green button is used to start the imaging automation; Figure S7: Time series imaging of a developing chicken embryo and chorioallantoic membrane, taken by the web-based camera system; Table S1: Optimized post-print processing for each resin used in this study. The reference [41–45] are cited in the supplementary materials.

**Author Contributions:** Conceptualization, G.L.; methodology, E.L., A.S., T.F. and G.A.; software, G.A.; formal analysis, E.L. and A.S.; investigation, G.L.; resources, G.L.; writing—original draft preparation, E.L., A.S., T.F. and G.L. writing—review and editing, G.L.; visualization, E.L., A.S. and G.A.; supervision, G.L.; project administration, G.L.; funding acquisition, G.L. All authors have read and agreed to the published version of the manuscript.

**Funding:** This research was funded by funds from Land Salzburg-Wirtschafts- und Forschungs-förderung (20102-P1900166-KZP01-2019; System Precision on Chips: Fertigungsprozesse und Applikationsforschung—SPOC 2.0).

**Institutional Review Board Statement:** Not applicable.

**Informed Consent Statement:** Not applicable.

**Data Availability Statement:** Not applicable.

**Conflicts of Interest:** The authors declare no conflict of interest.

# References


1.  Liaw, C.Y.; Guvendiren, M. Current and emerging applications of 3D printing in medicine. *Biofabrication* **2017**, *9*, 024102. [CrossRef] [PubMed]
2.  Au, A.K.; Huynh, W.; Horowitz, L.F.; Folch, A. 3D-Printed Microfluidics. *Angew. Chem. Int. Ed. Engl.* **2016**, *55*, 3862–3881. [CrossRef] [PubMed]
3.  Bhattacharjee, N.; Urrios, A.; Kang, S.; Folch, A. The upcoming 3D-printing revolution in microfluidics. *Lab. Chip.* **2016**, *16*, 1720–1742. [CrossRef] [PubMed]
4.  Warr, C.; Valdoz, J.C.; Bickham, B.P.; Knight, C.J.; Franks, N.A.; Chartrand, N.; van Ry, P.M.; Christensen, K.A.; Nordin, G.P.; Cook, A.D. Biocompatible PEGDA Resin for 3D Printing. *Acs. Appl. Bio. Mater.* **2020**, *3*, 2239–2244. [CrossRef] [PubMed]
5.  Gonzalez, G.; Baruffaldi, D.; Martinengo, C.; Angelini, A.; Chiappone, A.; Roppolo, I.; Pirri, C.F.; Frascella, F. Materials Testing for the Development of Biocompatible Devices through Vat-Polymerization 3D Printing. *Nanomaterials* **2020**, *9*, 1788. [CrossRef]
6.  Camara-Torres, M.; Sinha, R.; Mota, C.; Moroni, L. Improving cell distribution on 3D additive manufactured scaffolds through engineered seeding media density and viscosity. *Acta Biomater.* **2020**, *101*, 183–195. [CrossRef]
7.  le Guehennec, L.; van Hede, D.; Plougonven, E.; Nolens, G.; Verlee, B.; de Pauw, M.C.; Lambert, F. In vitro and in vivo biocompatibility of calcium-phosphate scaffolds three-dimensional printed by stereolithography for bone regeneration. *J. Biomed. Mater. Res. A* **2020**, *108*, 412–425. [CrossRef] [PubMed]
8.  Li, Y.; Jahr, H.; Zhou, J.; Zadpoor, A.A. Additively manufactured biodegradable porous metals. *Acta Biomater.* **2020**, *115*, 29–50. [CrossRef] [PubMed]
9.  Ingber, D.E. Human organs-on-chips for disease modelling, drug development and personalized medicine. *Nat. Rev. Genet.* **2022**, *23*, 467–491. [CrossRef]
10. Low, L.A.; Sutherland, M.; Lumelsky, N.; Selimovic, S.; Lundberg, M.S.; Tagle, D.A. Organs-on-a-Chip. *Adv. Exp. Med. Biol.* **2020**, *1230*, 27–42.
11. Rothbauer, M.; Rosser, J.M.; Zirath, H.; Ertl, P. Tomorrow today: Organ-on-a-chip advances towards clinically relevant pharmaceutical and medical in vitro models. *Curr. Opin. Biotechnol.* **2019**, *55*, 81–86. [CrossRef] [PubMed]
12. Tatara, A.M.; Wong, M.E.; Mikos, A.G. In vivo bioreactors for mandibular reconstruction. *J. Dent. Res.* **2014**, *93*, 1196–1202. [CrossRef] [PubMed]
13. Nowak-Sliwinska, P.; Segura, T.; Iruela-Arispe, M.L. The chicken chorioallantoic membrane model in biology, medicine and bioengineering. *Angiogenesis* **2014**, *17*, 779–804. [CrossRef]
14. Schimke, M.M.; Stigler, R.; Wu, X.; Waag, T.; Buschmann, P.; Kern, J.; Untergasser, G.; Rasse, M.; Steinmuller-Nethl, D.; Krueger, A.; et al. Biofunctionalization of scaffold material with nano-scaled diamond particles physisorbed with angiogenic factors enhances vessel growth after implantation. *Nanomedicine* **2016**, *12*, 823–833. [CrossRef] [PubMed]
15. Brantner, A.H.; Quehenberger, F.; Chakraborty, A.; Polliger, J.; Sosa, S.; della Loggia, R. HET-CAM bioassay as in vitro alternative to the croton oil test for investigating steroidal and non-steroidal compounds. *ALTEX* **2002**, *19*, 51–56.
16. Schrage, A.; Gamer, A.O.; van Ravenzwaay, B.; Landsiedel, R. Experience with the HET-CAM method in the routine testing of a broad variety of chemicals and formulations. *Altern. Lab. Anim.* **2010**, *38*, 39–52. [CrossRef]
17. Marshall, K.M.; Kanczler, J.M.; Oreffo, R.O. Evolving applications of the egg: Chorioallantoic membrane assay and ex vivo organotypic culture of materials for bone tissue engineering. *J. Tissue Eng.* **2020**, *11*, 2041731420942734. [CrossRef]
18. Schendel, K.U.; Erdinger, L.; Komposch, G.; Sonntag, H.G. Orthodontic materials studied in the HET-CAM test for mucosa-irritating effects. *Fortschr. Kieferorthop.* **1994**, *55*, 28–35. [CrossRef]
19. Martowicz, A.; Kern, J.; Gunsilius, E.; Untergasser, G. Establishment of a human multiple myeloma xenograft model in the chicken to study tumor growth, invasion and angiogenesis. *J. Vis. Exp.* **2015**, *99*, e52665. [CrossRef]
20. DeBord, L.C.; Pathak, R.R.; Villaneuva, M.; Liu, H.C.; Harrington, D.A.; Yu, W.; Lewis, M.T.; Sikora, A.G. The chick chorioallantoic membrane (CAM) as a versatile patient-derived xenograft (PDX) platform for precision medicine and preclinical research. *Am. J. Cancer Res.* **2018**, *8*, 1642–1660.
21. Altmann, B.; Grun, C.; Nies, C.; Gottwald, E. Advanced 3D Cell Culture Techniques in Micro-Bioreactors, Part II: Systems and Applications. *Processes* **2021**, *9*, 21. [CrossRef]
22. Miranda, I.; Souza, A.; Sousa, P.; Ribeiro, J.; Castanheira, E.M.S.; Lima, R.; Minas, G. Properties and Applications of PDMS for Biomedical Engineering: A Review. *J. Funct. Biomater.* **2021**, *13*, 2. [CrossRef] [PubMed]
23. Rosenbruch, M. The sensitivity of chicken embryos in incubated eggs. *ALTEX* **1997**, *14*, 111–113. [PubMed]
24. Zhang, J.; Hu, Q.; Wang, S.; Tao, J.; Gou, M. Digital Light Processing Based Three-dimensional Printing for Medical Applications. *Int. J. Bioprint.* **2020**, *6*, 242. [CrossRef]
25. Malas, A.; Isakov, D.; Couling, K.; Gibbons, G.J. Fabrication of High Permittivity Resin Composite for Vat Photopolymerization 3D Printing: Morphology, Thermal, Dynamic Mechanical and Dielectric Properties. *Materials* **2019**, *12*, 3818. [CrossRef] [PubMed]
26. Priyadarshini, B.M.; Dikshit, V.; Zhang, Y. 3D-printed Bioreactors for In Vitro Modeling and Analysis. *Int. J. Bioprint.* **2020**, *6*, 267. [CrossRef] [PubMed]

27. Gensler, M.; Leikeim, A.; Mollmann, M.; Komma, M.; Heid, S.; Muller, C.; Boccaccini, A.R.; Salehi, S.; Groeber-Becker, F.; Hansmann, J. 3D printing of bioreactors in tissue engineering: A generalised approach. *PLoS ONE* **2020**, *15*, e0242615. [CrossRef]
28. Achinas, S.; Heins, J.I.; Krooneman, J.; Euverink, G.J.W. Miniaturization and 3D Printing of Bioreactors: A Technological Mini Review. *Micromachines* **2020**, *11*, 853. [CrossRef]
29. Fan, D.; Li, Y.; Wang, X.; Zhu, T.; Wang, Q.; Cai, H.; Li, W.; Tian, Y.; Liu, Z. Progressive 3D Printing Technology and Its Application in Medical Materials. *Front. Pharmacol.* **2020**, *11*, 122. [CrossRef]
30. Hart, C.; Didier, C.M.; Sommerhage, F.; Rajaraman, S. Biocompatibility of Blank, Post-Processed and Coated 3D Printed Resin Structures with Electrogenic Cells. *Biosensors* **2020**, *10*, 152. [CrossRef]
31. Kraus, D.; Wolfgarten, M.; Enkling, N.; Helfgen, E.H.; Frentzen, M.; Probstmeier, R.; Winter, J.; Stark, H. In-vitro cytocompatibility of dental resin monomers on osteoblast-like cells. *J. Dent.* **2017**, *65*, 76–82. [CrossRef] [PubMed]
32. Marozin, S.; Simon-Nobbe, B.; Irausek, S.; Chung, L.W.K. Lepperdinger, Kinship of conditionally immortalized cells derived from fetal bone to human bone-derived mesenchymal stroma cells. *Sci. Rep.* **2021**, *11*, 1–13. [CrossRef] [PubMed]
33. Ghasemi-Mobarakeh, L.; Kolahreez, D.; Ramakrishna, S.; Williams, D. Key terminology in biomaterials and biocompatibility. *Curr. Opin. Biomed. Eng.* **2019**, *10*, 45–50. [CrossRef]
34. Kennedy, D.C.; Coen, B.; Wheatley, A.M.; McCullagh, K.J.A. Microvascular Experimentation in the Chick Chorioallantoic Membrane as a Model for Screening Angiogenic Agents including from Gene-Modified Cells. *Int. J. Mol. Sci.* **2021**, *23*, 452. [CrossRef] [PubMed]
35. Ribatti, D. The chick embryo chorioallantoic membrane in the study of tumor angiogenesis. *Rom. J. Morphol. Embryol.* **2008**, *49*, 131–135. [PubMed]
36. Ribatti, D.; Annese, T.; Tamma, R. The use of the chick embryo CAM assay in the study of angiogenic activiy of biomaterials. *Microvasc. Res.* **2020**, *131*, 104026. [CrossRef] [PubMed]
37. Roberts, M.J.; Bentley, M.D.; Harris, J.M. Chemistry for peptide and protein PEGylation. *Adv. Drug. Deliver. Rev.* **2012**, *64*, 116–127. [CrossRef]
38. Hussain, Z.; Khan, S.; Imran, M.; Sohail, M.; Shah, S.W.A.; de Matas, M. PEGylation: A promising strategy to overcome challenges to cancer-targeted nanomedicines: A review of challenges to clinical transition and promising resolution. *Drug Deliv. Transl. Re.* **2019**, *9*, 721–734. [CrossRef] [PubMed]
39. Browning, M.B.; Cereceres, S.N.; Luong, P.T.; Cosgriff-Hernandez, E.M. Determination of the in vivo degradation mechanism of PEGDA hydrogels. *J. Biomed. Mater. Res. A.* **2014**, *102*, 4244–4251.
40. BRatner, D. The Biocompatibility Manifesto: Biocompatibility for the Twenty-first Century. *J. Cardiovasc. Transl.* **2011**, *4*, 523–527. [CrossRef] [PubMed]
41. SHOPMEDVET.COM. Available online: https://www.shopmedvet.com/product/83705/wound-care-gauze-bandages (accessed on 5 June 2022).
42. Urrios, G.A.; Parra-Cabrera, C.; Bhattacharjee, N.; Gonzalez-Suarez, A.M.; Rigat-Brugarolas, L.G.; Nallapatti, U.; Samitier, J.; DeForest, C.A.; Posas, G.F.; Garcia-Cordero, J.L.; et al. 3D-printing of transparent bio-microfluidic devices in PEG-DA. *Lab Chip* **2016**, *16*, 2287–2294. [CrossRef] [PubMed]
43. Poly-(ethylene glycol)-diacrylate. Available online: https://www.sigmaaldrich.com/catalog/product/aldrich/455008 (accessed on 5 June 2022).
44. Detection of Sudan Dyes in Red Chilli Powder by Thin Layer Chromatography. Available online: https://www.omicsonline.org/scientific-reports/srep586.php (accessed on 5 June 2022).
45. MonoCure 3D Rapid UV Resin. Available online: https://www.fepshop.com/shop/uv-resin/monocure3d-resin/monocure-3d-rapid/ (accessed on 5 June 2022).

*Article*

# A Portable Microfluidic System for Point-of-Care Detection of Multiple Protein Biomarkers

Nan Li [†], Minjie Shen [†] and Youchun Xu *

State Key Laboratory of Membrane Biology, Department of Biomedical Engineering, School of Medicine, Tsinghua University, Beijing 100084, China; lin16@mails.tsinghua.edu.cn (N.L.); smj15@mails.tsinghua.edu.cn (M.S.)
* Correspondence: xyc2012@mail.tsinghua.edu.cn; Tel.: +86-10-62796071
† These authors contributed to this work equally.

**Abstract:** Protein biomarkers are indicators of many diseases and are commonly used for disease diagnosis and prognosis prediction in the clinic. The urgent need for point-of-care (POC) detection of protein biomarkers has promoted the development of automated and fully sealed immunoassay platforms. In this study, a portable microfluidic system was established for the POC detection of multiple protein biomarkers by combining a protein microarray for a multiplex immunoassay and a microfluidic cassette for reagent storage and liquid manipulation. The entire procedure for the immunoassay was automatically conducted, which included the antibody–antigen reaction, washing and detection. Alpha-fetoprotein (AFP), carcinoembryonic antigen (CEA) and carcinoma antigen 125 (CA125) were simultaneously detected in this system within 40 min with limits of detection of 0.303 ng/mL, 1.870 ng/mL, and 18.617 U/mL, respectively. Five clinical samples were collected and tested, and the results show good correlations compared to those measured by the commercial instrument in the hospital. The immunoassay cassette system can function as a versatile platform for the rapid and sensitive multiplexed detection of biomarkers; therefore, it has great potential for POC diagnostics.

**Keywords:** protein biomarker; microarray; microfluidic cassette; multiplex measurement; immunoassay; point-of-care testing

**Citation:** Li, N.; Shen, M.; Xu, Y. A Portable Microfluidic System for Point-of-Care Detection of Multiple Protein Biomarkers. *Micromachines* **2021**, *12*, 347. https://doi.org/ 10.3390/mi12040347

Academic Editor: Violeta Carvalho

Received: 10 March 2021
Accepted: 22 March 2021
Published: 24 March 2021

**Publisher's Note:** MDPI stays neutral with regard to jurisdictional claims in published maps and institutional affiliations.

## 1. Introduction

Protein biomarkers are indicators of many diseases, such as cancer, cardiovascular disorders, and infectious diseases, playing a critical role in the early diagnosis and treatment of diseases in clinical practice [1–4]. However, due to the complex molecular and pathogenic mechanisms of diseases, most cannot be diagnosed by only depending on a single biomarker [5–7]. For example, for the accurate diagnosis of cervical cancer at a curable stage, a combination of protein biomarkers is measured, including squamous cell carcinoma antigen (SCC-Ag), serum fragments of cytokeratin (CYFRA), carcinoma embryonic antigen (CEA) and soluble CD44 (sCD44) [8]. The simultaneous determination of a panel of protein biomarkers can significantly improve the specificity and accuracy of the diagnosis. Therefore, the establishment of a multiplexed analytical method with good specificity, sensitivity and speed for the determination of protein biomarkers is one of the most important needs in clinical diagnosis.

For the measurement of protein biomarkers, various promising technologies have been developed, among which the commercialized products mainly include the enzyme-linked immunosorbent assay (ELISA), chemiluminescent or electrochemiluminescent immunoassays, gel electrophoresis and mass spectrometry [9–11]. The routine method used for the multiplexed detection of protein biomarkers is ELISA or the ELISA-derived sandwich-type immunoassay in multi-well plates or multiple tubes. Since each well or tube can only be used to detect a single biomarker, multiplexed detection relies on tedious operations or bulky instruments and thus consumes large volumes of samples and reagents and requires a long analytical period. Recently, great efforts have been made to develop alternative methods for multianalyte immunoassays, which involve two strategies: multilabel and spatially resolved assay protocols [12,13]. The multilabel mode utilizes different labels to tag the corresponding analytes for simultaneous detection of various signals from biomarkers [14–16]. However, the multilabel strategy is limited by its poor quantification capability and throughput. The spatially resolved mode, which uses a single label, can simultaneously identify all targets at different spatial reaction locations [17–20]. Many kinds of spatially resolved array-based platforms have been developed in recent years with different signaling patterns, such as fluorescence [21], chemiluminescence [22] and giant magneto resistance [23]. In order to obtain better performances, researchers also explored other approaches, including electrochemical detection [24–27], surface plasmon resonance [28], cantilever [29], nanowire [30,31], electrical preconcentration [32,33] and various functional materials [34,35]. These approaches usually require microfabrication of the device or complicated material synthesis, which increased the cost of the detection system. Meanwhile, more rigorous operations of immunoassays are needed to ensure a good reproducibility, resulting in sophisticated instrument or intensive labor.

Although the sensitive detection methods and functional materials can improve the performance, the operation of immunoassays usually takes hours and requires highly skilled personnel, which makes it suitable only for non-emergency diagnoses in centralized labs. To overcome these shortcomings, microfluidic technology can be used for the automation and integration of multistep operations by using small cassettes, chips or disks. Several microfluidic platforms [36–39] have been developed to reduce the detection time and manual operations. On the other hand, protein microarrays have been tested in our previous studies to monitor and quantify multiple targets in human serum and urine [40–42]. The versatility and robustness of the microarray system has been demonstrated with thousands of real samples. Therefore, the combination of microarray and microfluidic devices provides a new solution to enhance the performance of multiple-biomarker detection and meet the needs for point-of-care (POC) diagnoses [43–45], whereas the majority of the existing microfluidic platforms are still semiautomatic, while fully automated platforms are burdened with large dead volumes and labor-intensive operations. In addition, platforms that rely on bulky off-chip pumps and valves to drive liquid movement result in cumbersome systems and increase the risk of contamination. Recently, the development of 'self-contained' microfluidic systems, which integrate all the necessary components for an entire assay, becomes a feasible solution to the aforementioned limitations [46,47]. Dai et al. [48] reported a pump-free microfluidic chip for the detection of multiple biomarkers. The driving of the samples and reagents is conducted by the capillary force along with the gravitational force, which enables the chip to be flux-adaptable and easy-to-use. However, this kind of driving manner cannot actively control the flow and the unidirectional flow also hinders its reaction efficiency. Hu et al. [49] described a self-contained and portable microfluidic chemiluminescence immune sensor for quantitative detection of biomarkers. The microfluidic device contains different reagents in individual reservoirs, which are isolated by on-chip mechanical valves. The reagents are driven by the combination control of a negative pressure and the valves. However, this negative pressure as the only driving force means that this microfluidic chip cannot achieve dynamic reaction, which otherwise needs more reaction time. Sinha et al. [50] developed a field-effect-transistor (FET) sensor array-based microfluidic system for the detection of multiple cardiovascular biomarkers. Although these FET-based biosensors are fast, reliable, compact and extremely sensitive, the fabrication can be complex and costly. Therefore, the demand for a rapid, automated, easy-to-use and portable system for multiplexed immunoassays in POC applications has not been fully satisfied.

In this study, a self-contained microfluidic cassette system is developed for POC detection of multiple protein biomarkers. By combining microarray and microfluidic technology, our system can perform a fully integrated immunoassay for the simultaneous detection of multiple biomarkers in serum samples. The self-contained microfluidic cassette is composed of a Luer syringe, which is used to drive liquid movement, and a rotary valve to regulate or switch the flow direction. The cassette can be divided into two functional parts: the reagent storage module and the immunoassay reaction module. The sample, reagents and waste can be stored in six chambers of the reagent storage module, which increases the practicality of the platform and allows the automation of the immunoassay. The immunoassay reaction module contains a protein microarray and a buffer chamber. Accordingly, a supporting device is constructed to conduct fluidic manipulation in the cassette and record the fluorescence signal. Once the serum sample is loaded into the cassette, multiple steps of the immunoassay are automatically executed, and the entire detection process can be completed within 40 min. Three tumor biomarkers that widely used in clinical diagnosis, alpha-fetoprotein (AFP), CEA and carcinoma antigen 125 (CA125), were chosen to validate the proposed system for the simultaneous detection of multiple biomarkers. Moreover, five clinical samples were tested on our system for further validation.

## 2. Materials and Methods

### 2.1. Design and Fabrication of the Cassette

The proposed microfluidic cassette (5 cm × 6.7 cm × 10.5 cm) designed for the immunoassay is schematically depicted in Figure 1A–C. The center of the cassette contains a Luer syringe (Kangdelai, Shanghai, China) sealed to one outlet of a channel of a polycarbonate (PC) rotary valve via a threaded interface. The other outlet of the channel is used to connect to each liquid chamber (Tiangen Biotech, Beijing, China) inserted in the main body of the cassette. The rotary valve functions as a switching valve inside the cassette to switch between the connection with the Luer syringe in the center and the liquid chambers located outside. When the outlet is precisely aligned to a certain inlet of the liquid chamber, liquid can be transferred between the Luer syringe and the liquid chamber. The polyurethane washer is sandwiched between the main body of the cassette and the rotary valve to avoid liquid leakage. The main body of the cassette and the PC base are tightly fixed in place by screws. Therefore, all of the fluids are guided to the appropriate inlets successively via cooperative control of the self-contained syringe and valve. On one side of the cassette, six chambers that can accommodate up to 1 mL of solution are connected to the main body of the cassette through interfaces. These chambers are covered by a PC lid with venting holes. On the other side of the cassette, a glass layer with a protein microarray printed on its surface is attached to the bottom of the reaction chamber. Double-sided adhesive tape (QL-9970-025, Wuxi Bright Technology, Wuxi, China) is used to bond the chamber and the glass. The reaction chamber is connected with the buffer chamber through a channel (0.3 mm in height and 0.5 mm in width). A pressure-sensitive adhesive (PSA) (90697, Adhesive Research, Beijing, China) cover slip is bonded to the main body of the cassette to seal the chamber and channels. The PC components are machined by a computer numerical control (CNC) machine (Hongyang Laser Co., Ltd., Beijing, China). All of the functional components used for fluid manipulation and reagent storage are integrated in the cassette.

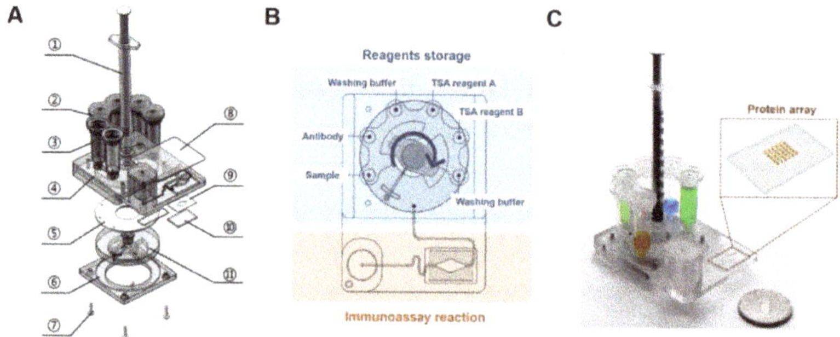



**Figure 1.** Schematic of the cassette used for the immunoassay. (**A**) The three-dimensional illustration of the cassette in an exploded view, 1. Luer syringe; 2. lid; 3. reagents and waste storage chambers; 4. main body of the cassette; 5. polyurethane washer; 6. cassette base; 7. screws; 8. pressure-sensitive adhesive (PSA) cover slip; 9. double adhesive tape; 10. glass; 11. rotary valve. (**B**) Schematic layout of the microfluidic cassette indicating various features. The blue area represents the reagent storage module, and the yellow area represents the immunoassay reaction module. (**C**) Photo of the cassette. The yellow box shows the glass with a protein microarray.

### 2.2. Design of the Supporting Device

As shown in Figure 2, the supporting device (21 cm × 18 cm × 24 cm) is designed for liquid manipulation and fluorescence imaging, which are controlled by an embedded microcontroller unit (MCU) (STM32F103RBT6, STMicroelectronics, Geneva, Switzerland). For liquid manipulation, the syringe piston and the rotary valve are precisely controlled by a stepper motor (28-T6, Shengsida Machinery Equipment Co., Ltd., Suzhou, China) and a digital servo (GDW DS945MG, Shenzhen Huaxiang World Technology Co. Ltd., Shenzhen, China) in a coordinated pattern to achieve basic liquid operations, including liquid transport and mixing. For fluorescence imaging, green light from a light-emitting diode (LED) (530 nm, Quadica Developments, Inc., Alberta, Canada) is filtered by a 510–540 nm bandpass excitation filter (Beijing Bodian Optical Tech. Co., Ltd., Beijing, China) through a microscope objective lens (5× objective lens, Nanjing Jiangnan Novel Optics Co., Ltd., Nanjing, China). The light is used to excite the fluorescent dye on the glass surface after performing the immunoassay. At the same time, the fluorescent images are captured by a digital camera (MVBS30U, CREVIS, Yongin-si, Korea) through a microscope objective lens (MRH00100, Nikon, Shanghai, China) and an optical filter (570 ± 15 nm, Beijing Bodian Optical Tech. Co., Ltd., Beijing, China). The system is powered by external 12 V DC power and communicates with the computer through a USB cable.

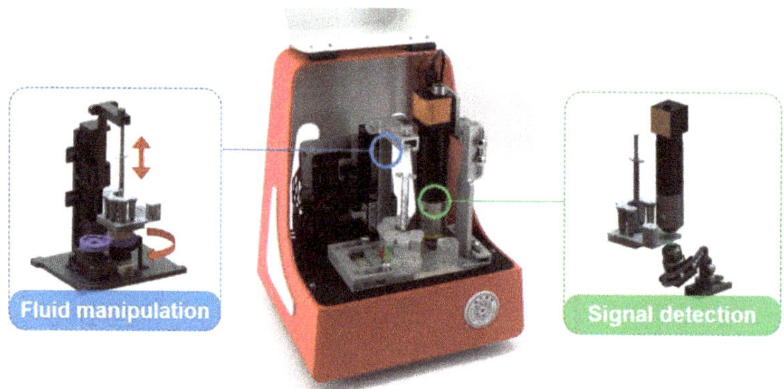



**Figure 2.** The completed prototype of the supporting device with fluid manipulation and signal detection modules.

### 2.3. Reagents and Materials

All antigens (AFP, CEA, CA125) and their antibodies were purchased from Fapon (Shenzhen, China). Cy3-labelled isotype IgG was purchased from Bioss (Beijing, China). Bovine serum albumin (BSA) and Tween-20 were provided by Sigma. Cy3-labelled tyramide was purchased from ApexBio (Houston, TX, USA). Glass slides were provided by Gous Optics (Shanghai, China) and modified with epoxy groups by CapitalBio (Beijing, China). The printing buffer was also purchased from CapitalBio. PBS (10 mM, pH = 7.4) was provided by Solarbio (Beijing, China). PBST (PBS with 0.1% Tween-20) was prepared as the washing buffer. The water used in the experiments was purified with a Milli-Q system (Millipore, Beijing, China).

### 2.4. Printing and Immobilization of the Antibodies

The antibodies were diluted in the printing buffer and then spotted onto epoxy-group-modified glass slides using a printing robot (Personal Arrayer 16, CapitalBio). After spotting, the glass slide was kept at room temperature for 8 h. The epoxy groups on the slides reacted with the free amino groups of the antibodies, covalently linking the antibodies to the glass surface. After that, the glass slide was blocked in 3% BSA solution for 1 h. The glass slides were washed 3 times and kept at 4 °C before use. There were two types of contact-printed arrays on the glass slide. The first one was used for optimizing the spotting concentrations (Figure S1A), and the second one was used to determine the spotting concentrations of antibodies for detection (Figure S1B).

### 2.5. Optimization of the Antibody Concentrations

The dosage of the capture antibody immobilized on the glass slide and the horseradish peroxidase (HRP)-labelled antibody were both optimized for each biomarker. For example, different concentrations of the capture antibody for AFP were spotted onto the glass slide, and different dilution ratios of HRP-labelled antibody were utilized to measure the AFP simulated solution (10 ng/mL). The optimal conditions were chosen according to the concentration, yielding a strong detection signal and low reagent consumption.

### 2.6. Immunoassays

The sample, reagents and buffer were loaded into the cassette. The immunoassay was conducted with the supporting device according to the programmed procedure. Simulated samples of AFP, CEA and CA125 at a series of concentrations were measured with our system. Calibration curves were established according to the results. Similarly, clinical serum samples were also tested on our system.

### 2.7. Data Acquisition and Analysis

Once the on-cassette immunoassay was completed, the fluorescent signal of the detection microarray was excited by an LED and captured by the camera. Differences in the median fluorescence intensity (MFI) between the signal and the blank were determined by LuxScan 3.0 software (CapitalBio). According to the data extracted from the images, standard curves for the three biomarkers were plotted, and the concentration of each target in the clinical samples was accurately quantified with the corresponding curves.

## 3. Results

### 3.1. Workflow of the Immunoassay Cassette

The cassette proposed in this study is designed to accomplish the simultaneous detection of multiple biomarkers in a "sample-in to answer-out" manner. The process of the immunoassay and the control program of the cassette are described below (Figure 3). A video is supplied to visualize this process (Supplementary Movie S1). Before the assay, the channel and reaction chamber of the cassette were blocked with PBST containing 2% BSA for 1 h to reduce nonspecific binding. The corresponding reagents for the immunoassay were preloaded into the storage chambers of the cassette. After adding the serum sample

(200 µL), the cassette was fixed to the supporting device. The automated manipulation procedure for the immunoassay was executed as follows:

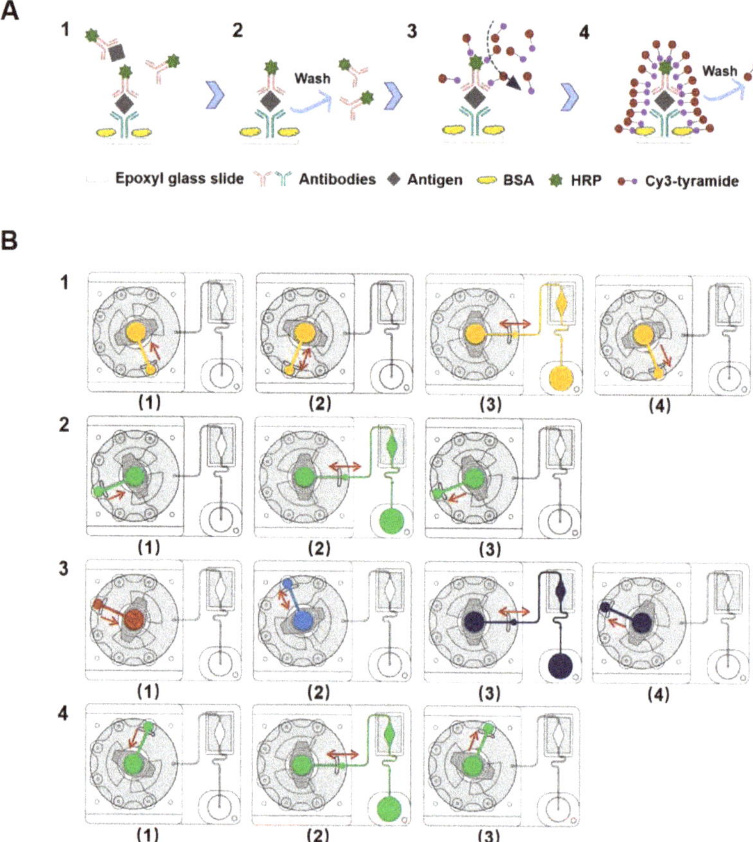



**Figure 3.** The cassette processing protocol. (**A**) Schematic showing the process of the immunoassay. (**1**) Antibody–antigen reaction; (**2**) washing; (**3**) horseradish peroxidase (HRP)-catalyzed signal amplification; (**4**) washing. The Cy3-tyramide molecule can be catalyzed by HRP under a $H_2O_2$ environment, and the activated tyramide can covalently bind to certain amino acid residues of the proximate proteins. (**B**) Illustration of the entire flow control system of the cassette: (**1**) the 200 µL serum sample is mixed with lyophilized antibody and then placed into the reaction chamber; (**2**) after completing the antibody–antigen reaction, the glass is washed with 800 µL washing buffer; (**3**) 200 µL TSA reagent A and 200 µL TSA reagent B are mixed prior to transfer into the reaction chamber; (**4**) the glass is washed with 800 µL washing buffer.

(1) Antibody–antigen reaction. The rotary valve was rotated to engage the inlet of the sample chamber. The serum sample (200 µL) was driven into the Luer syringe, and then the rotary valve was rotated clockwise to connect the inlet of the antibody chamber. The serum sample was dispensed into the antibody chamber and completely mixed with the lyophilized antibody for 1 min at a flow rate of 50 µL/s. After dissolving, the mixture was aspirated back into the Luer syringe, and the rotary valve was rotated clockwise to connect the reaction chamber. Subsequently, the mixture was transferred into the reaction chamber and gently pushed and pulled back and forth by the syringe to accelerate the reaction for 18 min at a flow velocity of 10 µL/s. After the affinity binding reaction, the

waste was transferred back to the sample chamber (Figure 3B1). The total time required for the antibody–antigen reaction step was 20 min.

(2) Washing. After the affinity binding step, the rotary valve was switched to connect the inlet of the first washing buffer chamber, and the buffer was driven into the Luer syringe. Then, the rotary valve was rotated clockwise to connect the reaction chamber again, and the reaction chamber was washed by pushing and pulling the washing buffer back and forth for 5 min (800 µL at 20 µL/s). After the washing process, the waste was discarded into the first washing chamber (Figure 3B2).

(3) HRP-catalyzed signal detection. Following the first washing process, the rotary valve was switched to connect the Luer syringe and the inlet of the tyramide signal amplification (TSA) reagent A chamber. After 200 µL TSA reagent A was driven into the Luer syringe, the rotary valve was rotated clockwise to connect to the TSA reagent B chamber. Then, TSA reagent A was transferred into the TSA reagent B chamber and mixed with 200 µL TSA reagent B for 30 s at a 50 µL/s flow velocity. Afterwards, the 400 µL volume of TSA reagent mixture was transferred into the reaction chamber to detect the signal from the protein microarray for 9 min at a 10 µL/s flow velocity. After the signaling reaction, the waste was discarded into the TSA reagent A chamber (Figure 3B3).

(4) Washing. The second washing step followed a process similar to that of the first washing step (Figure 3B4).

It is noticed that bubble trapping issues may sometimes occur in the reaction chamber. Though it has finite effect on the detection result as the bubbles are apt to attach to the top PSA in the reaction chamber, further upgrade of the cassette may eliminate these issues. Long and narrow chambers with rounded corners are preferred. In addition, injection molding of the cassette and hydrophilic modification on the surface of channels and chambers can also help to reduce the bubble formation.

The entire immunoassay process can be accomplished within 40 min. The TSA reaction steps were applied in this system to improve the sensitivity of the immunoassay [50]. Moreover, the above workflow can be easily reprogrammed to perform other experimental protocols.

### 3.2. Optimization of the Concentrations of Antigen and Antibody

First, the concentration of the capture antibody used for immobilization by printing on the glass slide along with the concentration of the HRP-labelled antibody were optimized to improve the performance of the immunoassay. As shown in Figure S2, a comparatively better signal was generated at a concentration of 0.6 mg/mL for the immobilization of the AFP capture antibody and at the dilution of 1:250 for the HRP-labelled AFP antibody. For CEA and CA125, the antibody concentrations used for printing were 0.6 and 0.4 mg/mL, respectively. The preferred dilution ratios of the HRP-labelled antibody for these three biomarkers were all 1:250. Second, according to the optimized results, the HRP-labelled antibodies for the three biomarkers were mixed and lyophilized in a tube in the cassette. During the immunoassay, the antibodies were dissolved for the automated detection of simulated and clinical samples.

### 3.3. Determination of the Specificity

Multitarget detection in a single chamber can result in cross-reactions between different antigens and antibodies during the immunoassay. Therefore, the specificity of the immunoassay was determined by evaluating the cross-reactivity (CR) of the reactions. For example, the CR between AFP and CEA-Ab was calculated by using Equation (1):

$$CR\ (\%) = MFI_{(AFP,\ CEA\text{-}Ab)}/MFI_{(AFP,\ AFP\text{-}Ab)} \times 100\% \tag{1}$$

The CR values were obtained by detecting three simulated samples containing only one target. The concentrations of the AFP, CEA and CA125 samples were 8 ng/mL, 8 ng/mL, and 32 U/mL, respectively. As shown in Figure 4, the values of the CR between unmatched antigens and antibodies were all below 2%, suggesting a high specificity of the binding between the antigens and their respective antibodies.

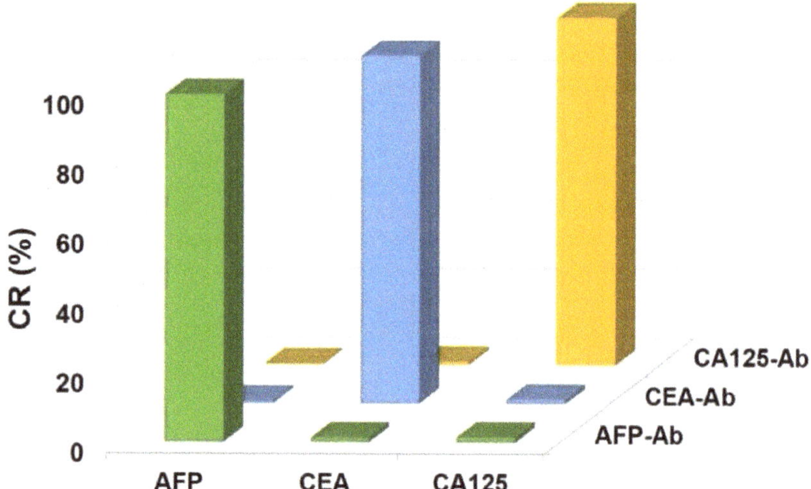


**Figure 4.** Cross-reactivities of the three pairs of antigens and antibodies.

*3.4. Calibration Curves of the Three Biomarkers*

Simulated samples with a series of different concentrations for each biomarker were prepared and analyzed by following the operation protocol to obtain the standard curves (Figure 5). The fitting equations and the limits of detection (LODs) are listed in Table 1. The LOD values were conservatively calculated by using the average MFI value of the blank sample plus three times the standard deviation, and the concentrations for AFP, CEA and CA125 were calculated to be 0.303 ng/mL, 1.870 ng/mL, and 18.617 U/mL, respectively. These concentrations are all within the normal range of those in a healthy individual. Therefore, our system is capable of disease monitoring and diagnosis. However, the Cy3-tyramide signaling method we applied in the immunoassay resulted in a smaller linear range than ELISA (0.01 to 1 ng/mL) [51], indicating that further optimization is needed in the future. In addition, the linear range could also be shifted according to clinical requirements by altering the concentration of the printing antibody and HRP-labelled antibody, the reaction time and other signal amplification parameters.

**Table 1.** Standard curves and the limits of detection (LODs).

| Target | Standard Curve | Adj. $R^2$ | LOD |
|--------|----------------|------------|-----|
| AFP | $y_{AFP} = 64{,}114.05 - 63{,}980.01/[1 + (x/7.10)^{2.4908}]$ | 0.9940 | 0.303 ng/mL |
| CEA | $y_{CEA} = 63{,}554.89 - 63{,}508.52/[1 + (x/21.50)^{2.5841}]$ | 0.9935 | 1.870 ng/mL |
| CA125 | $y_{CA125} = 62{,}549.65 - 62{,}485.99/[1 + (x/108.58)^{3.6293}]$ | 0.9979 | 18.617 U/mL |

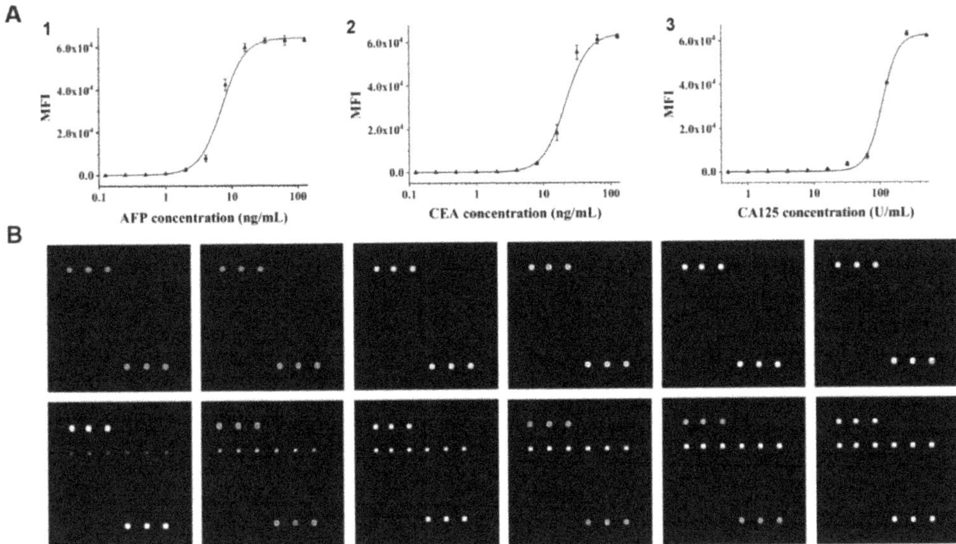

**Figure 5.** Establishment of the standard curves. (**A**) Calibration curves for the immunoassays of tumor markers. (**1**) Standard curve for alpha-fetoprotein (AFP); (**2**) standard curve for carcinoembryonic antigen (CEA); (**3**) standard curve for carcinoma antigen 125 (CA125). The error bars show the standard deviations of triplicate measurements. (**B**) Representative fluorescence images of microarrays during detection of AFP with simulated standard samples. The concentrations of AFP were 0, 0.125, 0.25, 0.5, 1, 2, 4, 8, 16, 32, 64 and 128 ng/mL (from left to right and top to bottom).

### 3.5. Detection of Clinical Samples

Clinical samples were also collected to further validate the use of the cassette system for diagnosis via protein biomarker detection. As shown in Table 2, the relative deviations of the concentrations of AFP, CEA, and CA125 in five clinical samples measured by our system and in the hospital were all within 20%, suggesting the potential of our system for clinical use. The accuracy of our system can be affected by the machining of the cassette compartments, the manual assembly of the cassette, and the signal extracted from the immunoassay array. Therefore, the accuracy of our system still has the potential for further improvement. In fact, there is still a gap between our system and a ready-to-use product in clinical applications. Additionally, we are still dedicated in the improvement of our system and hope for a better performance in the follow-up study.

**Table 2.** Detection of five clinical samples.

| No. | Measured in Hospital | | | Measured in Cassette System | | | | | |
|---|---|---|---|---|---|---|---|---|---|
| | AFP (ng/mL) | CEA (ng/mL) | CA125 (U/mL) | AFP (ng/mL) | RD [1] (%) | CEA (ng/mL) | RD (%) | CA125 (U/mL) | RD (%) |
| 1 | 2.75 | 2.39 | 7.51 | 2.66 | 3.4 | 2.62 | 9.5 | 7.82 | 4.1 |
| 2 | 3.57 | 2.43 | 27.9 | 3.39 | 5.0 | 2.58 | 6.3 | 30.89 | 10.7 |
| 3 | 1.63 | 2.52 | 8.65 | 1.64 | 0.5 | 2.24 | 11.0 | 10.02 | 15.8 |
| 4 | 6.08 | 1.97 | 19.14 | 6.49 | 6.7 | 2.08 | 5.4 | 15.52 | 18.9 |
| 5 | 8.59 | 2.14 | 12.2 | 7.52 | 12.5 | 2.30 | 7.5 | 11.91 | 2.4 |

[1] RD stands for the relative deviation between the concentration measured by the cassette system and that measured in hospital (by Roche cobas e601). RD (%) = $| C_{(Hospital)} - C_{(Cassette\ system)} | / C_{(Hospital)} \times 100\%$.

These results clearly demonstrated that multitarget detection of protein biomarkers can be achieved with our system. Fully automated detection of three protein biomarkers in clinical samples can be completed within 40 min in a "sample-in and answer-out" manner. In addition, our system is capable of functioning as a platform for the detection of more than three biomarkers according to the microarray-based immunoassay. Several previous methods for multi-target detections have been listed in Table S1. It is found that when combining immunoassay with microfluidic devices, they are apt to have several compromises in their sensitivity, simplicity, speed, and cost. Our method has comprehensive advantages regarding self-containment, detection time, sensitivity, integration and fabrication complexity. These features make our system suitable for POC applications.

## 4. Conclusions

In this study, a self-contained multiplexed immunoassay cassette system was developed by integrating a protein microarray into a microfluidic cassette for automated immunoassay. The entire process, including the antibody–antigen reaction, first wash, signal amplification and second wash, can be automatically accomplished within 40 min. The system is composed of a disposable cassette and a handheld device. There is no exchange of liquids between the cassette and the device, thus minimizing the possibility of carry-over contamination. The cassette system was applied to simultaneously detect AFP, CEA and CA125 in human serum, and good sensitivity and repeatability were obtained. Moreover, five clinical samples were detected with this system, and the obtained results were in acceptable agreement with those measured by the commercial instrument in the hospital. These results demonstrate that the presented immunoassay cassette system was sensitive, easy to use, rapid and could be easily expanded for the simultaneous detection of various kinds of protein biomarkers for POC applications.

**Supplementary Materials:** The following are available online at https://www.mdpi.com/2072-666 X/12/4/347/s1, Figure S1: (A) Printing design of the capture antibody on the optimization glass. Antibodies of AFP, Antibodies against AFP, CEA, and CA125 at different concentrations in printing buffer were printed on the surface of the epoxy group-modified glass. (B) The design for printing on the detection glass consisted of the optimal concentration of each antibody and controls (Cy3-IgG and printing buffer), Figure S2: Optimization of the antibody concentrations. (A) AFP, (B) CEA, (C) CA125, title, Table S1. Summary of POC multi-target immunoanalyzers, Movie S1. Automated processing with the cassette system. The video shows the on-cassette fluid control steps, and the sample and reagents are represented by different colored dyes.

**Author Contributions:** Conceptualization, N.L.; Funding acquisition, Y.X.; Investigation, Y.X.; Methodology, N.L. and M.S.; Validation, N.L. and M.S.; Writing—original draft, N.L. and M.S.; Writing—review and editing, Y.X. All authors have read and agreed to the published version of the manuscript.

**Funding:** This research was funded by the National Natural Science Foundation of China, grant number 22074078.

**Institutional Review Board Statement:** The study was approved by the Institutional Review Board (IRB) of Tsinghua University of China (No. 20180022). All blood samples were provided by donors following the IRB guidelines.

**Informed Consent Statement:** Informed consent was obtained from all subjects involved in the study.

**Conflicts of Interest:** The authors declare no conflict of interest.

## References

1. Allinson, J.L. Clinical biomarker validation. *Bioanalysis* **2018**, *10*, 957–968. [CrossRef] [PubMed]
2. Biomarkers Definitions Working Group; Atkinson, A.J., Jr.; Colburn, W.A.; DeGruttola, V.G.; DeMets, D.L.; Downing, G.J.; Hoth, D.F.; Oates, J.A.; Peck, C.C.; Schooley, R.T.; et al. Biomarkers and surrogate endpoints: Preferred definitions and conceptual framework. *Clin. Pharmacol. Ther.* **2001**, *69*, 89–95.
3. Califf, R.M. Biomarker definitions and their applications. *Exp. Biol. Med.* **2018**, *243*, 213–221. [CrossRef]
4. Wulfkuhle, J.D.; Liotta, L.A.; Petricoin, E.F. Proteomic applications for the early detection of cancer. *Nat. Rev. Cancer* **2003**, *3*, 267–275. [CrossRef]

5. Kingsmore, S.F. Multiplexed protein measurement: Technologies and applications of protein and antibody arrays. *Nat. Rev. Drug Discov.* **2006**, *5*, 310–320. [CrossRef]
6. Ludwig, J.A.; Weinstein, J.N. Biomarkers in cancer staging, prognosis and treatment selection. *Nat. Rev. Cancer* **2005**, *5*, 845–856. [CrossRef] [PubMed]
7. Wu, L.; Qu, X. Cancer biomarker detection: Recent achievements and challenges. *Chem. Soc. Rev.* **2015**, *44*, 2963–2997. [CrossRef] [PubMed]
8. Dasari, S.; Wudayagiri, R.; Valluru, L. Cervical cancer: Biomarkers for diagnosis and treatment. *Clin. Chim. Acta* **2015**, *445*, 7–11. [CrossRef]
9. Giljohann, D.A.; Mirkin, C.A. Drivers of biodiagnostic development. *Nature* **2009**, *462*, 461–464. [CrossRef]
10. Nimse, S.B.; Sonawane, M.D.; Song, K.S.; Kim, T. Biomarker detection technologies and future directions. *Analyst* **2016**, *141*, 740–755. [CrossRef] [PubMed]
11. Stolz, A.; Jooss, K.; Hocker, O.; Romer, J.; Schlecht, J.; Neususs, C. Recent advances in capillary electrophoresis-mass spectrometry: Instrumentation, methodology and applications. *Electrophoresis* **2019**, *40*, 79–112. [CrossRef]
12. Pei, X.; Zhang, B.; Tang, J.; Liu, B.; Lai, W.; Tang, D. Sandwich-type immunosensors and immunoassays exploiting nanostructure labels: A review. *Anal. Chim. Acta* **2013**, *758*, 1–18. [CrossRef]
13. Zong, C.; Wu, J.; Wang, C.; Ju, H.; Yan, F. Chemiluminescence imaging immunoassay of multiple tumor markers for cancer screening. *Anal. Chem.* **2012**, *84*, 2410–2415. [CrossRef]
14. Gill, R.; Zayats, M.; Willner, I. Semiconductor quantum dots for bioanalysis. *Angew. Chem. Int. Ed. Engl.* **2008**, *47*, 7602–7625. [CrossRef] [PubMed]
15. Joo, J.; Kwon, D.; Yim, C.; Jeon, S. Highly sensitive diagnostic assay for the detection of protein biomarkers using microresonators and multifunctional nanoparticles. *ACS Nano* **2012**, *6*, 4375–4381. [CrossRef] [PubMed]
16. Shikha, S.; Zheng, X.; Zhang, Y. Upconversion nanoparticles-encoded hydrogel microbeads-based multiplexed protein detection. *Nanomicro Lett.* **2018**, *10*, 31. [CrossRef]
17. Gupta, S.; Manubhai, K.P.; Kulkarni, V.; Srivastava, S. An overview of innovations and industrial solutions in Protein Microarray Technology. *Proteomics* **2016**, *16*, 1297–1308. [CrossRef] [PubMed]
18. Huang, W.; Whittaker, K.; Zhang, H.; Wu, J.; Zhu, S.W.; Huang, R.P. Integration of antibody array technology into drug discovery and development. *Assay Drug Dev. Technol.* **2018**, *16*, 74–95. [CrossRef]
19. Huang, Y.; Zhu, H. Protein array-based approaches forbiomarker discovery in cancer. *Genom. Proteom. Bioinform.* **2017**, *15*, 73–81. [CrossRef]
20. Spisák, S.; Guttman, A. Biomedical applications of protein microarrays. *Curr. Med. Chem.* **2009**, *16*, 2806–2815. [CrossRef]
21. Kwon, M.H.; Kong, D.H.; Jung, S.H.; Suh, I.B.; Kim, Y.M.; Ha, K.S. Rapid determination of blood coagulation factor XIII activity using protein arrays for serodiagnosis of human plasma. *Anal. Chem.* **2011**, *83*, 2317–2323. [CrossRef] [PubMed]
22. Zhao, Y.; Zhang, Y.; Lin, D.; Li, K.; Yin, C.; Liu, X.; Jin, B.; Sun, L.; Liu, J.; Zhang, A.; et al. Protein microarray with horseradish peroxidase chemiluminescence for quantification of serum alpha-fetoprotein. *J. Int. Med. Res.* **2015**, *43*, 639–647. [CrossRef]
23. Gao, Y.; Huo, W.; Zhang, L.; Lian, J.; Tao, W.; Song, C.; Tang, J.; Shi, S.; Gao, Y. Multiplex measurement of twelve tumor markers using a GMR multi-biomarker immunoassay biosensor. *Biosens. Bioelectron.* **2019**, *123*, 204–210. [CrossRef] [PubMed]
24. Qi, H.; Ling, C.; Ma, Q.; Gao, Q.; Zhang, C. Sensitive electrochemical immunosensor array for the simultaneous detection of multiple tumor markers. *Analyst* **2012**, *137*, 393–399. [CrossRef] [PubMed]
25. Sage, A.T.; Besant, J.D.; Lam, B.; Sargent, E.H.; Kelley, S.O. Ultrasensitive electrochemical biomolecular detection using nanostructured microelectrodes. *Acc. Chem. Res.* **2014**, *47*, 2417–2425. [CrossRef] [PubMed]
26. Khanmohammadi, A.; Aghaie, A.; Vahedi, E.; Qazvini, A.; Ghanei, M.; Afkhami, A.; Hajian, A.; Bagheri, H. Electrochemical biosensors for the detection of lung cancer biomarkers: A review. *Talanta* **2020**, *206*, 120251. [CrossRef]
27. Feng, J.; Li, Y.; Li, M.; Li, F.; Han, J.; Dong, Y.; Chen, Z.; Wang, P.; Liu, H.; Wei, Q. A novel sandwich-type electrochemical immunosensor for PSA detection based on PtCu bimetallic hybrid (2D/2D) rGO/g-$C_3N_4$. *Biosens. Bioelectron.* **2017**, *91*, 441–448. [CrossRef]
28. He, L.; Pagneux, Q.; Larroulet, I.; Serrano, A.Y.; Pesquera, A.; Zurutuza, A.; Mandler, D.; Boukherroub, R.; Szunerits, S. Label-free femtomolar cancer biomarker detection in human serum using graphene-coated surface plasmon resonance chips. *Biosens. Bioelectron.* **2017**, *89 Pt 1*, 606–611. [CrossRef]
29. Kartanas, T.; Ostanin, V.; Challa, P.K.; Daly, R.; Charmet, J.; Knowles, T. Enhanced Quality Factor Label-free Biosensing with Micro-Cantilevers Integrated into Microfluidic Systems. *Anal. Chem.* **2017**, *89*, 11929–11936. [CrossRef]
30. Gao, A.; Lu, N.; Dai, P.; Li, T.; Pei, H.; Gao, X.; Gong, Y.; Wang, Y.; Fan, C. Silicon-nanowire-based CMOS-compatible field-effect transistor nanosensors for ultrasensitive electrical detection of nucleic acids. *Nano Lett.* **2011**, *11*, 3974–3978. [CrossRef]
31. Wang, Z.; Lee, S.; Koo, K.; Kim, K. Nanowire-Based Sensors for Biological and Medical Applications. *IEEE Trans. Nanobiosci.* **2016**, *15*, 186–199. [CrossRef]
32. Ouyang, W.; Han, J. Universal amplification-free molecular diagnostics by billion-fold hierarchical nanofluidic concentration. *Proc. Natl. Acad. Sci. USA* **2019**, *116*, 16240–16249. [CrossRef] [PubMed]
33. Ouyang, W.; Han, J.; Wang, W. Enabling electrical biomolecular detection in high ionic concentrations and enhancement of the detection limit thereof by coupling a nanofluidic crystal with reconfigurable ion concentration polarization. *Lab Chip* **2017**, *17*, 3772–3784. [CrossRef]

34. Chang, H.; Zhang, H.; Lv, J.; Zhang, B.; Wei, W.; Guo, J. Pt NPs and DNAzyme functionalized polymer nanospheres as triple signal amplification strategy for highly sensitive electrochemical immunosensor of tumour marker. *Biosens. Bioelectron.* **2016**, *86*, 156–163. [CrossRef]

35. Lai, G.; Cheng, H.; Xin, D.; Zhang, H.; Yu, A. Amplified inhibition of the electrochemical signal of ferrocene by enzyme-functionalized graphene oxide nanoprobe for ultrasensitive immunoassay. *Anal. Chim. Acta* **2016**, *902*, 189–195. [CrossRef] [PubMed]

36. Chen, Y.; Meng, X.; Zhu, Y.; Shen, M.; Lu, Y.; Cheng, J.; Xu, Y. Rapid detection of four mycotoxins in corn using a microfluidics and microarray-based immunoassay system. *Talanta* **2018**, *186*, 299–305. [CrossRef] [PubMed]

37. Garcia-Cordero, J.L.; Maerkl, S.J. A 1024-sample serum analyzer chip for cancer diagnostics. *Lab Chip* **2014**, *14*, 2642–2650. [CrossRef]

38. Kloth, K.; Niessner, R.; Seidel, M. Development of an open stand-alone platform for regenerable automated microarrays. *Biosens. Bioelectron.* **2009**, *24*, 2106–2112. [CrossRef] [PubMed]

39. Knecht, B.G.; Strasser, A.; Dietrich, R.; Märtlbauer, E.; Niessner, R.; Weller, M.G. Automated microarray system for the simultaneous detection of antibiotics in milk. *Anal. Chem.* **2004**, *76*, 646–654. [CrossRef]

40. Chen, A.; Wang, G.; Cao, Q.; Wang, Y.; Zhang, Z.; Sun, Y.; Wang, H.; Xu, C.; Zhou, Q.; Han, P.; et al. Development of an antibody hapten-chip system for detecting the residues of multiple antibiotic drugs. *J. Forensic Sci.* **2009**, *54*, 953–960. [CrossRef]

41. Du, H.; Wu, M.; Yang, W.; Yuan, G.; Sun, Y.; Lu, Y.; Zhao, S.; Du, Q.; Wang, J.; Yang, S.; et al. Development of miniaturized competitive immunoassays on a protein chip as a screening tool for drugs. *Clin. Chem.* **2005**, *51*, 368–375. [CrossRef]

42. Du, H.; Yang, W.; Xing, W.; Su, Y.; Cheng, J. Parallel detection and quantification using nine immunoassays in a protein microarray for drug from serum samples. *Biomed. Microdevices* **2005**, *7*, 143–146. [CrossRef]

43. Baratchi, S.; Khoshmanesh, K.; Sacristan, C.; Depoil, D.; Wlodkowic, D.; McIntyre, P.; Mitchell, A. Immunology on chip: Promises and opportunities. *Biotechnol. Adv.* **2014**, *32*, 333–346. [CrossRef]

44. Mou, L.; Jiang, X. Materials for microfluidic immunoassays: A review. *Adv. Healthc. Mater.* **2017**, *6*, 1601403. [CrossRef] [PubMed]

45. Wang, P.; Kricka, L.J. Current and emerging trends in point-of-care technology and strategies for clinical validation and implementation. *Clin. Chem.* **2018**, *64*, 1439–1452. [CrossRef]

46. Boyd-Moss, M.; Baratchi, S.; Di Venere, M.; Khoshmanesh, K. Self-contained microfluidic systems: A review. *Lab Chip* **2016**, *16*, 3177–3192. [CrossRef] [PubMed]

47. Dai, B.; Yin, C.; Wu, J.; Li, W.; Zheng, L.; Lin, F.; Han, X.; Fu, Y.; Zhang, D.; Zhuang, S. A flux-adaptable pump-free microfluidics-based self-contained platform for multiplex cancer biomarker detection. *Lab Chip* **2021**, *21*, 143–153. [CrossRef]

48. Hu, B.; Li, J.; Mou, L.; Liu, Y.; Deng, J.; Qian, W.; Sun, J.; Cha, R.; Jiang, X. An automated and portable microfluidic chemiluminescence immunoassay for quantitative detection of biomarkers. *Lab Chip* **2017**, *17*, 2225–2234. [CrossRef] [PubMed]

49. Sinha, A.; Tai, T.Y.; Li, K.H.; Gopinathan, P.; Chung, Y.D.; Sarangadharan, I.; Ma, H.P.; Huang, P.C.; Shiesh, S.C.; Wang, Y.L.; et al. An integrated microfluidic system with field-effect-transistor sensor arrays for detecting multiple cardiovascular biomarkers from clinical samples. *Biosens. Bioelectron.* **2019**, *129*, 155–163. [CrossRef]

50. Meany, D.L.; Hackler, L.; Zhang, H.; Chan, D.W. Tyramide signal amplification for antibody-overlay lectin microarray: A strategy to improve the sensitivity of targeted glycan profiling. *J. Proteome Res.* **2011**, *10*, 1425–1431. [CrossRef] [PubMed]

51. Zhang, S.; Garcia-D'Angeli, A.; Brennan, J.P.; Huo, Q. Predicting detection limits of enzyme-linked immunosorbent assay (ELISA) and bioanalytical techniques in general. *Analyst* **2014**, *139*, 439–445. [CrossRef] [PubMed]

*micromachines*

*Article*

# Toward Development of a Label-Free Detection Technique for Microfluidic Fluorometric Peptide-Based Biosensor Systems

Nikita Sitkov [1,*], Tatiana Zimina [1], Alexander Kolobov [2], Vladimir Karasev [1], Alexander Romanov [1], Viktor Luchinin [1] and Dmitry Kaplun [3,*]

[1] Department of Micro- and Nanoelectronics, Saint Petersburg Electrotechnical University "LETI", 197376 Saint Petersburg, Russia; tmzimina@gmail.com (T.Z.); genetic-code@yandex.ru (V.K.); event-horizon@mail.ru (A.R.); cmid_leti@mail.ru (V.L.)
[2] Institute of Highly Pure Biopreparations, 197110 Saint Petersburg, Russia; indolicidin@mail.ru
[3] Department of Automation and Control Processes, Saint Petersburg Electrotechnical University "LETI", 197376 Saint Petersburg, Russia
* Correspondence: sitkov93@yandex.ru (N.S.); dikaplun@etu.ru (D.K.)

**Citation:** Sitkov, N.; Zimina, T.; Kolobov, A.; Karasev, V.; Romanov, A.; Luchinin, V.; Kaplun, D. Toward Development of a Label-Free Detection Technique for Microfluidic Fluorometric Peptide-Based Biosensor Systems. *Micromachines* **2021**, *12*, 691. https://doi.org/10.3390/mi12060691

Academic Editor: Violeta Carvalho

Received: 27 April 2021
Accepted: 11 June 2021
Published: 13 June 2021

**Publisher's Note:** MDPI stays neutral with regard to jurisdictional claims in published maps and institutional affiliations.

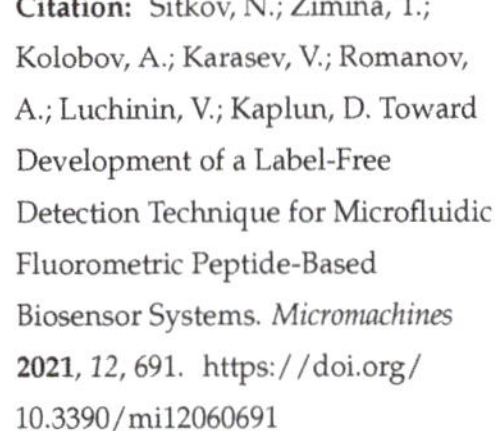



**Abstract:** The problems of chronic or noncommunicable diseases (NCD) that now kill around 40 million people each year require multiparametric combinatorial diagnostics for the selection of effective treatment tactics. This could be implemented using the biosensor principle based on peptide aptamers for spatial recognition of corresponding protein markers of diseases in biological fluids. In this paper, a low-cost label-free principle of biomarker detection using a biosensor system based on fluorometric registration of the target proteins bound to peptide aptamers was investigated. The main detection principle considered includes the re-emission of the natural fluorescence of selectively bound protein markers into a longer-wavelength radiation easily detectable by common charge-coupled devices (CCD) using a specific luminophore. Implementation of this type of detection system demands the reduction of all types of stray light and background fluorescence of construction materials and aptamers. The latter was achieved by careful selection of materials and design of peptide aptamers with substituted aromatic amino acid residues and considering troponin T, troponin I, and bovine serum albumin as an example. The peptide aptamers for troponin T were designed in silico using the «Protein 3D» (SPB ETU, St. Petersburg, Russia) software. The luminophore was selected from the line of ZnS-based solid-state compounds. The test microfluidic system was arranged as a flow through a massive of four working chambers for immobilization of peptide aptamers, coupled with the optical detection system, based on thick film technology. The planar optical setup of the biosensor registration system was arranged as an excitation-emission cascade including 280 nm ultraviolet (UV) light-emitting diode (LED), polypropylene (PP) UV transparent film, proteins layer, glass filter, luminophore layer, and CCD sensor. A laboratory sample has been created.

**Keywords:** peptide biosensor; lab-on-a-chip; label-free detection; peptide aptamers; protein biomarkers; microfluidic biochip; troponin T

## 1. Introduction

Noncommunicable diseases (NCDs) or chronic diseases kill around 40 million people each year, which is about 70 percent of all deaths worldwide [1]. NCDs include cardiovascular diseases, cancer, chronic respiratory diseases, diabetes, and many others. Such diseases require multiparametric combinatorial diagnostics for the selection of efficient treatment tactics [2]. This could be achieved by using biosensor systems based on peptide aptamers capable of selectively binding protein markers of diseases in biological fluids [3,4].

Currently, various instrumental methods are used in the diagnostics of NCDs. Thus, for example, in diagnostics of cardiovascular diseases, a vast range of techniques is applied: electrocardiography (ECG), echocardiography (echo-CG), computer tomography (CT), magnetic resonance imaging (MRI), stress tests (treadmill test, stress-echo), and laboratory diagnostic methods such as enzyme immunoassay, turbidimetry, etc. [5,6]. Instrumental techniques often require an accurate interpretation of the research results, which delays the diagnosis. Some are quite expensive, which limits their availability to a large number of patients [7]. Determination of molecular markers of cardiovascular diseases is an operative and informative diagnostic method, which is particularly important in treating acute conditions, enabling the early choice of a treatment strategy or surgery [7–9].

The biosensor according to the IUPAC, is "a device that uses specific biochemical reactions mediated by isolated enzymes, immunosystems, tissues, organelles or whole cells to detect chemical compounds usually by electrical, thermal or optical signals" [10].

Antibodies, nucleic acids, peptides, cells, etc. can be used as spatially complementary bio-recognition elements. The result of the bio-recognition process is converted into a measurable response due to various types of transducers. The most common are electronic and optical transducers. There are known examples of the implementation of biosensors using electrochemical, potentiometric, conductometric, impedimetric, voltammetric, capacitive methods, etc. [11,12]. The main feature of these methods is the modification of electrodes with various materials: graphene oxide [13], $MoS_2$ nanoflowers [14], mesoporous silicon [15], gold nanostructures [16,17], zinc oxide [18], various transistor structures [19–21], etc. In recent years, flexible electronics technologies for manufacturing biosensors have also gained in popularity [22]. Among them, electrochemical biosensors can achieve high selectivity and sensitivity, but their performance is limited by pH, ionic strength, and temperature of the biological fluid. In contrast, optical biosensors are more universal, since they are based on detecting changes of the incidental light frequency, phase, or polarization caused by biorecognition processes, and can be divided into categories such as photometric, fluorimetric, LIF (laser induced fluorescence), colorimetric, luminescence, light scattering, surface plasmon resonance (SPR), and other types of sensor principles, which could be implemented using fiber optics, nanomaterials, diffraction grids, and other designs [23]. To ensure high sensitivity in optical biosensors, special labels are often used, for example, organic dyes, quantum dots, or nanoclusters made of noble metals [24,25]. Label-free optical biosensor systems are also known [26–32]. In [28], microfluidic platforms capable of the quantitative detection of biomarkers in the format of point-of-care (POC) diagnosis are described. The versatility of the microfluidics-based system is demonstrated by implementation of the chemiluminescent immunoassay for quantitative detection of C-reactive protein (CRP) and testosterone in clinical samples. The concentration range for CRP determination was 50–300 $ng \times mL^{-1}$. The authors in [29] presented design and testing of a microfluidic platform for the immunoassay. The method is based on sandwiched ELISA, whereby the primary antibody is immobilized on nitrocellulose and, subsequently, magnetic beads are used as a label to detect the analyte. It takes approximately 2 h and 15 min to complete the assay. In [30], microfluidic platforms for disease biomarker detection are described, showing recent advances in biomarker detection using cost-effective microfluidic devices for disease diagnosis, with the emphasis on infectious disease and cancer diagnosis in low-resource settings for various biomarker detection using colorimetric, fluorescence, chemiluminescence, electrochemiluminescence (ECL), and electrochemical detection. In [31], a paper-based microfluidic device was described as an alternative technology for the detection of biomarkers by using affordable and portable devices for point-of-care testing (POCT). In [32], they describe a disposable electrochemical biosensor for the determination of cancer marker CA15-3 in blood. This procedure looks too complicated for express-testing, although a high sensitivity level has been achieved. All these articles describe methods with the involvement of immunoassays, which as has been pointed out above, makes them labor intensive and demanding for reagent storage conditions. It has been also noted by some authors that when antibodies are used as biorecognition elements in biosensors, maintaining high reproducibility can be problematic due to the cross-reactivity of the detection antibody and variability in the testing medium [33].

In this paper, a label-free biosensor system for the detection of protein biomarkers in biological fluids based on fluorometric registration of the target proteins bound to the peptide aptamer was proposed and investigated. The following goals were to be achieved: (1) Selection of materials for the design of the biosensor that do not emit secondary light in response to primary UV radiation, in the range of protein fluorescence (300–350 nm); (2) The study of emission and excitation spectra of luminophores (ZnS:X) with various activators, and the selection of luminophores re-emitting the fluorescence of proteins in the longer wavelength region, which is optimal for the receiving device; (3) Development of a technology to apply a luminophore layer to an optical window; (4) Design and testing of non-fluorescent peptide aptamers with substituted aromatic amino acid residues, spatially complementary to target protein biomarkers; and (5) Draft design of the biosensor with microfluidic transport element and casing with inlet and outlet. The proposed method of biosensor signal detection will make the diagnosis of NCDs more accessible and economical as well as convenient for use in point-of-care testing (POCT) systems and personalized medicine.

## 2. Materials and Methods

### 2.1. Protein-Markers

1. Cardiac troponin I (cTpI), T9924, Sigma-Aldrich (St. Louis, MO, USA).
2. Cardiac troponin T (cTpT), T0175, Sigma-Aldrich (St. Louis, MO, USA).

The spectral selection of the detection system was carried out using bovine serum albumin (BSA) obtained from Sigma-Aldrich (St. Louis, MO, USA), CAS Number: 9048-46-8.

### 2.2. Construction Materials

The following materials were tested: polyvinyl chloride (PVC), polypropylene (PP), polymethyl methacrylate (PMMA), and cover glass.

Polyvinyl chloride (PVC), Oracle Multi Color PVC Self Adhesive Film. PVC is a thermoplastic polymer with a melting temperature of 150–220 °C. At temperatures above 110–120 °C, it tends to decompose with the release of hydrogen chloride [34]. PVC is characterized by high biocompatibility. PVC is easily combined with practically all pharmaceutical products and also endures water and chemical reactions [35].

Polypropylene (PP) film, 40 μm thick. PE is a polymer material that has good dielectric properties and low absorbance. It is not soluble in most organic solvents. PP is a transparent polymer in a wide wavelength range from UV [36].

Polymethyl methacrylate (PMMA) is a thermoplastic transparent material. Disadvantages of PMMA are a tendency to surface damage (hardness 180–190 N/mm$^2$) and technological difficulties during thermoforming and vacuum forming, hence the appearance of internal stresses in places of bending during molding, which leads to the subsequent appearance of microcracks [37].

### 2.3. Inorganic Luminophores

The following inorganic photoluminophores were used—Green: ZnS:Cu, Blue: ZnS:Ag, and Orange: ZnS:CdS:Cu ("RPF "Luminofor" Corp.", Stavropol, Russian Federation) [38].

### 2.4. Apparatus

Spectrofluorimetric studies were carried out using an Agilent Cary Eclipse fluorescence spectrometer (Agilent Technologies, Santa Clara, CA, USA) [39].

Protein–peptide interaction analysis was studied using an Experion™ Automated Electrophoresis Station (Bio-Rad Laboratories, Hercules, CA, USA), which performs all of the steps of gel-based protein electrophoresis, microfluidic chips, and reagents to perform protein analyses.

The biosensor reader was equipped with a camera ToupCam 5.1 MP with a color CMOS sensor Aptina MT9P006 (C) with a resolution of 2592 × 1944 pixels and 5 MP photosensitive elements. The sensor size was 5.70 × 4.28 mm. Pixel size 2.2 × 2.2 μm.

Sensitivity at $\lambda$ = 550 nm was 0.53 W/lux-s. Dynamic range = 66.5 dB. ADC 12 bit parallel, 8-bit RGB. S/N = 40.5 dB. Spectral range: 380–650 nm.

UV-LED: 1 W, 275 nm, SMD 4545 Deep ultraviolet LG Chip (Korea): 5–9 V, 150 mA, angle 120–140 °, radiated power 8–10 mW.

### 2.5. Peptide Aptamers

The peptide aptamers were designed using data from Protein Data Bank [40] and "Protein 3D" software, developed at the Center of Microtechnology and Diagnostics of St. Petersburg Electrotechnical University «LETI» [41–43]. Two versions of peptides designed for troponin T were considered. The following requirements were put forward for the peptide aptamers being developed: complementarity to the target protein, selectivity, and absence of background fluorescence. Therefore, the presented sequences had differences in the substitution of aromatic amino acids. The sequences shown were synthesized and tested by capillary electrophoresis-on-a-chip. Preparation of peptides was carried out by solid state synthesis using «Applied Biosystems 430A» instrument and in situ method with N$\alpha$-Boc-protected amino acid residues.

### 2.6. Microfluidic Chip Fabrication Process

The biosensor was designed as a sandwich structure using thick film technologies. It consisted of a substrate made of a cover glass plate ($d \times l \times w$: 0.2 mm $\times$ 18 mm $\times$ 18 mm), on which a polymer pattern of microfluidic channels formed. The Ordyl Alpha 350 dry film photoresist layer ($d$ = 0.05 mm) was formed using standard photolithographic technology. The channels were hermitized with the polypropylene (PP) film ($d$ = 0.01 mm) using aa standard lamination technique. On the top of the PP layer, a diaphragm of PLA casing ($d$ = 0.4 mm) with a window for incidental UV light inlet was formatted using a 3D-printing technique and connected. The window was used to install the light filter to cut off the visible part of incidental light of the UV-LED. The other side of the glass substrate was covered with the luminophore layer for re-emission protein fluorescence in the range of visible light to which the CMOS-sensor is more sensitive. The luminophore layer was deposited using aerography technique from suspension.

The cover glass plates were treated with (3-aminopropyl)trimethoxysilane (APTMS) prior to the formation of microfluidic system topology relief in order to eliminate the nonspecific adsorption of test proteins on glass surfaces.

## 3. Results and Discussion

### 3.1. Designing Peptide Aptamers for Label-Free Detection

Peptide aptamers are combinatorial chain molecules consisting of about 15–35 amino acids, which have a high degree of affinity and specificity of attachment to the target protein molecule [44]. A design of peptide aptamers is a key problem in the development of a protein biosensor.

There are two main known groups of aptamers: oligonucleotides, most often obtained by the SELEX method [45], and peptide aptamers [46]. Since oligonucleotide aptamers do not always show sufficient selectivity for protein markers binding and demonstrate fluorescence at the UV excitation range, it seems reasonable to develop peptide aptamers, for which a number of modeling methods in silico [47,48] as well as experimental–combinatorial chemistry/high throughput analysis [49] methods are known.

The peptide aptamer search toolkit is growing and improving every year. As a resource for in vitro breeding, Ellington Lab developed the Aptamer Database, which collects catalogs of all published aptamers [49]. The database serves as a resource of information for creating aptamers for therapeutic, diagnostic, or scientific purposes.

Description of modern methods for the search and selection of peptide aptamers is presented in [50]. The described methods are mainly based on the improvement of SELEX technology and are generally very expensive and time-consuming [51].

An alternative to these methods are approaches based on computer modeling of protein structures and analysis of ligand interactions with target proteins. One of these approaches is molecular docking, which is performed by docking the ligand with the target protein binding site and moving it to determine the location and conformation that will be most beneficial for selective binding [47]. Computer simulation methods based on the principles of analysis of transient dynamic models are also used for this purpose. Based on time and length scales, these approaches can be divided into three main categories: quantum-mechanical modeling, atomic-molecular modeling, and mesoscale dynamics [52]. Monte Carlo methods and molecular dynamics are examples of atomic-molecular modeling [53]. On the other hand, the approach can be based on solving the Newtonian equation of motion in the conditions of the selected force fields, which determine the associated forces of molecular interactions. This approach was also effectively used to model biological structures and interactions between them [54]. A bead-based triple-overlay combinatorial strategy was proposed that can preserve inter-residue information during the screening process for a suitable complementary peptide to co-assemble target protein. The screening process commenced with a pentapeptide general library [55].

In this work, the peptide aptamers were designed in silico using the «Protein 3D» visualizer of supramolecular structures [41,56]. The software enables 3D structures of peptide chains to be visualized and the sites of mutual recognition to be identified. An important feature of the visualizer is the possibility of protein representation in the form of conjugated ion-hydrogen bond systems (CIHBS), considered as the basis for designing biostructures and energy transfer channels in these structures. The representation of proteins and peptides in the form of CIHBS has proved to be useful as a technology for in silico development of peptide aptamers. Based on this technology, a number of peptide sequences of 15–30 amino acid residues have been proposed and synthesized, and their ability of selective binding target proteins was experimentally tested using capillary electrophoresis [43,50].

To create the aptamer binding Troponin [57], a structural approach using the Protein 3D Supramolecular Structures Visualizer [44] was used to analyze the troponin complex. In addition, for the purpose of modifying the proposed sequence to replace aromatic amino acids that interfere with quantitative measurements, an attempt was made to use for this purpose the database of pentafragments developed at the CMID [42]. Protein Data Bank [40] contains files 1J1D and 1J1E of primary troponin complex structure, which is shown in Figure 1.

As can be seen in Figure 1a, troponin C is a globular molecule, and troponins T and I are extended $\alpha$-helices. The troponin C molecule has multiple contacts at the ends of two other troponin subunits. However, given the multiple nature of these contacts, the use of parts of the troponin C molecule as ligands is hardly advisable. Figure 1 shows that the troponins T and I molecules have a sharp bend of the helix structure, which is located in the region of 223–226 amino acid residues. In the area of bends between individual troponin molecules, a close contact begins and continues throughout a number of amino acids, and the helix of the troponin T molecule contacts two branches of troponin I, starting from 222 atoms and ending about 260. This can be well traced when analyzing the structure in the form of CIHBS in Protein 3D software (Figure 1b; selected by frame). The region of contact of troponin subunits is very specific, since hydrogen bonds are observed between the side chains of both molecules, contributing to their mutual recognition.

From the analysis of the CIHBS structure of the troponins complex, it follows that the troponin T region from amino acids 223 to 260 is promising for the creation of an aptamer on its basis, which would allow for the isolation of troponin T. The designed and synthesized peptides LETI-2 and LETI-7 are presented in Table 1. Peptide LETI-7 is designed as a non-fluorescent twin of LETI-2 by substituting aromatic amino acid residues Trp 237, Tyr 241 and 259, Phe 248, 257, and His 223. Sequence adjustments from 224 to 258 amino acids were performed using the computer modeling, in order to reduce the number of aromatic side chains. For these purposes, pentafragments containing aromatic amino acids were introduced into the program, and the program, based on the search for pentafragments, presented amino acid variants that can substitute a given once.

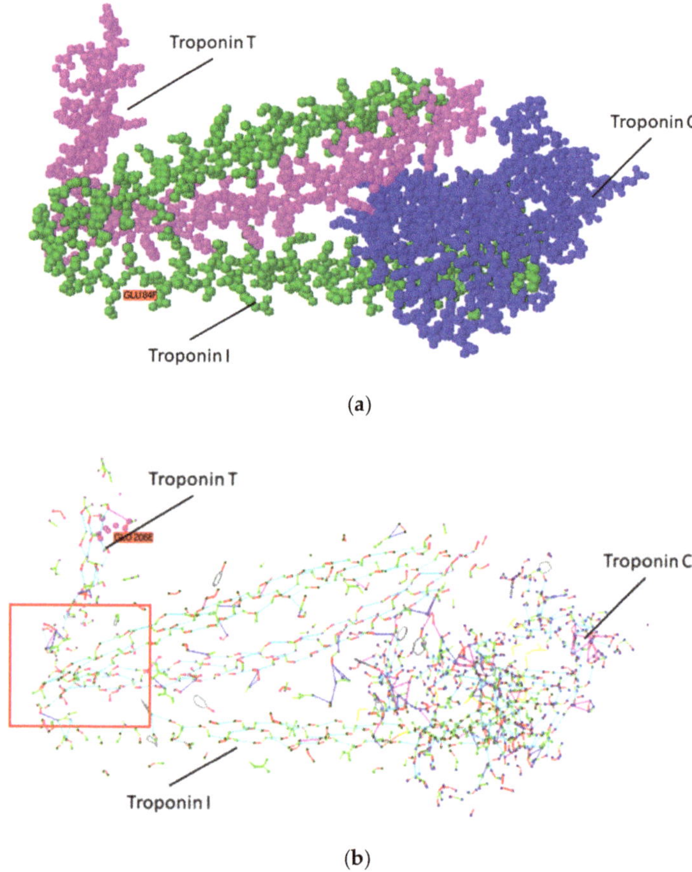


**Figure 1.** Structure of the troponin complex (**a**); troponin complex presented in the form of CIHBS rendering (**b**). Frame shows the area of contact between troponin T and troponin I, which is perspective for aptamer modeling.

**Table 1.** List of synthesized peptides.

| Name of Peptide | Structure | Number of Amino Acid Residues |
|---|---|---|
| LETI-2 | HLNEDQLREKAKELWQTIYNLEAEKFDLQEKFKQQKE | |
| LETI-7 | TLNEDQLREKAKELAQTIANLEAEKIDLQEKAKQQKYE | 38 |

Thus, the sample peptide aptamers were developed for troponin T (Figure 1) [50].

The peptide aptamer with substituted aromatic amino acids (LETI-7) was tested by capillary electrophoresis-on-a-chip with a Bio-Rad Experion station. Electropherograms of troponin T, peptide LETI-7, and their mixture are presented in Figure 2a. The data show that when peptides bind to the target protein, the elution time-shifts and the area of the aptamer zone decreases. Thus, the electrophoregrams presented in Figure 2a show the formation of a complex between LETI-7 and troponin T, which manifests itself in the change of migration time of the complex in comparison with pure troponin T. The electrophoretic study of interaction between the peptide aptamer LETI-7 and the off-target protein (troponin I) demonstrated (Figure 2b) that the elution time-shifts and the area of the aptamer zone did not change, which allows us to conclude that there is no binding between the peptide and the off-target protein (troponin I).

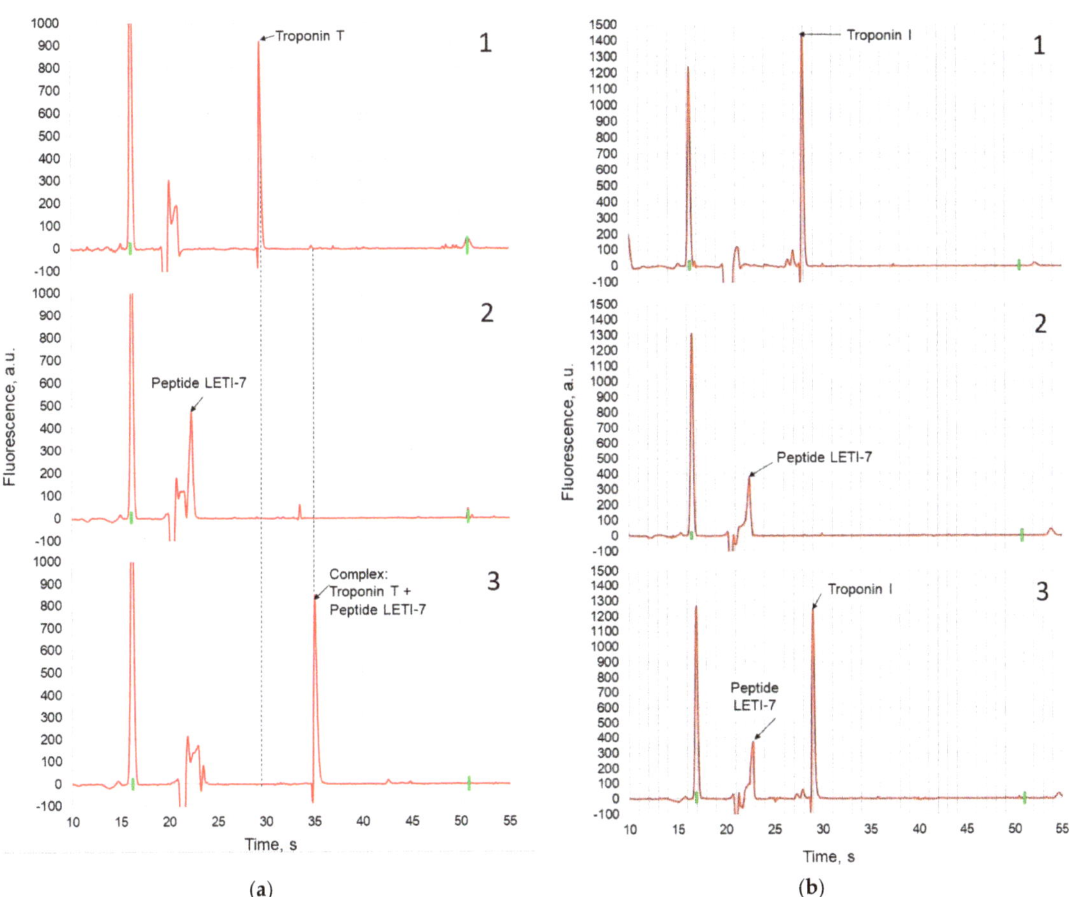


**Figure 2.** Capillary gel electrophoresis of peptide aptamer LETI-7 with troponin T (**a**) and Troponin I (**b**) using a Bio-Rad Experion station: 1—Protein, 2—Peptide LETI-7, 3—Mixture of peptide LETI-7 and protein.

### 3.2. Optical Scheme of Protein Detection in Biosensor Channels

In this work, we proposed to carry out direct detection of the fluorescence of biomarkers, selectively bound by peptide aptamers, without the use of special fluorescent labels (Figure 3a). The latter are labile and expensive substances and their exclusion from the protocol can increase the shelf life and ease the tests as well as reduce the cost of consumables. This is the novelty of the proposed technical solutions.

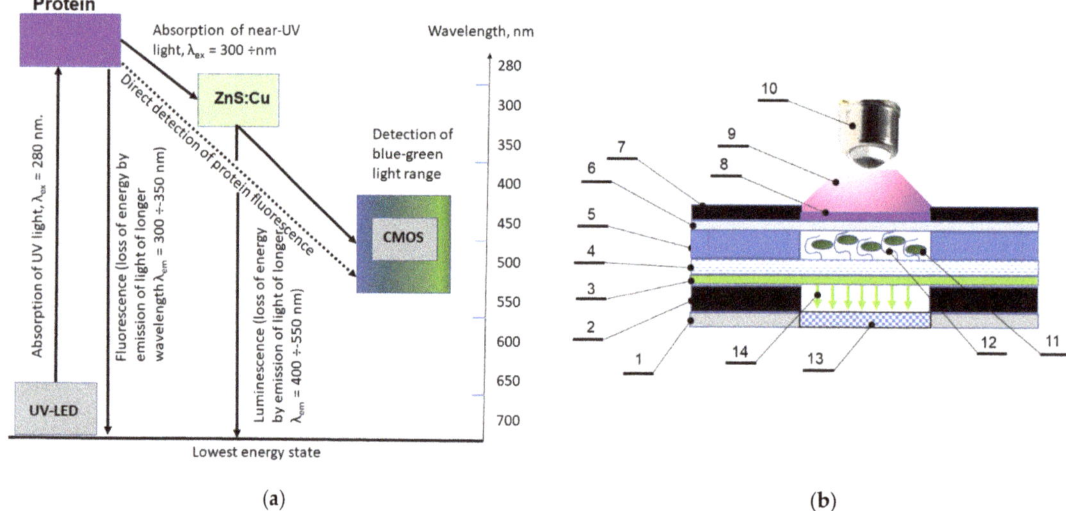

(a)                 (b)

**Figure 3.** Schematic presentation of the sensing principle (**a**) and relevant optical setup (**b**) optical setup for fluorimetric analysis using a biosensor connected to a video camera. 1—casing of CMOS camera; 2—PMMA base (d = 1.5 mm); 3—luminophore layer, 4—glass filter and base of microfluidic channel (d = 0.2 mm); 5—Ordyl Alpha 350 dry film photoresist casing of biosensor with microfluidic system relief, (d = 0.05 mm); 6 –PP film, hermitization layer, and UV window (d = 40 μm); 7—PLA (d = 0.4 mm) outer casing of biosensor; 8—filter; 9—UV light; 10—UV LED; 11—protein marker; 12—aptamer; 13—CMOS-sensor; 14—generated luminescence.

A design of the sandwich structure biosensor proposed is presented in Figure 4. The biosensor contains the luminophore layer (Figure 3b—layer 3), which is deposited on an outer surface of glass layer, thus sealing the microfluidic system of the biosensor, which is intended for the re-emission of near-UV light of protein fluorescence into the visible range of the spectrum in order to increase the sensitivity of the analytical method using a standard CMOS-sensor of the WEB camera. Three types of luminophores were tested (Figure 4) and ZnS:Cu (the green luminophore) was selected as a working component. The luminophore was deposited on the glass layer using acrylic lacquer. The spectral selection principle is presented in Figure 5 in which a coincidence of protein emission range and luminophore ZnS:Cu is demonstrated. The multilayer structure of the biosensor is presented in Figure 4, showing the topology of the 4-channel working area. The testing of the dynamic range of the biosensor was executed using BSA solution and the results are presented in Figure 5. At the present stage, the dynamic range of the sensor is about three orders.

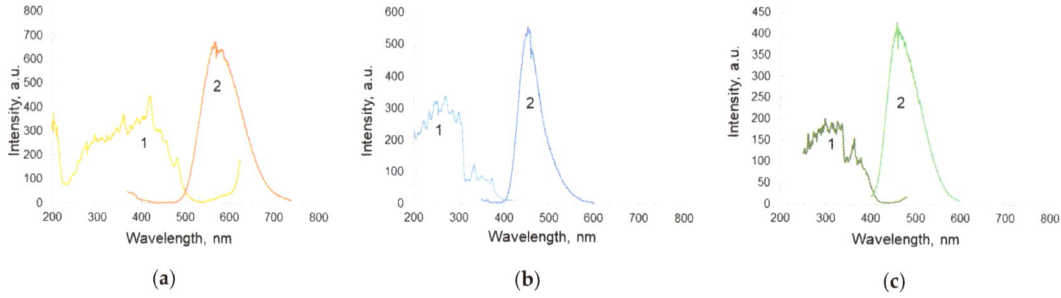

(a)                (b)                (c)

**Figure 4.** Excitation (**1**) and emission (**2**) spectra of ZnS:CdS:Cu (**a**); ZnS:Ag (**b**); and ZnS:Cu (**c**).

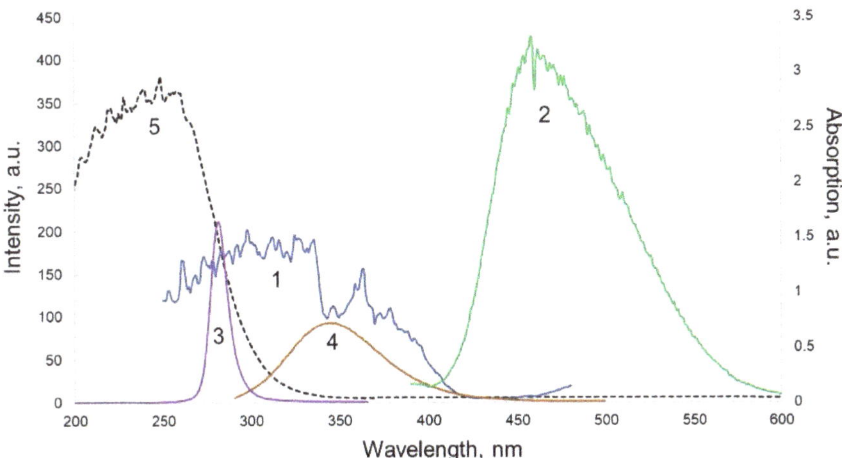

**Figure 5.** Excitation (**1**) and emission (**2**) spectra of the ZnS:Cu luminophore, UV-LED radiation spectrum (**3**) and fluorescence spectrum of BSA (**4**); (**5**) shows the range of spectrum absorbed by glass filter.

Fluorescence of most proteins is excited at the ultraviolet range of the spectrum of $\lambda$ = 220–280 nm. Proteins contain three amino acid residues that can contribute to fluorescence: tyrosine (Tyr), tryptophan (Trp), and phenylalanine (Phe). While tryptophan demonstrates the highest quantum yield of fluorescence, about 90% of all protein fluorescence is usually due to tryptophan residues. This natural fluorophore is extremely sensitive to the polarity of the environment [51]. Spectral shifts are often the result of several phenomena, among which ligand binding, protein–protein association, and denaturation can be distinguished. In addition, the emission maxima of proteins reflect the average availability of their tryptophan residues in the aqueous phase. Proteins absorb light in the vicinity of 280 nm, and the maxima of the fluorescence spectra lie in the region of $\lambda$ = 320–350 nm. The decay times of the fluorescence of tryptophan residues are in the range of 1–6 ns. Tyrosine fluoresces intensively in solution, but its fluorescence is much weaker in the composition of proteins. While maximum absorption of tyrosine and tryptophan occurs at a wavelength of 280 nm, phenylalanine is excited by shorter wavelengths, and the quantum yield of phenylalanine in proteins is low, so the fluorescence of this amino acid residue is negligible in the proposed detection scheme.

Since the purpose of this work was to study the possibility of implementing a highly sensitive analysis of protein markers of acute and chronic diseases on-a-chip without the use of fluorescent labels, and on the basis of intrinsic fluorescence of protein molecules in the ultraviolet range of radiation, it is obvious that to achieve this goal, biorecognition elements (peptide aptamers) immobilized on the chip should have minimal background fluorescence. These are not applied to nucleotide aptamers, since they fluoresce when excited by blue and ultraviolet radiation. Therefore, we use peptide aptamers, especially those not containing aromatic compounds.

Thus, in this study, luminophore was used to re-emit the fluorescence of proteins in the range of 300–350 nm into the visible range of 400–500 nm. As a model, BSA solution of various concentrations was used for calibration.

Luminophores are substances that luminesce under the influence of various kinds of excitations. The method of excitation is the basis for the classification of these substances, namely: photo luminophores—excited by light. The luminescent properties of inorganic substances are associated with the formation of structural and impurity defects in the crystal lattice of the compound during the synthesis process. The former appears as a result of thermal disproportionation of the luminophore lattice and represent vacancies and ions or atoms located in the interstices of the crystal lattice. The luminescence caused by such disturbances is called non-activated (or self-activated), since its formation does not require the introduction of an activating impurity. Impurity defects appear due to the introduction at high temperatures into the crystal lattice of ions or atoms of foreign elements. The luminescence caused by such disturbances is called activated, and the activating impurities are called activators. Most of the luminophores are activated. The concept of luminescence centers is associated with activators. The chemical state of activators in the crystal lattice and the structure of luminescence centers are still the subject of numerous studies. The excitation spectra of the luminophores show the dependence of the luminescence intensity on the wavelength of the excitation light. The emission spectra of luminophores show the distribution of the luminescence energy over wavelengths. According to Stokes law, the maximum of the emission spectrum is displaced with respect to the maximum of the absorption spectrum toward longer wavelengths. Sometimes the same activator in the same substrate produces emission bands located in different regions of the spectrum. For example, the ZnS:Cu emits in the blue, green, and red regions of the spectrum, depending on the concentration of copper and the conditions for preparation. The emission spectra can depend on the intensity and wavelength of the exciting light as well as on temperature [38].

Three types of luminophores are deposited on the glass in order to achieve a band shift from 350 nm to a visible band of the spectrum in order to receive the fluorescence using ordinary CMOS sensors. In Figure 4, the excitation/emission spectra of luminophores, namely ZnS:CdS:Cu, ZnS:Ag, and ZnS:Cu, are presented.

Furthermore, based on the data of the maximum fluorescence of the protein and taking into account that the range of effective reception of the CMOS sensor covers the region of 400–650 nm, it is necessary to choose the optimal luminophore for the biochip under study.

The maximum of the fluorescence spectrum of the ZnS:CdS:Cu luminophore corresponded to a wavelength of 580 nm, which is suitable for the CMOS registration of the signal. At the same time, the excitation spectrum (Figure 4a) had a maximum at 420 nm, which is of the fluorescence range of most proteins.

The maximum of the fluorescence spectrum of the ZnS:Ag luminophore corresponded to a wavelength of 450 nm, which is suitable for the range of possible wavelengths of effective detection with a CMOS sensor, while the excitation spectrum (Figure 4b) had a maximum at a wavelength of 430 nm, which is off the range of protein fluorescence.

The green luminophore ZnS:Cu corresponded to the research objectives to the greatest extent, since its excitation spectrum had a maximum precisely at the wavelength at which we observed the maximum emission from BSA (330 nm). The fluorescence spectrum of the ZnS:Cu luminophore (Figure 4c) had a maximum at a wavelength of 480 nm and covered the range of 400–550 nm, which is optimal for most CMOS sensors in webcams (Figure 6). Quantum efficiency of this luminophore is about 90%.

### 3.3. Preparation of Biosensor Model and Its Testing

The main components of the biosensor and its assembly are presented in Figure 6. The topology of microfluidic system, comprising inlet and outlet openings for sample load, and a main channel as a negative photomask is presented in Figure 6a. The corresponding structure formed using dry film photoresist Ordyl Alpha 350 is presented in Figure 6b. Assembly of the biochip in casing is presented in Figure 6c. In the biochip, the loaded sample should move in the channel of the microfluidic system to enter the active sites of the biochip, where the «target protein–peptide» complex is formed. An inlet opening is connected to a pump. The detection area (Figure 6d) was divided into four segments intended for the immobilization of the peptide aptamers. The chip casing was formed using a 3D printing technology. The luminophore layer was applied to the glass from the suspension liquid phase. Luminophore was grated in the mill, fractionated by sedimentation, and the fraction of 1–5 μm particles was suspended in acrylic lacquer and applied to a glass substrate using the airbrushing technique. All the glass surfaces were treated with (3-aminopropyl) trimethoxysilane, which at neutral pH ensures the sufficient neutrality of the glass surface.

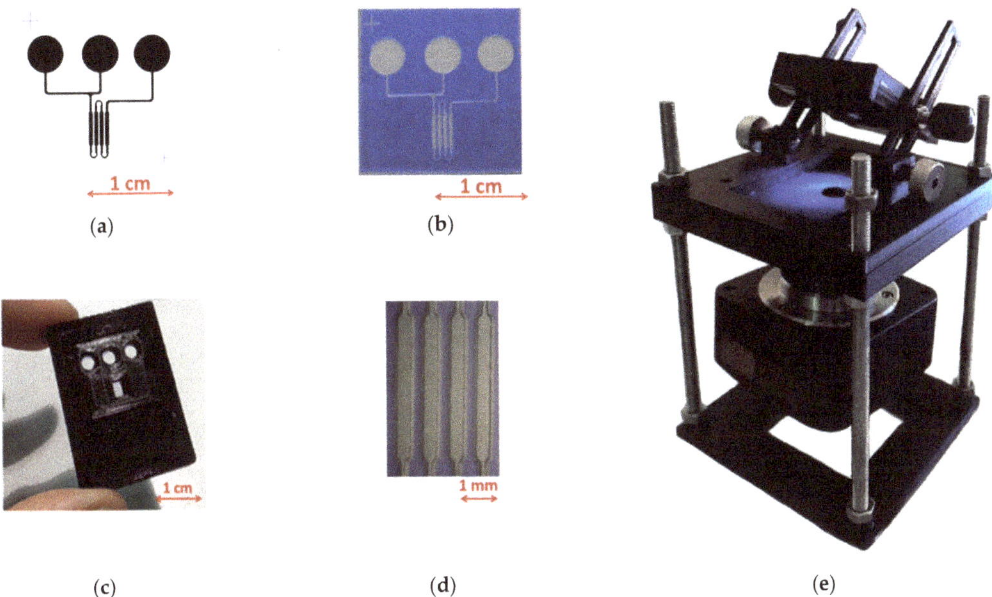

**Figure 6.** The biosensor design. Microfluidic system topology (**a**), SEM photograph of relief in the channel formed in photoresist film (**b**), view of biosensor (**c**), working channels (**d**), of the optical module for signal registration (**e**).

The biosensor assembly and reader were tested with the troponin T model. Channels 50 μm deep were filled with troponin T solutions in phosphate buffer at a concentration range of 6.4 ng/mL–800 ng/mL. The images of the working area for the channel filled with troponin T solution at a concentration of 8 μg/mL for assembly with clean glass (a) and with deposited luminophore layer (b) are presented in Figure 7a,b. The concentration relationships for the relative fluorescence units (RFU) for the signal recorded for clean glass (2) and for the biosensor with the deposited luminophore layer excited with differing concentrations of troponin T are presented in Figure 7c. The relative fluorescence values were calculated as follows: $RFU = (F_S - F_b)/(F_{Smax} - F_b)$, where $F_s$ is the fluorescence signal value for the sample; $F_b$ is the fluorescence signal value for the blank (buffer solution); and $F_{smax}$ is the fluorescence of the sample with the highest measured concentration. The relationship in Figure 7c has a linear character for concentrations and satisfies the linear equation: $y = 0.3769x - 0.1606$, with $R^2 = 0.9886$. The concentration limit of detection (LOD) for troponin T was 6.5 ng/mL, which was calculated by the standard procedure using standard deviation and slope values. The increase in sensitivity can be realized by reducing the background fluorescence, signal accumulation, and further spectral selection of the system.

Thus, Figure 7b shows the enhancement of the signal registration level due to the re-emission of fluorescence with luminophores into a longer wavelength range, where the CMOS sensor is more sensitive. The result was compared with direct detection of protein fluorescence without luminophore layer (Figure 7a) [58]. A relationship of the fluorescence intensity of the luminophore versus the concentration of protein in the working cell volume is presented in Figure 7c. The sensitivity achieved at this stage enables the quantitative detection of Troponin T, which is sufficient for the detection of most clinical levels of this biomarker in blood for diagnostics of AMI. The use of luminophore increases the sensitivity of the biosensor system by about an order of magnitude.

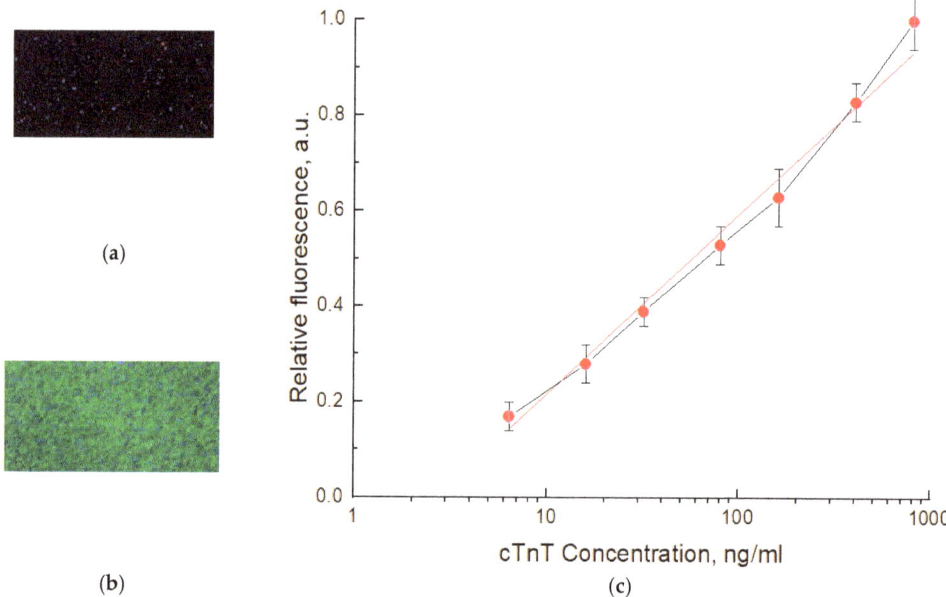

**Figure 7.** Images of the fragment of the biosensor working chamber filled with 800 ng/mL of troponin T solution recorded directly via glass window (**a**) and deposited onto the glass luminophore layer (**b**); relationship of relative fluorescence intensity versus concentration of cardiac troponin T for chips with the luminophore layer (**c**), where for curve, the red line indicates the linear approximation for the selected area.

## 4. Conclusions

In this paper, a spectral selection strategy was studied for direct label-free detection of the protein biomarkers of diseases in biological fluids for a polymer/glass microfluidic biosensor with peptide aptamers. In recent years, the interest is growing for the development of low-cost label-free methods for biomedical diagnostics. Thus, a number of papers have been presented describing the principles of such approaches [59–61]. In this work, we designed, in silico, the peptide aptamer for troponin T and its non-fluorescent digital twin. The main design of a biosensor with a polymer/glass microfluidic core element was developed and tested. The layer of solid state luminophore deposited on the glass window of the biosensor enabled more than one order of magnitude sensitivity increase to be achieved in the detection of the protein procedure. The spectral selection was targeted at defining the most efficient luminophore for which the excitation range of the spectrum coincides with the emission range of the protein, and luminescence of luminophore, in turn, fits the most sensitive range of a common web-cam CMOS sensor. Finally, the ZnS:Cu luminophore was selected and deposited using acrylic lacquer. The sensitivity of the system with re-emission was tested using troponin T solution. The concentration sensitivity of protein detection achieved at this stage was 6.5 ng/mL with the optical path length of 50 μm.

Thus, in accordance with the tasks set, based on the structural and technological features of the optical planar element, the following studies were completed:

1.  The fluorescence of a number of polymer materials was studied under excitation in the range of 260–280 nm, in order to select materials with minimal background fluorescence under radiation. The results showed that the selected materials meet the necessary requirements.
2.  A material for the preparation of the inlet window of the biosensor, transparent for UV radiation at 275 nm, was selected. The most fitting proved to be a PP film, 100 μm thick.

3. A peptide aptamer constructed using the «Protein 3D» software complementary to troponin T was designed. On its basis, its non-fluorescent digital twin was designed, which retained its three-dimensional complementarity to target protein, which has been shown with capillary electrophoresis-on-a-chip experiments.

4. For the outlet window transmitting fluorescence of troponin T for the channel depth of 50 μm in the range of 300–350 nm and absorbing incidental UV radiation at 275 nm, the most fitting material was glass. The cover glass plates were used for the preparation of this element.

5. Comparison of solid-state inorganic luminophores in order to select the optimal material for use in a planar optical element with an excitation range of 300–350 nm (Troponin T emission range) and an emission range of 450–550 nm (optimal for a receiving device) showed that a luminophore with a ZnS:Cu composition was the most consistent with the tasks.

6. A laboratory sample of the biosensor was designed and manufactured. Technology for applying a luminophore to glass was developed.

Further directions of the low-cost label-free system development are as follows: the development of improved miniature optical arrangements enabling a decrease in background light from the luminophore; improvement of software for signal capture and processing, and enabling signal accumulation during the capture process, etc.

**Author Contributions:** Conceptualization, N.S., T.Z., V.K. and A.K.; Methodology, T.Z., A.K., A.R., V.K., V.L. and N.S.; Software, D.K.; Validation, N.S. and T.Z.; Formal analysis, D.K. and V.L.; Investigation, N.S., T. Z., V.K. and A.K.; Resources, N.S. and T.Z.; Data curation, N.S., D.K. and T.Z.; Writing—original draft preparation, T.Z., N.S. and D.K.; Writing—review and editing, T.Z., N.S. and D.K.; Visualization, N.S., T.Z. and A.R.; Supervision, T.Z.; Project administration, V.L.; Funding acquisition, N.S. All authors have read and agreed to the published version of the manuscript.

**Funding:** The reported study was funded by RFBR, project number 19-32-90020.

**Acknowledgments:** The authors acknowledge with thanks A. P. Kablova for help with the preparation of illustrations.

**Conflicts of Interest:** The authors declare no conflict of interest.

# References


1. Daar, A.; Singer, P.; Leah Persad, D.; Pramming, S.K.; Matthews, D.R.; Beaglehole, R.; Bernstein, A.; Borysiewicz, L.K.; Colagiuri, S.; Ganguly, N.; et al. Grand challenges in chronic non-communicable diseases. *Nature* **2007**, *450*, 494–496. [CrossRef] [PubMed]

2. Hlapčić, I.; Belamarić, D.; Bosnar, M.; Kifer, D.; Dugac, A.V.; Rumora, L. Combination of systemic inflammatory biomarkers in assessment of chronic obstructive pulmonary disease: Diagnostic performance and identification of networks and clusters. *Diagnostics* **2020**, *10*, 1029. [CrossRef]

3. Reverdatto, S.; Burz, D.S.; Shekhtman, A. Peptide aptamers: Development and applications. *Curr. Med. Chem.* **2015**, *15*, 1082–1101. [CrossRef] [PubMed]

4. Madjid, M.; Safavi-Naeini, P.; Solomon, S.D.; Vardeny, O. Potential effects of coronaviruses on the cardiovascular system: A Review. *JAMA Cardiol.* **2020**, *5*, 831–840. [CrossRef] [PubMed]

5. Bax, J.J.; Baumgartner, H.; Ceconi, C.; Dean, V.; Fagard, R.; Funck-Brentano, C.; Kolh, P. Third universal definition of myocardial infarction. *J. Am. Coll. Cardiol.* **2012**, *60*, 1581–1598.

6. Johns Hopkins Medicine. Diagnosis and Screening for Cardiovascular Conditions. Available online: https://www.hopkinsmedicine.org/health/treatment-tests-and-therapies/diagnosis-and-screening-for-cardiovascular-conditions (accessed on 21 December 2020).

7. Deng, A.; Matloff, D.; Lin, C.E.; Probst, D.; Broniak, T.; Alsuwailem, M.; Belle, J.T.L. Development toward a triple-marker biosensor for diagnosing cardiovascular disease. *Crit. Rev. Biomed. Eng.* **2019**, *47*, 169–178. [CrossRef]

8. Aldous, S.J. Cardiac biomarkers in acute myocardial infarction. *Int. J. Cardiol.* **2013**, *164*, 282–294. [CrossRef]

9. Mythili, S.; Mythili, S. Diagnostic markers of acute myocardial infarction. *Biomed. Rep.* **2015**, *3*, 743–748. [CrossRef]

10. Nagel, B.; Dellweg, H.; Gierasch, L.M. Glossary for chemists of terms used in biotechnology (IUPAC Recommendations 1992). *Pure Appl. Chem.* **2013**, *64*, 143–168. [CrossRef]

11. Szunerits, S.; Mishyn, V.; Grabowska, I.; Boukherroub, R. Electrochemical cardiovascular platforms: Current state of the art and beyond. *Biosens. Bioelectron.* **2019**, *131*, 287–298. [CrossRef]

12. Dhara, K.; Mahapatra, D.R. Review on electrochemical sensing strategies for C-reactive protein and cardiac troponin I detection. *Microchem. J.* **2020**, *156*, 104857. [CrossRef]

13. Sharma, A.; Bhardwaj, J.; Jang, J. Label-free, highly sensitive electrochemical aptasensors using polymer-modified reduced graphene oxide for cardiac biomarker detection. *ACS Omega* **2020**, *5*, 3924–3931. [CrossRef]

14. Vasudevan, M.; Tai, M.J.; Perumal, V.; Gopinath, S.C.; Murthe, S.S.; Ovinis, M.; Mohamed, N.M.; Joshi, N. Highly sensitive and selective acute myocardial infarction detection using aptamer-tethered MoS2 nanoflower and screen-printed electrodes. *Biotechnol. Appl. Biochem.* **2020**, 1–10. [CrossRef]

15. Mi, X.; Li, H.; Tan, R.; Tu, Y. Dual-modular aptasensor for detection of cardiac troponin I based on mesoporous silica films by electrochemiluminescence/electrochemical impedance spectroscopy. *Anal. Chem.* **2020**, *92*, 14640–14647. [CrossRef]

16. Negahdary, M.; Heli, H. An electrochemical troponin I peptisensor using a triangular icicle-like gold nanostructure. *Biochem. Eng. J.* **2019**, *151*, 107326. [CrossRef]

17. Negahdary, M.; Behjati-Ardakani, M.; Heli, H. An electrochemical troponin T aptasensor based on the use of a macroporous gold nanostructure. *Microchim. Acta* **2019**, *186*, 377. [CrossRef]

18. Shanmugam, N.R.; Muthukumar, S.; Tanak, A.S.; Prasad, S. Multiplexed electrochemical detection of three cardiac biomarkers cTnI, cTnT and BNP using nanostructured ZnO-sensing platform. *Future Cardiol.* **2018**, *14*, 131–141. [CrossRef] [PubMed]

19. Kirste, R.; Rohrbaugh, N.; Bryan, I.; Bryan, Z.; Collazo, R.; Ivanisevic, A. Electronic biosensors based on III-nitride semiconductors. *Ann. Rev. Anal. Chem.* **2015**, *8*, 149–169. [CrossRef] [PubMed]

20. Sarangadharan, I.; Regmi, A.; Chen, Y.W.; Hsu, C.-P.; Chen, P.C.; Chang, W.H.; Lee, G.Y.; Chyi, J.I.; Shiesh, S.C.; Lee, G.B.; et al. High sensitivity cardiac troponin I detection in physiological environment using AlGaN/GaN high electron mobility transistor (HEMT) biosensors. *Biosens. Bioelectron.* **2018**, *100*, 282–289. [CrossRef] [PubMed]

21. Sinha, A.; Tai, T.Y.; Li, K.H.; Gopinathan, P.; Chung, Y.D.; Sarangadharan, I.; Ma, H.P.; Huang, P.C.; Shiesh, S.C.; Wang, Y.L.; et al. An integrated microfluidic system with field-effect-transistor sensor arrays for detecting multiple cardiovascular biomarkers from clinical samples. *Biosens. Bioelectron.* **2019**, *129*, 155–163. [CrossRef] [PubMed]

22. Sharma, A.; Jang, J. Flexible electrical aptasensor using dielectrophoretic assembly of graphene oxide and its subsequent reduction for cardiac biomarker detection. *Sci. Rep.* **2019**, *9*, 1–10.

23. Damborský, P.; Švitel, J.; Katrlík, J. Optical biosensors. *Essays Biochem.* **2016**, *60*, 91–100.

24. Chen, J.; Ran, F.; Chen, Q.; Luo, D.; Ma, W.; Han, T.; Wang, C.; Wang, C. A fluorescent biosensor for cardiac biomarker myoglobin detection based on carbon dots and deoxyribonuclease I-aided target recycling signal amplification. *RSC Adv.* **2019**, *9*, 4463–4468. [CrossRef]

25. Khan, S.; Hasan, A.; Attar, F.; Sharifi, M.; Siddique, R.; Mraiche, F.; Falahati, M. Gold nanoparticle-based platforms for diagnosis and treatment of myocardial infarction. *ACS Biomater. Sci. Eng.* **2020**, *6*, 6460–6477. [CrossRef] [PubMed]

26. El-Said, W.A.; Fouad, D.M.; El-Safty, S.A. Ultrasensitive label-free detection of cardiac biomarker myoglobin based on surface-enhanced Raman spectroscopy. *Sens. Actuators B Chem.* **2016**, *228*, 401–409. [CrossRef]

27. Diware, M.S.; Cho, H.M.; Chegal, W.; Cho, Y.J.; Kim, D.S.; Kim, K.S.; Paek, S.H. Ultrasensitive, label-free detection of cardiac biomarkers with optical SIS sensor. *Biosens. Bioelectron.* **2017**, *87*, 242–248. [CrossRef] [PubMed]

28. Hu, B.; Li, J.; Mou, L.; Liu, Y.; Deng, J.; Qian, W.; Sun, J.; Cha, R.; Jiang, X. An automated and portable microfluidic chemiluminescence immunoassay for quantitative detection of biomarkers. *Lab Chip* **2017**, *17*, 2225–2234. [CrossRef]

29. Bhalla, N.; Chung, D.W.Y.; Chang, Y.J.; Uy, K.J.S.; Ye, Y.Y.; Chin, T.Y.; Pijanowska, D.G. Microfluidic platform for enzyme-linked and magnetic particle-based immunoassay. *Micromachines* **2013**, *4*, 257–271. [CrossRef]

30. Sanjay, S.T.; Fu, G.; Dou, M.; Xu, F.; Liu, R.; Qi, H.; Li, X. Biomarker detection for disease diagnosis using cost-effective microfluidic platforms. *Analyst* **2015**, *140*, 7062–7081. [CrossRef] [PubMed]

31. Soum, V.; Park, S.; Brilian, A.I.; Kwon, O.S.; Shin, K. Programmable paper-based microfluidic devices for biomarker detections. *Micromachines* **2019**, *10*, 516. [CrossRef]

32. Oliveira, R.A.D.; Materon, E.M.; Melendez, M.E.; Carvalho, A.L.; Faria, R.C. Disposable microfluidic immunoarray device for sensitive breast cancer biomarker detection. *ACS Appl. Mater. Interfaces* **2017**, *9*, 27433–27440. [CrossRef] [PubMed]

33. Honikel, M.M.; Lin, C.E.; Probst, D.; Belle, J.T. Facilitating earlier diagnosis of cardiovascular disease through point-of-care biosensors: A review. *Crit. Rev. Biomed. Eng.* **2018**, *46*, 53–82. [CrossRef]

34. Aziz, S.B.; Brza, M.A.; Nofal, M.M.; Abdulwahid, R.T.; Hussen, S.A.; Hussein, A.M.; Karim, W.O. A comprehensive review on optical properties of polymer electrolytes and composites. *Materials* **2020**, *13*, 3675. [CrossRef]

35. Maheshwari, N.; Chatterjee, G.; Rao, V.R. A technology overview and applications of Bio-MEMS. *J. ISSS* **2014**, *3*, 39–59.

36. Chilukoti, G.R.; Periyasamy, A.P. Ultra High Molecular Weight Polyethylene for Medical Applications. 2012. Available online: https://www.researchgate.net/profile/Govardhan-Rao-Chilukoti/publication/298626009_Ultra_high_molecular_weight_polyethylene_for_medical_applications/links/589038ee92851c9794c4ecd3/Ultra-high-molecular-weight-polyethylene-for-medical-applications.pdf (accessed on 21 December 2020).

37. Umar, A.; Khairil, J.B.A.K.; Nor, A.B. A review of the properties and applications of poly (methyl methacrylate) (PMMA). *Polym. Rev.* **2015**, *55*, 678–705.

38. Chanturia, V.A.; Dvoichenkova, G.P.; Morozov, V.V.; Koval'chuk, O.E.; Podkamenny, Y.A.; Yakovlev, V.N. Experimental justification of luminophore composition for indication of diamonds in X-ray luminescence separation of kimberlite ore. *J. Min. Sci.* **2018**, *54*, 458–465. [CrossRef]

39. Cary Eclipse Fluorescence Spectrophotometer. Available online: https://www.agilent.com/en/product/molecular-spectroscopy/fluorescence-spectroscopy/fluorescence-systems/cary-eclipse-fluorescence-spectrophotometer (accessed on 22 December 2020).
40. Protein Data Bank. Available online: https://www.rcsb.org/ (accessed on 22 December 2020).
41. Visualizer of Supramolecular Biostructures «Protein 3D». Available online: http://protein-3d.ru/index_e.html (accessed on 21 December 2020).
42. Karasev, V. Data on the application of the molecular vector machine model: A database of protein pentafragments and computer software for predicting and designing secondary protein structures. *Data Brief* **2020**, *28*, 104815. [CrossRef]
43. Sitkov, N.O.; Zimina, T.M.; Karasev, V.A.; Lemozerskii, V.E.; Kolobov, A.A. Design of Peptide Ligands (Aptamers) for Determination of Myeloperoxidase Level in Blood Using Biochips. In Proceedings of the 2020 IEEE Conference of Russian Young Researchers in Electrical and Electronic Engineering, ElConRus, St. Petersburg and Moscow, Russia, 27–30 January 2020; Volume 2020, pp. 1599–1603.
44. Li, J.; Tan, S.; Chen, X.; Zhang, C.Y.; Zhang, Y. Peptide aptamers with biological and therapeutic applications. *Curr. Med. Chem.* **2011**, *18*, 4215–4222. [CrossRef]
45. Tuerk, C.; Gold, L. Systematic evolution of ligands by exponential enrichment: RNA ligands to bacteriophage T4 DNA polymerase. *Science* **2009**, *249*, 505–510. [CrossRef] [PubMed]
46. Groner, B.; Borghouts, C.; Kunz, C. Peptide aptamer libraries. *Comb. Chem. High Throughput Screen.* **2008**, *11*, 135–145. [CrossRef]
47. Antes, I. DynaDock: A new molecular dynamics-based algorithm for protein-peptide docking including receptor flexibility. *Proteins* **2010**, *78*, 1084–1104. [CrossRef]
48. Obarska-Kosinska, A.; Iacoangeli, A.; Lepore, R.; Tramontano, A. PepComposer: Computational design of peptides binding to a given protein surface. *Nucleic Acids Res.* **2016**, *44*, 522–528. [CrossRef]
49. Lee, J.F.; Hesselberth, J.R.; Meyers, L.A.; Ellington, A.D. Aptamer database. *Nucleic Acids Res.* **2004**, *32*, D95–D100. [CrossRef]
50. Ali, M.H.; Elsherbiny, M.E.; Emara, M. Updates on aptamer research. *Int. J. Mol. Sci.* **2019**, *20*, 2511. [CrossRef] [PubMed]
51. Agnew, H.D.; Coppock, M.B.; Idso, M.N.; Lai, B.T.; Liang, J.X.; McCarthy-Torrens, A.M.; Warren, C.M.; Heath, J.R. Protein-catalyzed capture agents. *Chem. Rev.* **2019**, *119*, 9950–9970. [CrossRef] [PubMed]
52. Gompetrs, R.; Renner, E.; Mehta, M. Enabling technologies for innovative new materials. *Am. Lab.* **2005**, *37*, 12–14.
53. Sim, A.Y.L.; Minary, P.; Levitt, M. Modelling nucleic acids. *Curr. Opin. Struct. Biol.* **2012**, *22*, 273–278. [CrossRef] [PubMed]
54. Karplus, M.; McCammon, J.A. Molecular dynamics symulations of biomolecules. *Nat. Struct. Biol.* **2002**, *9*, 646–652. [CrossRef] [PubMed]
55. Chan, K.H.; Lim, J.; Jee, J.E.; Aw, J.H.; Lee, S.S. Peptide-peptide co-assembly: A design strategy for functional detection of C-peptide, a biomarker of diabetic neuropathy. *Int. J. Mol. Sci.* **2020**, *21*, 9671. [CrossRef] [PubMed]
56. Sitkov, N.O.; Karasev, V.A.; Luchinin, V.V.; Zimina, T.M. Development of biosensors for express-detection of protein markers of diseases in blood using peptide biorecognition elements. *AIP Conf. Proc.* **2019**, *2140*, 020072.
57. Takeda, S.; Yamashita, A.; Maeda, K.; Maeda, Y. Structure of the core domain of human cardiac troponin in the $Ca^{2+}$-saturated form. *Nature* **2003**, *424*, 35–41. [CrossRef]
58. Ghisaidoobe, A.B.; Chung, S.J. Intrinsic tryptophan fluorescence in the detection and analysis of proteins: A focus on Förster resonance energy transfer techniques. *Int. J. Mol. Sci.* **2014**, *15*, 22518–22538. [CrossRef]
59. Jang, D.; Chae, G.; Shin, S. Analysis of surface plasmon resonance curves with a novel sigmoid-asymmetric fitting algorithm. *Sensors* **2015**, *15*, 25385–25398. [CrossRef] [PubMed]
60. Oliveira, N.; Souza, E.; Ferreira, D.; Zanforlin, D.; Bezerra, W.; Borba, M.A.; Arruda, M.; Lopes, K.; Nascimento, G.; Martins, D.; et al. A sensitive and selective label-free electrochemical DNA biosensor for the detection of specific dengue virus serotype 3 sequences. *Sensors* **2015**, *15*, 15562–15577. [CrossRef] [PubMed]
61. Wang, F.; Anderson, M.; Bernards, M.T.; Hunt, H.K. PEG functionalization of whispering gallery mode optical microresonator biosensors to minimize non-specific adsorption during targeted, label-free sensing. *Sensors* **2015**, *15*, 18040–18060. [CrossRef] [PubMed]

 *micromachines*

Article

# Development of an Automated, Non-Enzymatic Nucleic Acid Amplification Test

Zackary A. Zimmers [1], Alexander D. Boyd [1], Hannah E. Stepp [1], Nicholas M. Adams [1] and Frederick R. Haselton [1,2,*]

1   Department of Biomedical Engineering, Vanderbilt University, Nashville, TN 37212, USA;
    zackary.a.zimmers@vanderbilt.edu (Z.A.Z.); alexander.d.boyd@vanderbilt.edu (A.D.B.);
    hannah.e.stepp@vanderbilt.edu (H.E.S.); n.adams@vanderbilt.edu (N.M.A.)
2   Department of Chemistry, Vanderbilt University, Nashville, TN 37212, USA
*   Correspondence: rick.haselton@vanderbilt.edu

**Abstract:** Among nucleic acid diagnostic strategies, non-enzymatic tests are the most promising for application at the point of care in low-resource settings. They remain relatively under-utilized, however, due to inadequate sensitivity. Inspired by a recent demonstration of a highly-sensitive dumbbell DNA amplification strategy, we developed an automated, self-contained assay for detection of target DNA. In this new diagnostic platform, called the automated Pi-powered looping oligonucleotide transporter, magnetic beads capture the target DNA and are then loaded into a microfluidic reaction cassette along with the other reaction solutions. A stepper motor controls the motion of the cassette relative to an external magnetic field, which moves the magnetic beads through the reaction solutions automatically. Real-time fluorescence is used to measure the accumulation of dumbbells on the magnetic bead surface. Left-handed DNA dumbbells produce a distinct signal which reflects the level of non-specific amplification, acting as an internal control. The autoPiLOT assay detected as little as 5 fM target DNA, and was also successfully applied to the detection of *S. mansoni* DNA. The autoPiLOT design is a novel step forward in the development of a sensitive, user-friendly, low-resource, non-enzymatic diagnostic test.

**Keywords:** automation; non-enzymatic; DNA amplification; L-DNA; microfluidic; fluorescence

**Citation:** Zimmers, Z.A.; Boyd, A.D.; Stepp, H.E.; Adams, N.M.; Haselton, F.R. Development of an Automated, Non-Enzymatic Nucleic Acid Amplification Test. *Micromachines* **2021**, *12*, 1204. https://doi.org/10.3390/mi12101204

Academic Editor: Nam-Trung Nguyen

Received: 25 August 2021
Accepted: 30 September 2021
Published: 30 September 2021



## 1. Introduction

Nucleic acids are among the most important biomarkers of disease, and a large variety of diagnostic nucleic acid tests (NATs) have been developed to detect their presence in diagnostic settings, including polymerase chain reaction (PCR), loop-mediated isothermal amplification (LAMP), and rolling circle amplification [1–4]. Although very powerful, these tests rely on the use of enzymes, which poses several challenges for use at the point of care or in low-resource settings. For example, enzymes are typically the most expensive reagent in a NAT, and they require robust low-temperature storage conditions and labor-intensive sample preparation methods to achieve high sensitivity [5,6].

In response to these challenges, non-enzymatic NATs have been developed which amplify a target-induced signal using only thermodynamically-driven DNA hybridization reactions. Examples include hybridization chain reaction, entropy-driven catalysis, and catalyzed hairpin assembly [7–10]. These tests overcome the obstacles posed by enzymes, but typically have sub-optimal sensitivity due to their poor amplification. There have been several interesting demonstrations of non-enzymatic NATs that achieve increased sensitivity via exponential growth, including cascaded catalyzed hairpin assembly [11], branched or hyperbranched hybridization chain reaction [12,13], and dendritic amplification [14], but perhaps the most promising is the recent demonstration of a dumbbell DNA amplification scheme [15]. Each DNA dumbbell, shown in Figure 1, has four binding domains which bind the opposite dumbbell. The original authors performed the amplification assay by capturing target DNA on magnetic beads and sequentially incubating the beads with U1, then U2, then U1, etc. Over the course of 35 dumbbell incubations, a limit of detection of 5 copies/reaction was reported; this far surpasses limits of detection typical of other non-enzymatic NATs.

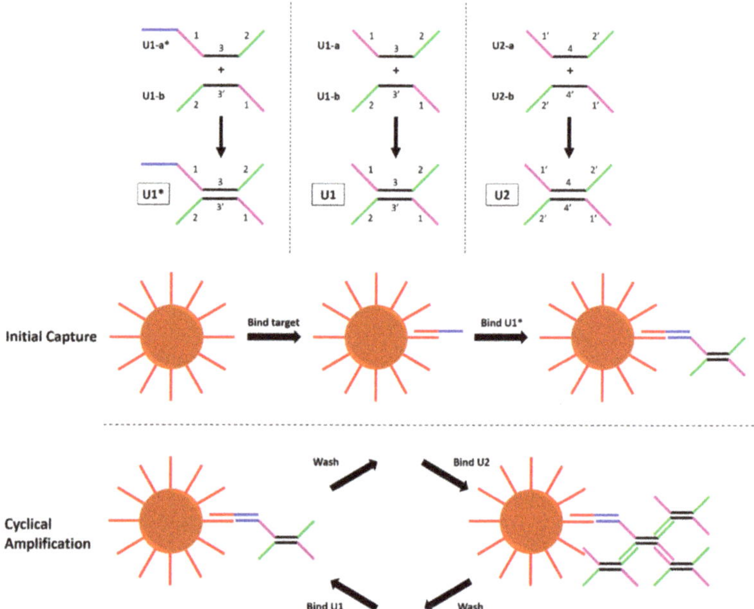

**Figure 1.** Overview of the dumbbell amplification assay. (**Top**) The DNA dumbbells used in this study. Complementary binding domains are denoted as 1, 1′. (**Bottom**) Binding steps in the dumbbell amplification assay. After binding target and U1*, the beads cycle between incubations of U1 and U2, with intermediate wash steps.

A serious drawback of this work which was not discussed is the amount of time, labor, and reagents required. Due to the single-stranded complementary binding domains, combining the two dumbbells in solution will cause uncontrolled binding to one another, rather than controlled accumulation on the surface of the magnetic beads. Therefore, the dumbbells must be added to the beads one at a time, with wash steps in between each incubation. To perform 35 successive dumbbell incubations, as well as the initial target capture step, approximately 18 h were required. Over the course of those 18 h, several pipetting steps were required every 30 min, each one a new opportunity for human error. These challenges are exacerbated by the lack of an internal negative control; a second parallel reaction was required to measure non-specific amplification, doubling the total amount of labor, time, and reagents required. These two obstacles, the undesirably high number of hands-on steps and the lack of an internal negative control, make the dumbbell DNA assay impractical for diagnostic applications. If the difficulties of performing this assay can be alleviated, it would have all the qualities of a powerful NAT appropriate for use in low-resource settings.

Previous works have pioneered the use of pre-arrayed reaction cassettes as a simplified and effective method of magnetic sample processing [16–18]. In this approach, different solutions are loaded into microfluidic tubing and separated by air gaps. The surface tension of these solutions maintains the integrity of the air gaps, and sealing the ends of the tubing immobilizes the contents. Magnetic beads are then transported through the different solutions via movement of an external magnet. Inspired by these methods, we designed an automated reaction processor to perform the dumbbell amplification assay without the need for repeated human pipetting steps. After the initial target capture step, the magnetic beads are loaded into a reaction cassette along with each dumbbell, and automatic movement of the tubing relative to a stationary magnetic field performs the steps previously performed manually. Fluorescent labels on the dumbbells create a real-time optical readout that can be measured throughout the reaction.

To eliminate the need for a second control reaction, left-handed dumbbells with their own distinct fluorescent labels are included as a built-in negative control. Left-handed DNA (L-DNA) is the chiral enantiomer of right-handed DNA (D-DNA). Although only D-DNA is found in nature, advancements in chemical synthesis techniques have enabled the commercial production of synthetic L-DNA. L-DNA is identical in chemical composition, and exhibits identical solubility, thermodynamic properties, and binding behavior as its right-handed counterpart [19–21]. Having previously demonstrated the utility of L-DNA as a built-in control for non-enzymatic DNA circuits [10], we used them here as a measure of non-specific amplification in the dumbbell amplification assay.

Finally, as a proof of principle of the potential diagnostic applications for this automated, non-enzymatic DNA amplification reaction, we demonstrated the detection of *Schistosoma mansoni* DNA. *S. mansoni* is the most widespread member of the family of parasitic flatworms which cause schistosomiasis, a leading neglected tropical disease responsible for the loss of 4.5 million disability-adjusted life years (DALYs) [22–24]. The target sequence is part of a 121 bp tandem repeat sequence which comprises approximately 12% of the *S. mansoni* genome, and has previously been targeted with PCR assays [25–27]. The PCR limit of detection was reported to be 1.28 pg/mL [27]; given the highly-repeated nature of the target sequence in the genome, this corresponds to approximately 790,000 copies/mL of the target sequence.

## 2. Materials and Methods

### 2.1. Oligonucleotides

All D-DNA sequences were purchased from Integrated DNA Technologies (Skokie, IL, USA), and all L-DNA sequences were purchased from Biomers (Ulm, Germany). Each dumbbell component was modified with a 5' FAM for D-DNA or a 5' Texas Red (TXR) for L-DNA. All fluorescently-labeled oligonucleotides were HPLC purified. A complete list of DNA sequences is given in Table 1, with distinct binding domains separated by underscores. Oligonucleotides were suspended in tris/EDTA (TE) buffer at a concentration of 100 μM and stored at $-20$ °C for long-term storage. For short-term storage, subsequent aliquots were created by diluting to 4 μM in 2X saline sodium citrate (SSC) buffer, and stored at 4 °C.

The dumbbell sequences (first 7 in Table 1) were adapted from a previous work by Xu et al. [15]. Sequences original to this work (last 5 in Table 1) are those designed to detect *S. mansoni* DNA, as well as 'U1-a* tag' and 'U1* removal', which were designed for a toehold-mediated strand displacement system to separate the dumbbells from the magnetic beads.

### 2.2. Design of the AutoPiLOT Reaction Processor

The automated Pi-powered looping oligonucleotide transporter (autoPiLOT) is an automated reaction processing device controlled by a Raspberry Pi B 3+ microcomputer. The Pi is connected via USB to two Arduino Unos, called the sensor Arduino and motor Arduino. Both Arduinos are connected to a 6 mm tactile button that, when pressed by the user, activates different segments of their onboard code. A circuit diagram is shown in Supplementary Materials. The master code for the Raspberry Pi, as well as the onboard code for both Arduinos, are available upon request.

Two infrared reflective object sensors (Digi-Key, QRD1114) are connected to the sensor Arduino for detection of liquid/air interfaces. A stepper motor (Applied Motion Products, STR4 drive, HT23 motor) is connected to the motor Arduino, which turns gears controlling the movement of the microfluidic tubing. Fluorinated ethylene propylene (FEP) tubing—

also commonly called PTFE tubing—with 1/8″ outer diameter and 3/32″ inner diameter (McMaster-Carr, 9369T24) was used to house the reaction contents. 5/8″ × 5/8″ × 1/4″ neodymium magnets (K&J Magnetics Inc, BAA4) were used to control the magnetic beads. A Qiagen ESElog fluorometer was used for fluorescence measurements. Housing for the tubing, magnets, and fluorometer was printed using a MarkForged Mark 2 3D printer, and the 3D drawing is also available upon request.

**Table 1.** Oligonucleotide sequences used in this work.

| Name | 5′ Mod | Sequence (5′–3′) |
| --- | --- | --- |
| U1-a | FAM/TXR | CTAGCTCATACATC_ATCCTATCTATCCAGAC_TCTCACACGTACTC |
| U1-a* | | TCGCTCTTACAAGGCA_CTAGCTCATACATC_ATCCTATCTATCCAGAC_TCTCACACGTACTC |
| U1-b | FAM/TXR | CTAGCTCATACATC_GTCTGGATAGATAGGAT_TCTCACACGTACTC |
| U2-a | FAM/TXR | GATGTATGAGCTAG_GAGATGCAATCGACTGT_GAGTACGTGTGAGA |
| U2-b | FAM/TXR | GATGTATGAGCTAG_ACAGTCGATTGCATCTC_GAGTACGTGTGAGA |
| Capture | Biotin | TTTTTTTTTT_CTCATTCACCTACG |
| Development target | | TGCCTTGTAAGAGCGA_CGTAGGTGAATGAG |
| U1-a* tag | | GTCAGTGA_TCGCTCTTACAAGGCA_CTAGCTCATACATC_ATCCTATCTATCCAGAC_TCTCACACGTACTC |
| U1* removal | | TGCCTTGTAAGAGCGA_TCACTGAC |
| *S. mansoni* capture | Biotin | TTTTTTTTTT_ATATTAACGCCCACG |
| *S. mansoni* target | | GATTATTTGCGAGAG_CGTGGGCGTTAATAT |
| *S. mansoni* U1-a* | | CTCTCGCAAATAATC_CTAGCTCATACATC_ATCCTATCTATCCAGAC_TCTCACACGTACTC |

The autoPiLOT is shown in Figure 2. The housing was 3D printed to hold the magnets, IR sensors, fluorometer, and microfluidic tubing. Not pictured is the Raspberry Pi microcomputer and Arduinos which control the device. The low cost and portability of these components make this device easily transportable and independent of a laptop or other computer for operation. The microfluidic tubing is held between the gears, which are controlled by a stepper motor. Rotation of the gears therefore moves the microfluidic tubing which contains the various reaction fluids, as previously described for one-directional sample prep devices [28,29]. The reaction cassette was moved at a speed of 0.2 cm/s to transport the magnetic beads, and 7.5 cm/s to break the beads out of the magnetic field.

To account for variations in air gap size or fluid volume during the preparation of the reaction cassette, a pair of infrared sensor/receivers was interfaced with the system to detect the locations of the liquid/air interfaces in the tubing. Before performing the cyclical amplification reaction, a "reconnaissance run" is performed which detects changes in IR transmittance as the tubing moves from one end to the other. These changes in transmission indicate a change from liquid to air, and the locations and sizes of each fluid chamber are mapped using this information. Because it uses these measurements to guide the movement of the tubing, the autoPiLOT reaction is adaptable to variations in the pre-loaded reaction cassettes.

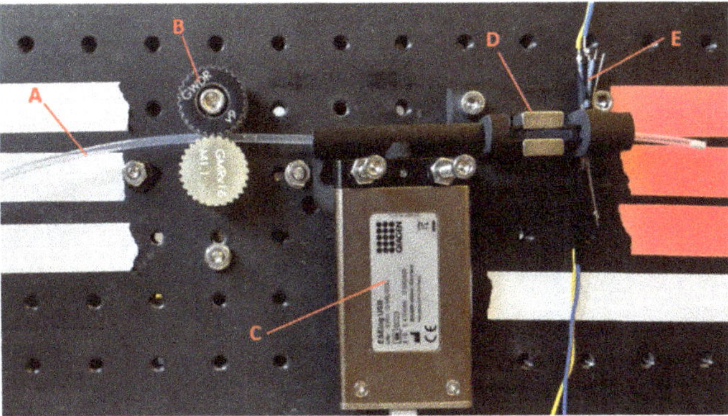

**Figure 2.** Photograph of the autoPiLOT reaction processor. Microfluidic polymer tubing (A) contains the reaction solutions. Two gears (B) controlled by a stepper motor move the tubing. A fluorometer (C) measures the fluorescence on the beads inside a light-tight chamber. Two neodynium magnets (D) hold the magnetic beads in their field. An IR sensor/receiver pair (E) detects the liquid/air interfaces in the tubing.

The movement of the magnetic beads between chambers is a result of the stationary magnetic field holding the beads as the tubing is moved by the gears. Tween 20 decreases the surface tension of the fluid chambers, which allows the beads to break through the interface. Several different concentrations of Tween were tested to see if they impacted the binding of the dumbbells to one another, and the results (shown in Supplementary Materials) show that none inhibited binding. Since dumbbell binding appeared equally efficient for all tested buffer conditions, 0.05% Tween in 5X SSC buffer was used in accordance with the protocol for the original dumbbell assay [15].

The movement of the magnetic beads through the four reaction chambers (shown in Figure 3) is as follows: from their starting place in the beads chamber, they are carried into U2. Here, the beads are mixed and then removed from the magnetic field (via a fast movement speed) and then incubated for 30 min. The beads are then recollected in the magnetic field and carried into the wash chamber, where they are dispersed across the length of the chamber. The wash chamber is moved across the optical path of the fluorometer for fluorescence readings, and then the beads are transported into U1. Again, they are mixed and incubated to bind dumbbells. The beads are then moved in the reverse direction to U2, and the cycle begins again. Fluorescence measurements are collected in the wash chamber after dumbbell incubations number 1, 3, 5, etc. All reaction steps after the incubation with dumbbell U1* are therefore automated.

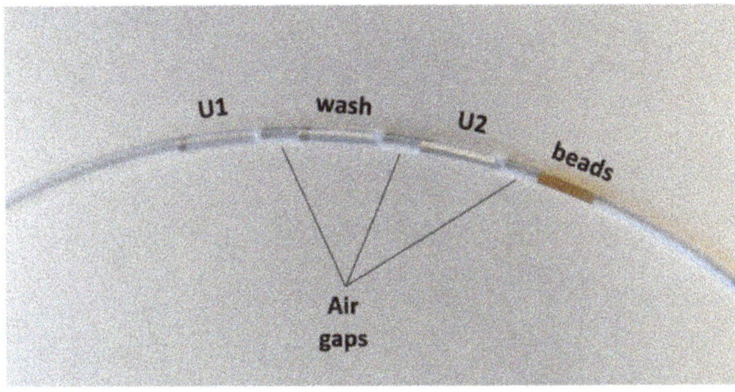


**Figure 3.** Picture of a pre-arrayed reaction cassette for use in the autoPiLOT reaction processor.

### 2.3. Gel Electrophoresis Studies

A series of experiments were conducted using agarose gel electrophoresis to investigate whether the predicted DNA hybridization events were taking place. Gels contained 3% ultra-pure agarose (Thermo Fisher) in 0.5X tris/borate/EDTA (TBE) buffer, and were stained with 1X GelRed nucleic acid stain (GoldBio, originally 10,000X in water). An ultra-low range DNA ladder (ThermoFisher) was included in each gel at a concentration of 2 ng/µL. Gels were run at 60V for approximately 1 h, then imaged using a Bio-Rad Gel Doc EZ Imager.

### 2.4. Dumbbell Formation

To form the double-stranded dumbbell structures, the two oligonucleotides (for example, U1-a and U1-b form dumbbell U1) were combined at equal molar ratios to the desired final concentration and then heated with the following thermal profile: 95 °C for 5 min, followed by 50 °C for 10 min, and finally by 37 °C for 10 min. This process was performed in a Qiagen Rotor-Gene PCR thermal cycler. The dumbbells were then stored at room temperature until use. For experiments using both D-DNA and L-DNA dumbbells, both D-DNA and L-DNA were combined in the same tube. For example, D-DNA U1-a and U1-b were combined with L-DNA U1-a and U1-b in the same tube for a final solution of right- and left-handed U1.

### 2.5. Magnetic Bead Functionalization

Dynabeads MyOne Streptavidin T1 magnetic beads (ThermoFisher) were functionalized with a biotinylated capture probe designed to bind the DNA target of interest. First, the beads were washed three times using hybridization buffer (5X SSC buffer + 0.05% Tween 20). Each wash consists of magnetically separating the beads from solution and, while separated, removing the supernatant and adding fresh buffer. After washing, the beads were resuspended in the biotinylated capture probe at a ratio of 2.5 nanomoles DNA per mg beads, briefly vortexed, and incubated on a rotisserie for 20 min. The beads were then washed three times with hybridization buffer and resuspended in blocking buffer (1X PBS + 16 µM free biotin + 0.05% Tween 20) and incubated on the rotisserie for 30 min. Experiments investigating other blocking methods used blocking buffers in which the free biotin was replaced with 2% bovine serum albumin (BSA) by volume, 200 µg/mL salmon sperm DNA, or removed altogether. The beads were then washed three times with hybridization buffer and stored in stocking buffer (1X PBS + 0.05% Tween 20) at a final concentration of 1 mg/mL. D-Biotin and BSA were purchased from Thermo Fisher, and Tween 20 was purchased from Sigma Aldrich.

### 2.6. Dumbbell Amplification Reactions

An amount of 50 µg of functionalized magnetic beads was used in each reaction. The beads were washed three times in hybridization buffer and resuspended in 100 µL of sample solution. The solution was vortexed to disperse the beads throughout and incubated on a rotating rotisserie. Sequence-specific nucleic acid capture via functionalized magnetic beads has previously been shown to be a slow process, with low efficiency for incubation times shorter than 30 min [17]. Therefore, 30 min was chosen as the incubation time. Next, 10 µL of U1* at 1 µM was added, and the beads incubated for another 30 min on the rotisserie. The beads were then washed three times in hybridization buffer, and resuspended in 75 µL hybridization buffer. At this point, the functionalized beads have bound target, which has in turn bound the first dumbbell U1*. Incubations in dumbbells U1 and U2 were then performed automatically using the autoPiLOT reaction processor.

The microfluidic FEP tubing was loaded with the following fluid chambers, each separated by a small air gap approximately 1 cm in length: 100 µL U1 at 250 nM, 100 µL hybridization buffer, 100 µL U2 at 250 nM, 100 µL magnetic bead solution. To load a fluid chamber into the tubing, the liquid was pipetted directly into the tube with the pipette tip flush against the opening. The tubing was then gently tipped to move the fluid chamber further along, clearing room for the next solution to be inserted. After the four fluid chambers were loaded in the tubing, both ends were sealed with Cha-seal tube sealing compound. The loaded tube (shown in Figure 3) was run on the autoPiLOT to move the magnetic beads back and forth between the two dumbbell solutions, reading fluorescence on the beads in the wash chamber after every other dumbbell incubation. The total reaction time was a function of the number of dumbbell incubations; 15 incubations on the autoPiLOT took approximately 8 h (15 incubations × 0.5 h/incubation, plus the time to move the beads back and forth between chambers).

## 3. Results

### 3.1. Validation of DNA Hybridization Events

Each dumbbell consists of two partially-complementary oligonucleotides; Figure 1 shows a full breakdown of each dumbbell and its components. Briefly, dumbbells U1 and U2 consist of U1-a, U1-b, U2-a, and U2-b, where a and b are used to denote the two strands in each dumbbell. U1* is a modified version of U1 with an extra binding domain on the 5′ end of U1-a to bind to the target DNA. U1* serves as the bridge between the target DNA and the network of dumbbells that forms during the amplification assay. Gel electrophoresis was used to validate that the individual dumbbell strands bind to form dumbbells as predicted.

For each of the three dumbbells, the respective single-stranded components were combined at equal molar ratios and visualized on a gel to confirm hybridization into double-stranded dumbbells; these results are shown in Figure 4A. With the exception of U1-a*, which is 60 bases, all of the single-stranded components are 45 bases in length. These can be seen as the lower bands on the gel. For dumbbells U1 and U1* (Figure 4A, left lanes 3 and 6), the lower band disappears when the two single-stranded components are combined and a new band is formed corresponding to 90 bases, showing that the two strands have hybridized to form the double-stranded dumbbell. The formation of dumbbell U2 appears incomplete, as there is still a prominent band at 45 bases indicative of remaining single-stranded DNA (Figure 4A, right lane 3). Even when different ratios of U2-a and U2-b were used, this extra band persisted (Supplementary Materials). The prominent band at 90 bases, however, is evidence that the dumbbell U2 is still forming, even if not quite as efficiently as U1 and U1*.

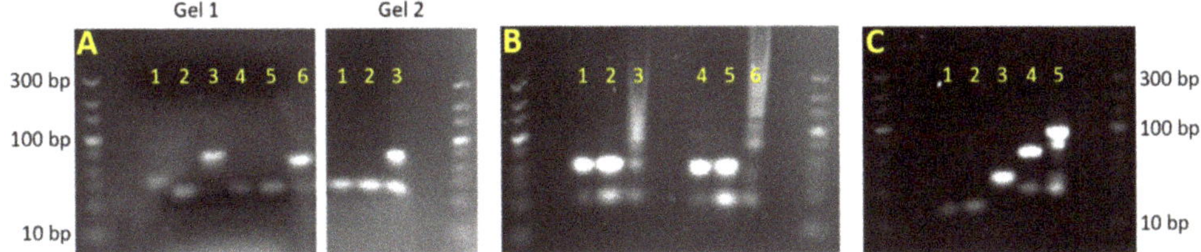


**Figure 4.** Gel electrophoresis results for D-DNA dumbbell binding studies. The same DNA ladder is used throughout, and shown at the right and left sides of each gel. (**A**) Gel 1, lanes 1-3: U1-a*, U1-b, and U1*. Gel 1, lanes 4-6: U1-a, U1-b, and U1. Gel 2, lanes 1-3: U2-a, U2-b, and U2. (**B**) Lanes 1–3: U1, U2, and U1+U2. Lanes 4–6: identical, but with 5′ FAM fluorescent labels. (**C**) Lane 1: capture probe. Lane 2: target. Lane 3: capture+target. Lane 4: U1*. Lane 5: capture + target + U1*.

Next, the binding of the dumbbells U1 and U2 to one another was examined. Figure 4B shows that once combined at equal ratios, the dumbbells form products of varying sizes. Distinct bands corresponding to complexes of 2, 3, and 4 dumbbells can be seen; the bands resulting from even larger complexes blend together to form a smear on the gel. This test was duplicated using FAM-labeled DNA to examine whether the presence of fluorescent labels had any apparent effect on dumbbell binding. The results (Figure 4B, right) show that the FAM-labeled dumbbells hybridized in a similar fashion as the unlabeled dumbbells. These findings validate three key assumptions moving forward: (1) the dumbbells form as expected from their single-stranded components, (2) the dumbbells bind to each other as expected to form large dumbbell complexes, and (3) attaching fluorescent labels to these dumbbells does not inhibit their affinity for one another.

After validating the D-DNA dumbbell binding events, the previous experiments were repeated using left-handed DNA. The L-DNA sequences are completely identical, the only differences are the chirality of the nucleotide bases and the replacement of FAM with Texas Red. As expected, the results (Supplementary Materials) show that the L-DNA dumbbells exhibit the same behavior as the D-DNA; the dumbbells assemble from their single-stranded components and bind to each other to form large dumbbell networks.

### 3.2. Validation of Target Capture by Magnetic Beads

There are two other key DNA binding steps in the dumbbell DNA amplification assay which were not examined in the previous gel experiments: the binding of target DNA to the capture probe-functionalized beads, and the binding of U1* to the target. Both of these binding events were also examined via gel electrophoresis. The results, shown in Figure 4C, show that the capture and target strands (lanes 1 and 2) hybridize to form a capture-target complex (lane 3). The addition of U1* (lane 4) then creates a capture-target-U1* complex (lane 5). This suggests that the magnetic beads which have been modified with capture probe will capture the target DNA out of solution, and subsequently bind the modified dumbbell U1*. The presence of U1* provides a starting point for cyclical dumbbell amplification. Microscopy was also used to confirm that the fluorescently-labeled U1* was indeed attached to the magnetic beads. Supplementary Materials show beads which have been incubated in target DNA, followed by the FAM-labeled U1*, using brightfield imaging and fluorescence imaging. The fluorescence image shows that FAM (in green) has attached to the beads, which confirms that the binding steps shown in Figure 4C also occur when bound to the surface of magnetic beads.

### 3.3. Fluorescence Measurements Reflect the Amount of DNA on the Beads

One of the features that makes quantitative PCR a gold standard NAT is the quantitative information gained by real-time fluorescence measurements. Fluorescence is read during each thermal cycle to create fluorescence vs. cycle data, and these data can be used to estimate the starting amount of target DNA. To achieve real-time fluorescence measurements in the autoPiLOT assay, fluorescence was measured directly on the surface of the beads. Capture probe-functionalized magnetic beads were incubated with varying amounts of FAM-labeled target DNA, shown in black in Figure 5A. The resulting fluorescence curve, shown in Figure 5B, demonstrates that the autoPiLOT fluorescence readings are directly proportional to the amount of DNA attached to the beads. Next, the same experiment was performed using unlabeled target DNA and followed by incubating with FAM-labeled U1*, as shown in Figure 5B. Just as each target had one FAM fluorophore in the previous design, each target should bind one FAM-labeled U1* in this design. The results (Figure 5B, red points) show that the two fluorescence curves are approximately equal at lower concentrations of target DNA. The U1* fluorescence tapers off at higher concentrations compared to that of the target. One possible explanation for this discrepancy is that when the beads already have a large amount of their surfaces covered in target DNA, the binding of additional DNA is inhibited by steric hindrance [30]. If true, this would suggest that the amplification observed in the autoPiLOT assay will also taper off as the bead surfaces become too overcrowded.

### 3.4. Performance of Parallel L-DNA Dumbbells

The amplification behavior of L-DNA control dumbbells was measured to validate that it was an accurate measure of the non-specific amplification of the D-DNA dumbbells. The fluorometer in the autoPiLOT reaction processor has two fluorescent channels, one that detects FAM on the D-DNA and one that detects Texas Red on the L-DNA. The first step was to ensure that the intensities of these two channels were adjusted such that D-DNA dumbbells produced a signal equal to that of L-DNA dumbbells. The gain of the Texas

Red channel was adjusted until the fluorescent signal was equal across multiple dumbbell concentrations (see Figure 6A). Moving forward, the magnitudes of fluorescent measurements are assumed to be directly comparable between D-DNA and L-DNA; matching measurements indicate matching amounts of dumbbells.

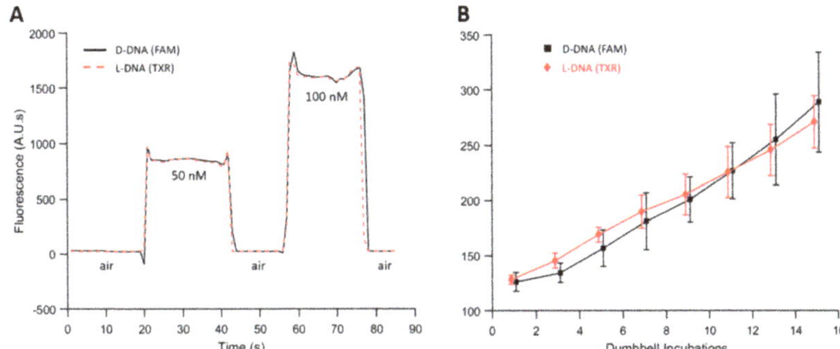

**Figure 5.** Matching D-DNA and L-DNA fluorescent signals in the autoPiLOT. (**A**) Reaction chambers containing D-DNA and L-DNA dumbbells at 50 and 100 nM were passed through the optical path of the autoPiLOT fluorometer. (**B**) No-target control reactions performed on the autoPiLOT measuring both D-DNA (black) and L-DNA (red) fluorescence. The mean of three trials ± one standard deviation is shown.

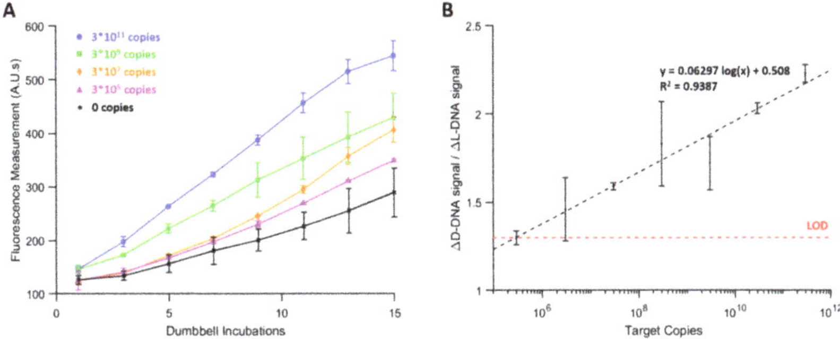

**Figure 6.** (**A**) D-DNA amplification curves for autoPiLOT trials for a range of target copy numbers. (**B**) Signal ratio, or the ratio of change in D-DNA signal to change in L-DNA signal after 15 dumbbell incubations, is plotted (black dots) as a function of target copy number. A logarithmic line best fit is overlaid as a black dashed line. The red dashed line represents the threshold signal ratio used to determine the limit of detection.

Next, negative control reactions were performed using the autoPiLOT. No target was bound by the beads, so only non-specific amplification was observed. The fluorescent measurements for both D-DNA and L-DNA dumbbells over the course of 15 dumbbell incubations are shown in Figure 6B. The amplification of both enantiomers is virtually identical; analysis using two-way analysis of variance (ANOVA) found that the chirality of the dumbbells had no significant effect on the observed variance ($\alpha < 0.05$). These findings confirm the hypothesis that the amplification of the left-handed dumbbells matches the non-specific amplification of the right-handed dumbbells.

Finally, autoPiLOT amplification was compared between samples with and without L-DNA dumbbells included. Samples contained $3*10^{11}$ copies of target to ensure that amplification was high, and any differences in performance caused by the presence of the L-DNA would be exaggerated. Over the course of 15 dumbbell incubations, the D-DNA amplification (shown in Figure 7) was found to be identical according to two-way ANOVA, $\alpha < 0.05$, regardless of whether or not L-DNA dumbbells have been included. The parallel

L-DNA dumbbells exhibit two key traits: they amplify at a rate equal to the non-specific amplification of the D-DNA dumbbells, and they do not impact the amplification of the D-DNA dumbbells. These findings demonstrate the utility of a dual-chirality design as a means to measure both target-induced amplification and non-specific amplification in a single reaction.

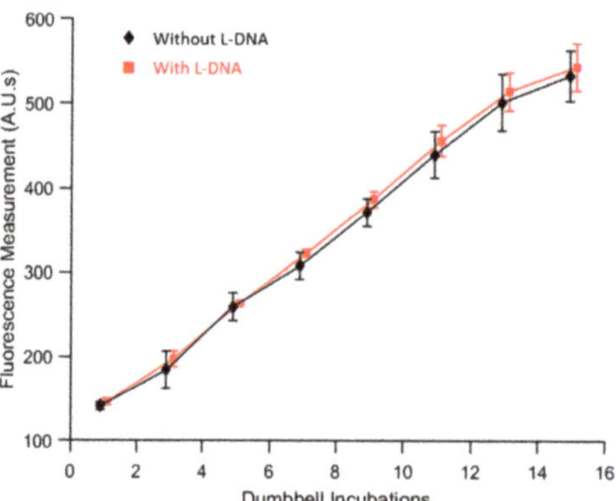


**Figure 7.** Comparison of D-DNA amplification with (red) and without (black) parallel L-DNA dumbbells included. The mean of three trials ± one standard deviation is shown.

### 3.5. AutoPiLOT Performance and Limit of Detection Studies

To determine the analytical limit of detection for the autoPiLOT reaction, serial dilutions of target DNA ranging from $3 \times 10^{11}$ to $3 \times 10^5$ copies per reaction were amplified for 15 cycles of 30 min incubations. The D-DNA amplification curves for several of these target concentrations are shown in Figure 8. Analysis using two-way ANOVA revealed that both target copy number and number of dumbbell incubations had significant effects on the resulting fluorescence measurements. The slopes of the amplification curves increase with increasing target DNA, resulting in increased fluorescent intensity, especially at later cycles.

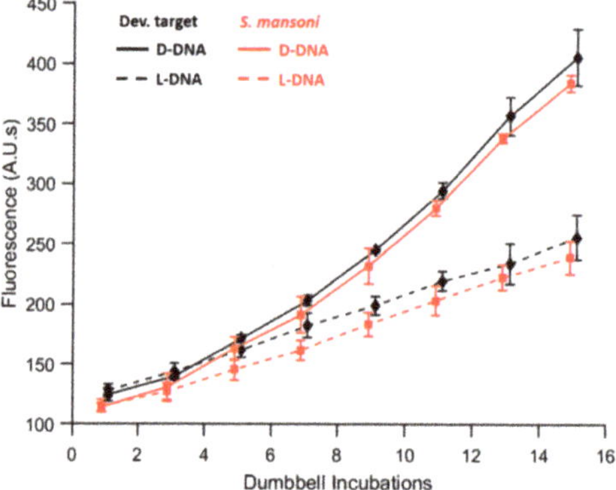


**Figure 8.** Comparison of autoPiLOT performance for detection of the development target sequence (black) and the *S. mansoni* target sequence (red). D-DNA signals are shown as solid lines, and L-DNA signals as dashed lines. The mean of three trials ± one standard deviation is shown.

To interpret the results of a dual-chirality dumbbell amplification reaction, the signals from both the D-DNA and L-DNA dumbbells were compared. This eliminates the need to perform no-target control reactions. The change in fluorescent signal can be calculated by simply subtracting the first fluorescent measurement from each subsequent measurement;

the ratio of change in D-DNA to change in L-DNA was used to compare the two signals. Let this ratio be called the signal ratio. The signal ratio changes with each new fluorescent measurement, but in general tends to increase with additional incubations. Figure 8 shows the signal ratio after 15 dumbbell incubations as a function of target copy number.

Diagnostically, to determine whether a given sample is positive, there must be a threshold signal ratio; an observed signal ratio greater than this threshold is interpreted as a positive result. A ratio of 1 may seem like a reasonable threshold, since D-DNA and L-DNA amplification are expected to be equal in the absence of target DNA. However, there is a certain level of expected noise that means a negative sample may still exceed a signal ratio of 1. To determine a threshold, three autoPiLOT reactions were performed on separate days, and the formula $\mu_0 + 3\sigma_0$ was used, where $\mu_0$ and $\sigma_0$ represent the mean and standard deviation of the signal ratio. Using this calculation, a signal ratio of 1.30 was determined as the threshold to determine positive results. This value is plotted as a red dashed line in Figure 8, and intersects with the linear line of best fit at $3 \times 10^5$ copies. Thus, the limit of detection of this reaction was determined to be approximately $3 \times 10^5$ copies/reaction, or 5 fM (pink line in Figure 8).

### 3.6. Application of the AutoPiLOT to Detection of S. mansoni DNA

The previous experiments have all used the originally-published development target sequence (see Table 1, target) to test the autoPiLOT platform on well-established DNA sequences. To apply this platform in a diagnostic setting, several DNA elements must be redesigned to integrate a new target sequence. As an example, DNA from the genome of *Schistosoma mansoni* was chosen as a new target. The stepwise binding motif of the dumbbell DNA assay means that dumbbells U1 and U2 do not need to be redesigned to integrate a new target sequence; only the target-binding domain on U1-a* and the capture sequence need to be changed to bind the desired target. New magnetic beads were functionalized using the *S. mansoni* capture probe, and the autoPiLOT assay was performed on samples containing $3 \times 10^7$ copies of the *S. mansoni* target. The results are shown in Supplementary Materials alongside the results for the original target sequence. The signal ratios after 15 cycles were calculated as $2.21 \pm 0.09$ for the original target and $2.18 \pm 0.23$ for the *S. mansoni* target, which are not significantly different based on a two-way *t*-test, $\alpha < 0.05$. These results suggest that the autoPiLOT assay can be applied to new targets without sacrificing performance. Although the limit of detection was not determined for the new target, the identical performance at a concentration of $3 \times 10^7$ copies/reaction suggests that it would be comparable to previously-determined limit of $3 \times 10^5$ copies/reaction.

## 4. Discussion

Real-time fluorescence measurements were shown to reflect the amount of DNA which had accumulated on the magnetic beads. Signal amplification was approximately linear, and the slope increased with increasing amounts of target DNA in the sample, as shown in Figure 7. The autoPiLOT assay exhibited a limit of detection of 5 fM, and showed promising results when applied to the detection of *S. mansoni* DNA, demonstrating the potential for a sensitive, low-cost, point-of-care non-enzymatic NAT.

The observed linear amplification is not in agreement with the previously-reported exponential dumbbell DNA amplification. This discrepancy is likely due to a combination of factors. When the beads are transported into a new chamber in the autoPiLOT, they are briefly mixed back and forth to disperse them across the length of the chamber. Despite this, the beads quickly settle to the bottom of the tubing where they remain for most of the 30 min. Dumbbell binding efficiency would likely improve if the beads were homogenously dispersed throughout the chamber; the effective exposed surface area of the beads would be much larger. Another potential cause of decreased amplification is steric hinderance from the surrounding dumbbells. The more dumbbells are attached to the bead surface, the more chaotic and intertwined the growing network of dumbbells becomes. This would result in amplification which becomes less efficient the more dumbbells are bound; this behavior is suggested in the amplification curves shown in Figure 8. At lower fluorescence values (in the 100–250 A.U. range), the amplification has some concavity indicative of exponential amplification. This is most visible for the $3 \times 10^5$ and $3 \times 10^7$ copy number curves. As the reaction progresses and more dumbbells attach to the beads, the curve becomes linear, and even begins to taper off at high enough values. Future work should focus on optimization of the autoPiLOT magnetic bead control strategies, in an effort to more efficiently bind dumbbells and achieve exponential amplification.

Future research should also work toward the implementation of a lower-cost fluorescence imaging system. The fluorometer used in the autoPiLOT is the only expensive component, with a price tag of several thousand dollars. Recently, smartphone-based fluorescence microscopes have grown in popularity and sophistication, bringing high quality digital cameras and multifunctionality to the point of care [31,32]. It is estimated that approximately 80% of the world population uses smartphones, making a smartphone optical readout highly applicable, even in low-resource settings [33].

An alternative strategy to avoid use of a costly fluorometer would be a colorimetric readout. The original demonstration of the dumbbell amplification assay utilized a colorimetric readout, in which dumbbells were labeled with biotin, and avidin-labeled horseradish peroxidase (HRP) was added to the beads at the end of the end of the reaction [15]. The drawback of this strategy is that it prohibits the use of two distinct readouts, one for D-DNA and one for L-DNA. The use of a dual-chirality design allowed for the monitoring of not only target signal, but also non-specific amplification in each reaction. Switching to a colorimetric readout simplifies the autoPiLOT components, but sacrifices the innovative dual-chirality functionality.

Reaction automation in the autoPiLOT removes the need for many hands-on pipetting steps, but does not by default decrease the total reaction time. Future research may also investigate shortening the overall autoPiLOT reaction time. We have shown (in Supplementary Materials) that the incubation period in each dumbbell can be greatly reduced while still observing signal amplification. The exact effect on signal-to-noise ratio, however, remains unknown. Further testing of several incubation times to determine a balance between reaction time and sensitivity would be useful for diagnostic applications in which time is limited. Additionally, a more flexible design in which the reaction runs only as long as necessary could be developed; samples with higher target concentrations would amplify quicker, terminating the reaction quicker than samples with less target.

Finally, the performance of the autoPiLOT diagnostic platform should be evaluated using real urine samples. The preliminary data presented here suggest that the autoPiLOT is an effective strategy to detect *S. mansoni* DNA, which has previously been detected in urine samples via PCR. The autoPiLOT demonstrated a limit of detection of 3,000,000/mL target copies, nearly as sensitive as previously described PCR assays. It is possible that reaction sensitivity will decrease when applied to DNA extracted from urine, since urine is known to contain inhibitors such as nucleases. In the case of nucleases, heating the urine to 75 °C for 10 min has been shown to deactivate nucleases in urine [10]. Sample-prep steps such as heating are likely required to maximize extraction efficiency from urine. Losses due to extraction efficiency may be countered by increasing sample volume; large volumes are easily obtained. These characteristics suggest that the autoPiLOT has the potential to be a highly-sensitive, low-resource, non-enzymatic NAT, and the first of its kind.

## 5. Conclusions

The dumbbell DNA amplification scheme has previously been shown to be a highly-sensitive, non-enzymatic method of detecting target DNA. The largest obstacles to its use in diagnostic applications were the extremely high number of manual pipetting steps and hands-on time required, and the need for an additional control reaction. The autoPiLOT platform has overcome these obstacles through automation of the reaction in a self-contained, microfluidic reaction cassette. The components are primarily 3D-printed parts and low-cost electronics, making the autoPiLOT a low-resource compatible diagnostic platform.

The autoPiLOT assay also demonstrates the utility of left-handed L-DNA as an internal control. Whereas a traditional assay format would require the performance of two parallel reactions, one with and one without target, for interpretation of results, the dual-chirality design used here discretely measured both specific and non-specific signals in the same reaction. In addition to saving time and reagents, this opens the door to testing unpurified biological samples which may exhibit varying rates of non-specific amplification that cannot be accurately simulated with a parallel control reaction.

**Supplementary Materials:** The following are available online at https://www.mdpi.com/article/10.3390/mi12101204/s1. Figure S1: Circuit diagram for the autoPiLOT device. Figure S2: The effect of buffer and mixing conditions on dumbbell binding. Figure S3: Gel electrophoresis results for U2 formation with varying U2-a U2-b ratios. Figure S4: Gel electrophoresis results for L-DNA dumbbell binding studies. Figure S5: Microscopic images of magnetic beads bound with fluorescently-labeled DNA. Figure S6: The effect of magnetic beads on fluorescence measurements. Figure S7: Overview of the TMSD strategy to remove DNA from magnetic beads. Figure S8: Comparison of different strategies to remove DNA from magnetic beads. Figure S9: Comparison of different bead blocking protocols effect on non-specific binding and amplification. Figure S10: The effect of dumbbell incubation time on signal amplification.

**Author Contributions:** Conceptualization, Z.A.Z., N.M.A. and F.R.H.; formal analysis, Z.A.Z. and H.E.S.; investigation, Z.A.Z., A.D.B. and H.E.S.; methodology, Z.A.Z.; software, A.D.B. and Z.A.Z.; supervision, F.R.H.; validation, Z.A.Z. and H.E.S.; visualization, Z.A.Z.; writing—original draft, Z.A.Z.; writing—review and editing, Z.A.Z., N.M.A. and F.R.H. All authors have read and agreed to the published version of the manuscript.

**Funding:** This research was funded by NIH/NHGRI, grant number R42HG009470, and by NIH/NIAID, grant number R01AI57827.

**Conflicts of Interest:** The authors declare no conflict of interest. The funders had no role in the design of the study; in the collection, analyses, or interpretation of data; in the writing of the manuscript, or in the decision to publish the results.

# References


1. Musso, D.; Roche, C.; Nhan, T.-X.; Robin, E.; Teissier, A.; Cao-Lormeau, V.-M. Detection of Zika virus in saliva. *J. Clin. Virol.* **2015**, *68*, 53–55. [CrossRef] [PubMed]
2. Vallejo, A.F.; Martínez, N.L.; González, I.J.; Arévalo-Herrera, M.; Herrera, S. Evaluation of the loop mediated isothermal DNA amplification (LAMP) kit for malaria diagnosis in P. vivax endemic settings of Colombia. *PLoS Negl. Trop. Dis.* **2015**, *9*, e3453. [CrossRef] [PubMed]
3. Watanabe, M.; Kawaguchi, T.; Isa, S.-I.; Ando, M.; Tamiya, A.; Kubo, A.; Saka, H.; Takeo, S.; Adachi, H.; Tagawa, T. Ultra-sensitive detection of the pretreatment EGFR T790M Mutation in non–small cell lung cancer patients with an EGFR-activating mutation using droplet digital PCR. *Clin. Cancer Res.* **2015**, *21*, 3552–3560. [CrossRef] [PubMed]
4. Yao, Q.; Wang, Y.; Wang, J.; Chen, S.; Liu, H.; Jiang, Z.; Zhang, X.; Liu, S.; Yuan, Q.; Zhou, X. An Ultrasensitive Diagnostic Biochip Based on Biomimetic Periodic Nanostructure-Assisted Rolling Circle Amplification. *ACS Nano* **2018**, *12*, 6777–6783. [CrossRef]
5. Chandrasekaran, A.R.; Punnoose, J.A.; Zhou, L.; Dey, P.; Dey, B.K.; Halvorsen, K. DNA nanotechnology approaches for microRNA detection and diagnosis. *Nucleic Acids Res.* **2019**, *47*, 10489–10505. [CrossRef]
6. Yun, W.; Jiang, J.; Cai, D.; Zhao, P.; Liao, J.; Sang, G. Ultrasensitive visual detection of DNA with tunable dynamic range by using unmodified gold nanoparticles and target catalyzed hairpin assembly amplification. *Biosens. Bioelectron.* **2016**, *77*, 421–427. [CrossRef] [PubMed]
7. Jung, C.; Ellington, A.D. Diagnostic applications of nucleic acid circuits. *Acc. Chem. Res.* **2014**, *47*, 1825–1835. [CrossRef] [PubMed]
8. Dirks, R.M.; Pierce, N.A. Triggered amplification by hybridization chain reaction. *Proc. Natl. Acad. Sci. USA* **2004**, *101*, 15275–15278. [CrossRef]
9. Fan, Z.; Yao, B.; Ding, Y.; Zhao, J.; Xie, M.; Zhang, K. Entropy-driven amplified electrochemiluminescence biosensor for RdRp gene of SARS-CoV-2 detection with self-assembled DNA tetrahedron scaffolds. *Biosens. Bioelectron.* **2021**, *178*, 113015. [CrossRef]
10. Zimmers, Z.A.; Adams, N.M.; Haselton, F.R. Addition of mirror-image L-DNA elements to DNA amplification circuits to distinguish leakage from target signal. *Biosens. Bioelectron.* **2021**, *188*, 113354. [CrossRef] [PubMed]
11. Chen, X.; Briggs, N.; McLain, J.R.; Ellington, A.D. Stacking nonenzymatic circuits for high signal gain. *Proc. Natl. Acad. Sci. USA* **2013**, *110*, 5386–5391. [CrossRef]

12. Liu, L.; Liu, J.-W.; Wu, H.; Wang, X.-N.; Yu, R.-Q.; Jiang, J.-H. Branched hybridization chain reaction circuit for ultrasensitive localizable imaging of mRNA in living cells. *Anal. Chem.* **2018**, *90*, 1502–1505. [CrossRef]

13. Bi, S.; Chen, M.; Jia, X.; Dong, Y.; Wang, Z. Hyperbranched hybridization chain reaction for triggered signal amplification and concatenated logic circuits. *Angew. Chem. Int. Ed.* **2015**, *54*, 8144–8148. [CrossRef]

14. Xuan, F.; Hsing, I.-M. Triggering hairpin-free chain-branching growth of fluorescent DNA dendrimers for nonlinear hybridization chain reaction. *J. Am. Chem. Soc.* **2014**, *136*, 9810–9813. [CrossRef] [PubMed]

15. Xu, G.; Zhao, H.; Reboud, J.; Cooper, J.M. Cycling of Rational Hybridization Chain Reaction To Enable Enzyme-Free DNA-Based Clinical Diagnosis. *ACS Nano* **2018**, *12*, 7213–7219. [CrossRef]

16. Bordelon, H.; Adams, N.M.; Klemm, A.S.; Russ, P.K.; Williams, J.V.; Talbot, H.K.; Wright, D.W.; Haselton, F.R. Development of a low-resource RNA extraction cassette based on surface tension valves. *ACS Appl. Mater. Interfaces* **2011**, *3*, 2161–2168. [CrossRef] [PubMed]

17. Adams, N.M.; Creecy, A.E.; Majors, C.E.; Wariso, B.A.; Short, P.A.; Wright, D.W.; Haselton, F.R. Design criteria for developing low-resource magnetic bead assays using surface tension valves. *Biomicrofluidics* **2013**, *7*, 014104. [CrossRef] [PubMed]

18. Adams, N.M.; Bordelon, H.; Wang, K.-K.A.; Albert, L.E.; Wright, D.W.; Haselton, F.R. Comparison of three magnetic bead surface functionalities for RNA extraction and detection. *ACS Appl. Mater. Interfaces* **2015**, *7*, 6062–6069. [CrossRef]

19. Urata, H.; Ogura, E.; Shinohara, K.; Ueda, Y.; Akagi, M. Synthesis and properties of mirror-image DNA. *Nucleic Acids Res.* **1992**, *20*, 3325–3332. [CrossRef] [PubMed]

20. Hauser, N.C.; Martinez, R.; Jacob, A.; Rupp, S.; Hoheisel, J.D.; Matysiak, S. Utilising the left-helical conformation of L-DNA for analysing different marker types on a single universal microarray platform. *Nucleic Acids Res.* **2006**, *34*, 5101–5111. [CrossRef] [PubMed]

21. Young, B.E.; Kundu, N.; Sczepanski, J.T. Mirror-Image Oligonucleotides: History and Emerging Applications. *Chem. Eur. J.* **2019**, *25*. [CrossRef] [PubMed]

22. Steinmann, P.; Keiser, J.; Bos, R.; Tanner, M.; Utzinger, J. Schistosomiasis and water resources development: Systematic review, meta-analysis, and estimates of people at risk. *Lancet Infect. Dis.* **2006**, *6*, 411–425. [CrossRef]

23. Hotez, P.J.; Brindley, P.J.; Bethony, J.M.; King, C.H.; Pearce, E.J.; Jacobson, J. Helminth infections: The great neglected tropical diseases. *J. Clin. Investig.* **2008**, *118*, 1311–1321. [CrossRef] [PubMed]

24. Colley, D.G.; Bustinduy, A.L.; Secor, W.E.; King, C.H. Human schistosomiasis. *Lancet* **2014**, *383*, 2253–2264. [CrossRef]

25. Sarhan, R.M.; Kamel, H.H.; Saad, G.A.; Ahmed, O.A. Evaluation of three extraction methods for molecular detection of Schistosoma mansoni infection in human urine and serum samples. *J. Parasit. Dis.* **2015**, *39*, 499–507. [CrossRef] [PubMed]

26. Hamburger, J.; Turetski, T.; Kapeller, I.; Deresiewicz, R. Highly repeated short DNA sequences in the genome of Schistosoma mansoni recognized by a species-specific probe. *Mol. Biochem. Parasitol.* **1991**, *44*, 73–80. [CrossRef]

27. Enk, M.J.; e Silva, G.O.; Rodrigues, N.B. A salting out and resin procedure for extracting Schistosoma mansoni DNA from human urine samples. *BMC Res. Notes* **2010**, *3*, 115. [CrossRef]

28. Creecy, A.; Russ, P.K.; Solinas, F.; Wright, D.W.; Haselton, F.R. Tuberculosis biomarker extraction and isothermal amplification in an integrated diagnostic device. *PLoS ONE* **2015**, *10*, e0130260. [CrossRef]

29. Russ, P.K.; Karhade, A.V.; Bitting, A.L.; Doyle, A.; Solinas, F.; Wright, D.W.; Haselton, F.R. A Prototype Biomarker Detector Combining Biomarker Extraction and Fixed Temperature PCR. *J. Lab. Autom.* **2016**, *21*, 590–598. [CrossRef] [PubMed]

30. Mahshid, S.S.; Camire, S.; Ricci, F.; Vallee-Belisle, A. A Highly Selective Electrochemical DNA-Based Sensor That Employs Steric Hindrance Effects to Detect Proteins Directly in Whole Blood. *J. Am. Chem. Soc.* **2015**, *137*, 15596–15599. [CrossRef]

31. Knowlton, S.; Joshi, A.; Syrrist, P.; Coskun, A.F.; Tasoglu, S. 3D-printed smartphone-based point of care tool for fluorescence-and magnetophoresis-based cytometry. *Lab A Chip* **2017**, *17*, 2839–2851. [CrossRef] [PubMed]

32. Sung, Y.; Campa, F.; Shih, W.-C. Open-source do-it-yourself multi-color fluorescence smartphone microscopy. *Biomed. Opt. Express* **2017**, *8*, 5075–5086. [CrossRef] [PubMed]

33. Ong, D.; Poljak, M. Smartphones as mobile microbiological laboratories. *Clin. Microbiol. Infect.* **2020**, *26*, 421–424. [CrossRef]

 *micromachines*

*Article*

# Osmotically Enabled Wearable Patch for Sweat Harvesting and Lactate Quantification

Tamoghna Saha, Jennifer Fang, Sneha Mukherjee, Charles T. Knisely, Michael D. Dickey * and Orlin D. Velev *

Department of Chemical and Biomolecular Engineering, North Carolina State University, Raleigh, NC 27695-7905, USA; tsaha@ncsu.edu (T.S.); jfang7@ncsu.edu (J.F.); smukhe22@ncsu.edu (S.M.); ctknisel@ncsu.edu (C.T.K.)
* Correspondence: mddickey@ncsu.edu (M.D.D.); odvelev@ncsu.edu (O.D.V.)

**Abstract:** Lactate is an essential biomarker for determining the health of the muscles and oxidative stress levels in the human body. However, most of the currently available sweat lactate monitoring devices require external power, cannot measure lactate under low sweat rates (such as in humans at rest), and do not provide adequate information about the relationship between sweat and blood lactate levels. Here, we discuss the on-skin operation of our recently developed wearable sweat sampling patch. The patch combines osmosis (using hydrogel discs) and capillary action (using paper microfluidic channel) for long-term sweat withdrawal and management. When subjects are at rest, the hydrogel disc can withdraw fluid from the skin via osmosis and deliver it to the paper. The lactate amount in the fluid is determined using a colorimetric assay. During active sweating (e.g., exercise), the paper can harvest sweat even in the absence of the hydrogel patch. The captured fluid contains lactate, which we quantify using a colorimetric assay. The measurements show the that the total number of moles of lactate in sweat is correlated to sweat rate. Lactate concentrations in sweat and blood correlate well only during high-intensity exercise. Hence, sweat appears to be a suitable biofluid for lactate quantification. Overall, this wearable patch holds the potential of providing a comprehensive analysis of sweat lactate trends in the human body.

**Keywords:** paper microfluidics; sweat; sensing; hydrogels; lactate; osmotic pumping; evaporation; capillary; wicking; biochemical assay

**Citation:** Saha, T.; Fang, J.; Mukherjee, S.; Knisely, C.T.; Dickey, M.D.; Velev, O.D. Osmotically Enabled Wearable Patch for Sweat Harvesting and Lactate Quantification. *Micromachines* **2021**, *12*, 1513. https://doi.org/10.3390/mi12121513

Academic Editors: Violeta Carvalho, Senhorinha Teixeira and João Eduardo P. Castro Ribeiro

Received: 28 October 2021
Accepted: 2 December 2021
Published: 4 December 2021

**Publisher's Note:** MDPI stays neutral with regard to jurisdictional claims in published maps and institutional affiliations.



## 1. Introduction

Lactate is a by-product of glycolysis, which is generated as an outcome of anaerobic glucose metabolism due to the high energy demand of the body. It is a pivotal biomarker for determining oxidative stress levels, muscle health, and tissue hypoxia [1–4]. Lactate is predominantly present in eccrine sweat [5], blood [3,6], and to a certain extent in tears [7] and saliva [8]. The net amount of lactate present in these biofluids is highly dependent on the physiological state and the site of anaerobic metabolism in the human body. Lactate monitoring is of prime importance in certain groups of individuals, especially those who are frequently prone to oxygen-deficit conditions such as athletes and military personnel [3] (for improving training and performance endurance). Critical lactate levels in the body can abruptly alter the fluid pH and cause several detrimental effects on human health such as seizures, sepsis, renal failure, tumors [9], cerebral stroke [10], and panic disorders [11].

Lactate content in eccrine sweat depends on several parameters such as the exertion type, duration, intensity, and sweat rate [11]. Medium-intensity exercise (60–70% of maximum heart rate) for a short duration, or heat-induced sweating can generate a lactate concentration of up to 25 mM, while exhaustive outdoor exercises could generate up to 115 mM [12–16]. On a contrary note, it is also proven that the sweat lactate concentration varies inversely with sweat rate [17]. This suggests that with rigorous physical activity, the rate of sweat release is greater than the rate of lactate generation. There is also evidence that suggests sweat lactate levels vary minimally with age [15], sex [15,18], and the sampling location (due to varying sweat gland density) [6,19]. Apart from eccrine sweat glands, there two other types of sweat glands in the body have been identified: apocrine and apoeccrine [5]. However, there is less evidence of lactate being present in the sweat secreted by these glands, and our focus here is on eccrine glands in the forearms as a commonly accepted lactate testing location.

Similar to most other biomarkers, researchers have looked upon the correlation between the lactate concentration in sweat and blood. Alam et al. have evaluated that blood lactate concentration ranges between 0.5 and 2 mM for a healthy subject at rest [3]. During a low to medium-intensity exercise workout, the blood lactate concentration neither shows a significant change in its magnitude nor reflects a good correlation with sweat lactate concentration [13,20]. Goodwin et al. have shown that once the body crosses its lactate threshold (LT), blood lactate concentration rises exponentially and can peak up to 11 mM [21]. Such high blood lactate can inhibit muscle contractions and is undesirable. A study by Karpova et al. has shown a positive correlation between blood and sweat lactate levels during high-intensity exercise [1]. Thus, the extent of lactate generation in both sweat and blood are well understood under active sweating. However, to the best of our knowledge, there are no literature reports of studies that simultaneously monitor sweat and blood lactate levels under low sweating conditions or after a sufficiently long-time post exercise.

Researchers have been able to quantify sweat and blood lactate levels using different sensing and quantification techniques [7,11–54]. Established commercial and laboratory-based sensing platforms function in the presence of an enzyme (either lactate oxidase ($LO_x$) or lactate dehydrogenase (LDH)). Earlier skin-interfacing wearable sensor prototypes developed on colorimetric principles typically operate by transporting sweat through complex serpentine microfluidic channels to a sensing zone, which contains a lactate specific enzymatic assay. The incoming lactate in the sweat reacts and generates a colored product whose intensity is quantified via phone-based image analysis. This intensity directly correlates to the concentration of lactate in sweat [22–25].

Another class of wearable device prototypes involve electrochemical measurements where lactate concentration is determined based on the magnitude of the current generated at the functionalized electrode surface [26–51]. Such sensing platforms usually comprise a three-electrode system where the enzyme is immobilized on the working electrode either via cross-linking, adsorption, electrostatics, or covalent bonding [3]. Lactate in sweat can be also quantified using chemiluminescence [52,53], HPLC [30,55], or enzyme free colorimetry [47].

Despite delivering on-spot sweat lactate detection with high sensitivity, wide linear dynamic range (LDR), and low limit of detection (LOD), the current wearable platforms face the following three key challenges: they (1) remain inoperative under the conditions of normal human activity—which rarely involves profuse sweating [7,8,12,22–25,40–42, 44,48,52]; (2) do not provide a comprehensive analysis of the human lactate metabolism and its correlation to blood lactate levels [27,30,34,36,40–42,46–48,50,56]; and (3) require an external power source for continuous operation [8,12,36,40–42,47,48,50–52,54]. Thus, most of the current sweat lactate sensors function only under "active sweating"; i.e., they either require the subjects to undergo strenuous physical exertion, ingest sweat stimulating agents [36,57], or elevate body temperature [13] prior to testing. As a result, these prototypes can quantify the lactate concentration in sweat only during exercise and not under any other physiological condition (such as rest or post-exercising conditions). Furthermore, conducting measurements via active sweating is not only inconvenient and tedious but also raises questions concerning its clean capture, external contamination, low sweat volumes due to uncontrolled evaporation, and dilution of biomarkers because of excessive sweating. Some prototypes have quantified sweat lactate release from rested subjects [34,36,58]. However, these prototypes are not a wearable platform, have a low sampling duration, do not report their sweat lactate results post-exercise, and do not study the correlation between blood and sweat lactate levels. Hence, there is a necessity of an efficient sweat lactate monitoring device, which is capable of delivering information under all physiological conditions (rest, during exercise, and post-exercise), and providing a comprehensive understanding of the human lactate metabolism remains unaddressed.

This article discusses the on-skin performance of our recently developed zero-powered, non-invasive wearable sweat sampling patch [14,59,60]. We use three different pumping materials (based on different osmotic strength) in our patch for sampling sweat lactate from the forearm region of the body. We chose the forearm, since it is standard for such measurements and has one of the highest sweat gland densities in the body [5]. The patches were tested under five different physiological conditions (rest, medium-intensity exercise, high-intensity exercise, post-medium-intensity exercise, and post-high-intensity exercise). We initially present a comparative analysis of the amount of lactate sampled by each pumping source. Then, we discuss the operating mechanism of each osmotic on-skin pumping material toward facilitating long-term lactate withdrawal and estimate how much lactate is sampled via osmosis. Next, we study the dependency of sweat lactate on the sweat rate. Finally, we seek to understand the correlation between sweat and blood lactate levels.

## 2. Experimental Section

### 2.1. Materials

The experiments were performed using Whatman filter paper (Grade 542, GE Healthcare Life Sciences, Waukesha, WI, USA); Sylgard-184 elastomer (Dow Corning, Midland, MI, USA); phosphate buffer saline (PBS), acrylamide (Sigma, St. Louis, MO, USA), N-N$'$ methylenebisacrylamide (Sigma, St. Louis, MO, USA), 2-hydroxy-4$'$-(2-hydroxyethoxy)-2-methylpropiophenone (Sigma, St. Louis, MO, USA), D-glucose, alcohol wipes (medical grade), L-lactic acid quick test strips (QQLLAC10, BioAssay Systems, Hayward, CA, USA), and a lactate plus meter (Nova Biomedical, Waltham, MA, USA).

### 2.2. Patch Fabrication

Sylgard-184 silicone elastomer and its curing agent were mixed in a 10:1 *w/w* ratio and cured for 12 h at 70 °Cto make the base for the PDMS sheet. The patch was prepared by attaching two PDMS sheets together—38 mm × 15 mm × 2 mm (bottom) and 30 mm × 15 mm × 1 mm (top). A single hole matching the hydrogel diameter was punched at 6 mm from one of the edges on the bottom sheet to encase the hydrogel. A section of Whatman 542 paper was cut out using a $CO_2$ laser cutter (Universal Laser Systems VLS 3.5, Scottsdale, AZ, USA) and was sandwiched between the top and bottom PDMS sheets. Our previous research on capillary-assisted evaporative pumping determined the dimension of the paper channel for this study [59]. The PDMS sheets were attached together using additional silicone as a binder (Sylgard-184), making sure that it did not contact the paper channel. The whole patch was treated in an oven at 40 °C overnight to achieve firm adhesion between the sheets.

### 2.3. Hydrogel Synthesis and PDMS Disk Fabrication

The hydrogels for the osmotic interface were made using acrylamide monomer, N, N$'$-methylenebisacrylamide as the crosslinker and 2-hydroxy-4$'$-(2-hydroxyethoxy)-2-methylpropiophenone as the photo-initiator. The monomer solution contained 22% (*w/w*) acrylamide, 0.48% (*w/w*) crosslinker, and 0.15% (*w/w*) photo-initiator. The solution was cured inside a circular Petri dish (47 mm diameter) under a 175 mW/cm$^2$ UV lamp (Sunray 400-SM, Uvitron International, West Springfield, MA, USA) for three minutes. Disks of 6 mm diameter were punched out and stored in either 4 M glucose or PBS solution for 24 h. These solutions served as the osmolytes. After 24 h, the infused disks were transferred to a fresh solution in a vial and stored for further usage. A single disk was taken out, blotted with a paper napkin (to remove the residual osmolyte on the hydrogel surface), and fitted inside the circular space in the patch before usage.

Disks of 6 mm diameter were punched out from a 2 mm thick PDMS sheet. This was referred as the PDMS disk. The PDMS disk was fitted inside the circular space in the patch before usage.

*2.4. On-Skin (In Vivo) Sweat Analysis*

Human trials were conducted on individuals in five sets: (a) NEX (i.e., without exposing them to any physical activity) for 2 h; (b) during MEX for an hour; (c) PMEX for 2 h; (d) during HEX for 30 min; and (e) PHEX for 2 h. All trials were conducted under normal ambient conditions (22 °C, 45% relative humidity (RH)) and strictly as per two approved IRB protocols (UNC-18-1959 and UNC-19-3065). Eight healthy subjects (5 females and 3 males), aged 20–28, were recruited from the NC State University campus. All 8 subjects were tested under NEX, MEX, and PMEX trials, while 3 subjects were also tested under MEX and PMEX trials. All subjects gave written, informed consent before participation in the study.

We used three types of patches for our human trials, which differ by the type of their pumping material: 4 M glucose hydrogel, PBS hydrogel, and PDMS disk. Both subjects' forearms were initially washed with alcohol and DI water. Then, these patches were snapped on both forearms of every individual using a Velcro strap. The subjects resided on a bench for 2 h when testing under NEX. After 2 h, the patches were taken off and stored for analysis. A paper strip was also gently rubbed (dry rub test) on the skin to check the base lactate level after 2 h. Blood lactate was measured using the conventional finger-pricking test. This concluded the NEX trials. The skin was rewashed using alcohol and DI water, and a fresh new batch of all three patch types was interfaced onto the skin again during the MEX trials. After an hour, the patches were taken off, and a dry rub test was performed. Blood lactate was also measured. This concluded the MEX trials. The skin was washed again with alcohol and DI water. A fresh new batch of all three patch types was interfaced onto the skin again for 2 h. After 2 h, the patches were taken off and stored for analysis. Additionally, a dry rub test and blood lactate test were also conducted. This concluded the PMEX trials.

The HEX and PHEX trials were conducted 24 h post PMEX trial completion. The skin was initially washed with alcohol and DI water. Then, the patches were snapped on both forearms. The subjects exercised under HEX for 30 min. After the test, both dry rub and blood lactate tests were conducted. The protocol followed for the PHEX trial was the same as that for the PMEX trial. The subjects refrained from any food or drinks throughout the testing period. The paper channel in the patch was treated with our previously developed colorimetric lactate assay (lactate dehydrogenase based) to determine the lactate levels after the completion of each trial [14]. A further detailed description of the testing protocol is provided in the Supplementary Information.

*2.5. Statistical Analysis*

We performed a two-sample t-test for statistical significance. Significance is denoted as $p < 0.05$.

## 3. Results and Discussion

### 3.1. Sweat Sampling Mechanism of the Wearable Patch

Our wearable patch functions based on the simultaneous action of three basic effects to deliver long-term sweat and biomarker sampling: osmosis, capillary wicking, and evaporation [14]. Osmosis is the main pumping mechanism for sweat extraction. A hydrogel infused with a highly concentrated solute (osmolyte) generates the osmotic driving force with the skin. The hydrogel is hosted inside the circular chamber of the patch, as shown in Figure 1. The hydrogel disk and the circular end of the paper directly interface the skin. The hydrogel withdraws sweat and its associated biomarkers due to its higher osmotic strength (with respect to sweat) (Figure 1). These withdrawn sweat and biomarkers are sampled on a paper microfluidic channel. The paper channel has two sections: (a) a long rectangular strip (sandwiched between two PDMS sheets), with one end in contact with the hydrogel and the other end connected to (b) a pad with a large surface area (evaporation pad), which is left open to allow sweat evaporation. The sweat, biomarkers, and the osmolyte travel through the rectangular section of the paper channel via capillary action toward the evaporation pad. The incoming sweat evaporates from the pad, leaving behind all biomarkers and non-volatile osmolytes. These biomarkers can be quantified using different analytical techniques. Since osmosis is a colligative property, the patch does not require any external power source for long-term operation. The patch will continue to operate (withdraw sweat from skin) as long as the chemical potential difference between the skin and hydrogel is maintained. In our previous work, we have analyzed the in vitro performance of the patch and studied all its components [14,59,60]. We found that hydrogel equilibrated in 4 M glucose solution functions best in delivering long-term model biomarker collections on paper. Hence, we decided to use a patch with 4 M glucose hydrogel and dry paper for all our human subject studies.

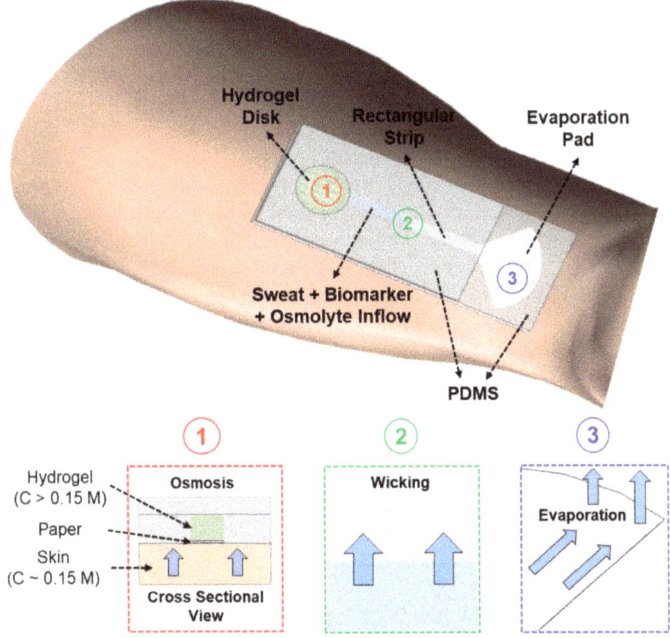


**Figure 1.** Schematic of the wearable, zero-powered, and non-invasive sweat sensing platform on skin. The platform comprises of a hydrogel pad and a paper microfluidic channel. Simultaneous action of osmosis, capillary wicking, and evaporation leads to long-term sweat withdrawal from skin even at resting perspiration rates.

## 3.2. Sampling of Sweat Lactate Using Different Pumping Materials in the Patch

We characterized the efficiency of three different pumping sources in our patch for sampling sweat lactate from the human skin: 4 M glucose hydrogel, PBS hydrogel, and a PDMS hydrogel substitute disk. The three osmolytes were chosen since they have varying levels of osmotic strength and are benign to the human skin upon contact. The 4 M glucose hydrogel has the highest osmotic strength. The PBS hydrogel is isotonic with sweat, while the PDMS disk is inert and has no osmotic action. We embedded each of these materials in the patch and estimated the amount of lactate that they sample on paper under five different physiological conditions: rest (NEX), medium-intensity exercise (MEX, 60–70% of maximum heart rate), post-medium-intensity exercise (PMEX), high-intensity exercise

(HEX, >85 % maximum heart rate), and post-high-intensity exercise (PHEX) (Figure S1). The base lactate level under each physiological condition was measured by mechanically rubbing the pad of the paper channel on skin (without hydrogel). This method of lactate collection was referred as the "dry rub" test. Such a comprehensive study would provide a better understanding regarding how effectively lactate is sampled via osmosis under varying physiological conditions, unlike our previous work, where we only used the 4 M glucose hydrogel patch to sample lactate during rest and MEX condition. The exercise protocol for the human trials and the calibration plot of the lactate assay protocol is presented in the Supplementary Information (Figures S2 and S3). The methodology of calibration has been reported in our previous paper [14]. The sensitivity and LOD of our colorimetric lactate assay are 0.18 a.u./mM and 750 μM, respectively.

Figures 2 and 3 present a comparative analysis of sweat lactate and volume intake by different pumping materials in the patch under varying physiological conditions, respectively. Figure 2a summarizes the set of all the physiological conditions in which the subjects were tested. After 2 h of rest (NEX) trials, the 4 M glucose hydrogel, PDMS disk, and the PBS hydrogel patch collected $\approx$17 $\pm$ 11 nmoles, $\approx$17.5 $\pm$ 12 nmoles, and $\approx$13.3 $\pm$ 7.5 nmoles of lactate, respectively. The dry rub test resulted in $\approx$23.63 $\pm$ 9.5 nmoles of lactate on paper after 2 h (Figure 2b). The results show that all three pumping materials in the patch can sample lactate from the skin during rest. The 4 M glucose hydrogel patch shows visible fluid flow on the paper channel (Figures 2b and 3). In contrast, the PBS hydrogel patch did not show any visible fluid collection but did collect lactate on paper (Figure 3). As PBS hydrogel is isotonic to sweat, it is evident that the lactate on paper originates from the natural lactate generation rate of the body ($\approx$0.3–1.3 nmole.cm$^{-2}$ min $^{-1}$) [36]. The same reasoning also holds for the PDMS disk patch (Figures 2b and 3). The lactate readouts of all three pumping materials matched well with the dry rub test (Figure 2b). This justifies that lactate readout of our patch matches well with base lactate levels after 2 h and shows that it can function efficiently under low sweating conditions.

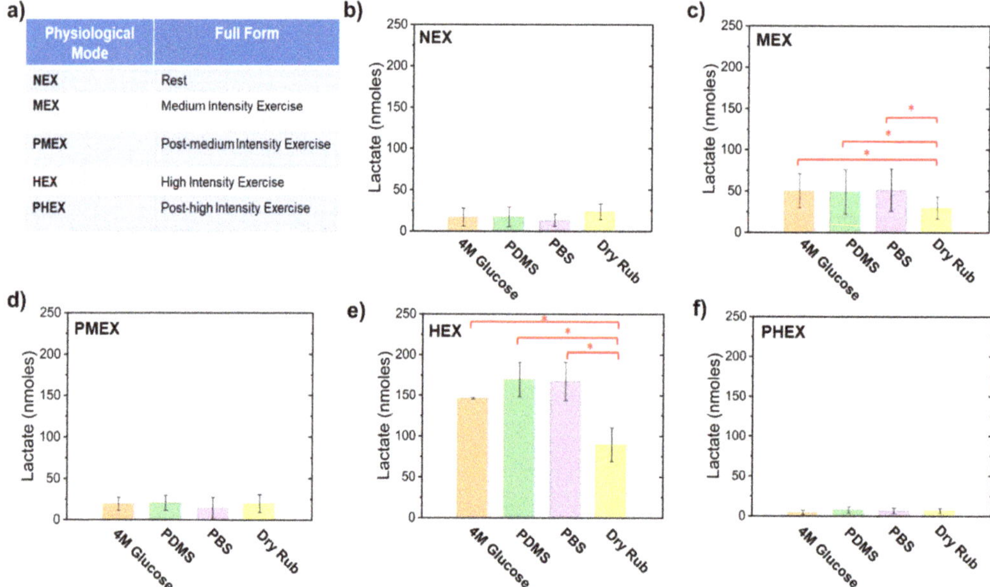

**Figure 2.** Plots presenting a comparative analysis of the sampled average sweat lactate quantity with different pumping materials in the patch under varying physiological conditions. (**a**) Table highlighting the full forms of all the physiological conditions in which the human subjects were tested. Plot showing the amount of lactate sampled during (**b**) NEX trial, (**c**) MEX trial, (**d**) PMEX trial, (**e**) HEX trial, and (**f**) PHEX trial. These plots show that all pumping materials in the patch can sample lactate on paper. Error bars denote the standard deviation obtained from all subjects. * $p < 0.05$.

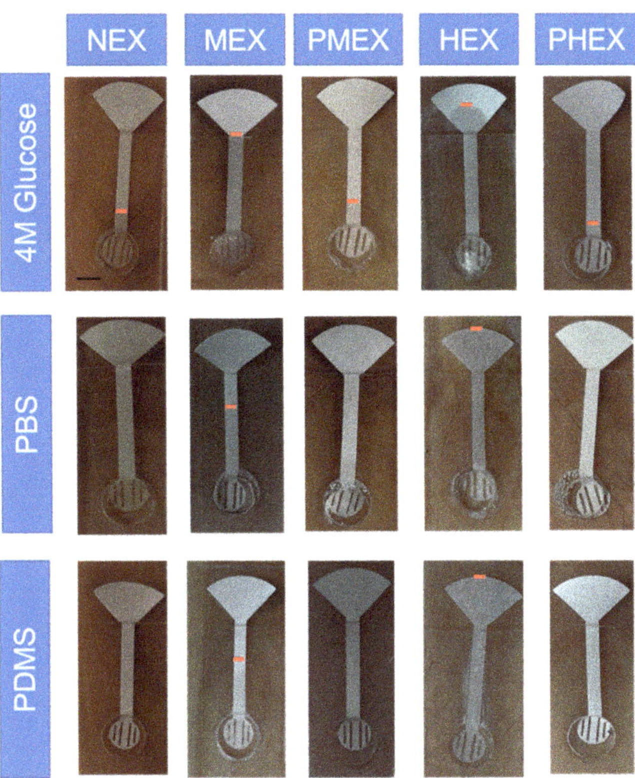



**Figure 3.** Optical images of the patch on skin showing the levels of sweat collection on the paper channel of patches with different pumping materials under different physiological conditions. The red mark on the paper channel indicates the position of sweat after each trial. Scale bar = 5 mm.

The patches were also tested under conditions of medium-intensity exercise (MEX). The 4 M glucose hydrogel, PDMS disk, and PBS hydrogel patches sampled $\approx$50.55 $\pm$ 20.6 nmoles, $\approx$49.32 $\pm$ 26.8 nmoles, and $\approx$51.56 $\pm$ 25.61 nmoles of lactate, respectively, after 1 h of exercise. The dry rub test resulted in $\approx$30 $\pm$ 13.4 nmoles of lactate collection on paper after 1 h (Figure 2c). The amounts of lactate sampled by all three pumping materials were significantly larger ($p < 0.05$) than the NEX trial results and dry rub test during MEX. All three patch types also showed greater sweat collection on paper than NEX trials.

Exercise increases the energy demand of the body. A part of this high-energy demand is compensated via anaerobic glycolysis, which leads to the generation of lactate in sweat [1–3,13,14]. Our patch was able to detect this increase with all three pumping materials. Exercise also leads to "active sweat" generation. Hence, lactate sampled by a 4 M glucose patch is released via both osmosis and active sweat under exercise. The release of active sweat makes the 4 M glucose patch collect more lactate than the dry rub test. PBS hydrogel and PDMS disk patches show elevated lactate levels solely due to the inflow of active sweat, since they do not create osmotic gradients with inherent fluid withdrawing power.

The patches were tested under post-medium intensity exercise (PMEX) conditions for 2 h. The 4 M glucose hydrogel, PDMS disk, and PBS hydrogel collected samples containing $\approx$19.50 $\pm$ 8.0 nmoles, $\approx$21.0 $\pm$ 9.0 nmoles, and $\approx$14.6 $\pm$ 12.6 nmoles of lactate, respectively. The dry rub test resulted in $\approx$19.8 $\pm$ 10.8 nmoles of lactate on paper after 2 h (Figure 2d).

The 4 M glucose patch showed visible fluid flow on paper, while the PBS hydrogel patch and PDMS patch did not sample any fluid on paper. During PMEX, the aerobic glycolysis rate increases, and the lactate in sweat gets converted to pyruvate [2]. Hence, the amount of lactate in sweat decreases as the body cools down after exercise. Our patch was able to detect this change with all three pumping materials. Additionally, similar lactate readouts with the NEX trials also indicate that base lactate levels were reached within 2 h after exercise completion in all subjects.

High-intensity exercise (HEX) trials were also conducted with all three patch types. The 4 M glucose hydrogel, PDMS disk, and PBS hydrogel sampled $\approx$143.76 $\pm$ 10.7 nmoles, $\approx$170.30 $\pm$ 14.9 nmoles, and $\approx$168.10 $\pm$ 16.8 nmoles of lactate, respectively. The dry rub test resulted in $\approx$89.30 $\pm$ 14.6 nmoles of lactate on paper after 30 min (Figure 2e). All three patch types sampled $\approx$5 times larger sweat volume on paper than MEX trials. The amounts of lactate sampled by all three pumping materials and dry rub were significantly larger ($p < 0.05$) than the lactate sampled during MEX trials. Additionally, the lactate readouts of all three pumping materials were significantly higher than the readout of the dry rub test. HEX rapidly induces anaerobic glycolysis and ramps up the lactate levels in sweat [2,13,17]. HEX also leads to greater active sweat release than MEX, which makes all three types of patches sample larger samples of lactate than the dry rub test. This confirms that the amount of lactate in sweat depends on the sweat rate [1,17,18].

The patches were next tested under PHEX conditions for 2 h after the MEX trials. The 4 M glucose hydrogel, PDMS disk, and PBS hydrogel sampled $\approx$5.05 $\pm$ 2.1 nmoles, $\approx$9.45 $\pm$ 3.3 nmoles, and $\approx$10.09 $\pm$ 4.7 nmoles of lactate, respectively. The dry rub test resulted in $\approx$6.03 $\pm$ 5.8 nmoles of lactate on paper after 2 h (Figure 2f). All three patches and dry rub tests showed significantly lower lactate levels ($p < 0.05$) than HEX and PMEX trials. The 4 M glucose patch showed visible fluid flow on paper, while the PBS hydrogel patch and PDMS patch did not sample any fluid on paper (Figure 3). Significantly lower levels of lactate than HEX trials prove that most of the lactate in sweat had been converted to pyruvate within 2 h. Moreover, significantly lower levels of lactate than PMEX trials show that the rate of aerobic metabolism (conversion of lactate to pyruvate) during PHEX is higher than PMEX. Hence, less lactate appears in sweat during the PHEX trials. Our patch was able to detect this change with all three pumping materials. Additionally, similar lactate readouts with the NEX trials also prove that base lactate levels were also reached within 2 h.

### 3.3. Working Mechanism of Different Pumping Materials on-skin toward the Sampling of Sweat Lactate

We interpret the transport effects of each pumping media based on the amount of lactate withdrawn by them through the paper–skin interface. Figure 4a highlights how a 4 M glucose hydrogel facilitates lactate withdrawal from skin during NEX, PMEX, and PHEX trials. Once the hydrogel interfaces the skin, osmosis is initiated due to the chemical potential difference between the hydrogel and sweat inside the skin. The glucose solution (from the hydrogel) and the osmotically withdrawn sweat and lactate (from skin) collectively appear in the form of a colorless liquid that wicks onto the rectangular section of the paper channel over time. Since no active sweat is released at rest, the lactate registered on paper is derived completely from the osmotically withdrawn sweat (Figure 2b,d,f) [14]. The PBS hydrogel is isotonic with sweat and has no fluid withdrawing power. Hence, the patch does not show any visible sweat collection at the paper-skin interface (Figure 4b). However, the paper registers a considerable amount of lactate from the skin (Figure 2b,d,f). This lactate amount is an outcome of the wet paper absorbing some of the naturally generated lactate from the human body [14]. The PDMS disk collects lactate on the paper in contact with skin but without any traces of fluid on the paper channel (Figures 2b,d,f and 4c). The absence of fluid on the paper channel is expected, since PDMS has no intrinsic fluid-withdrawing capacity. The lactate appears completely due to its natural production rate in the body.

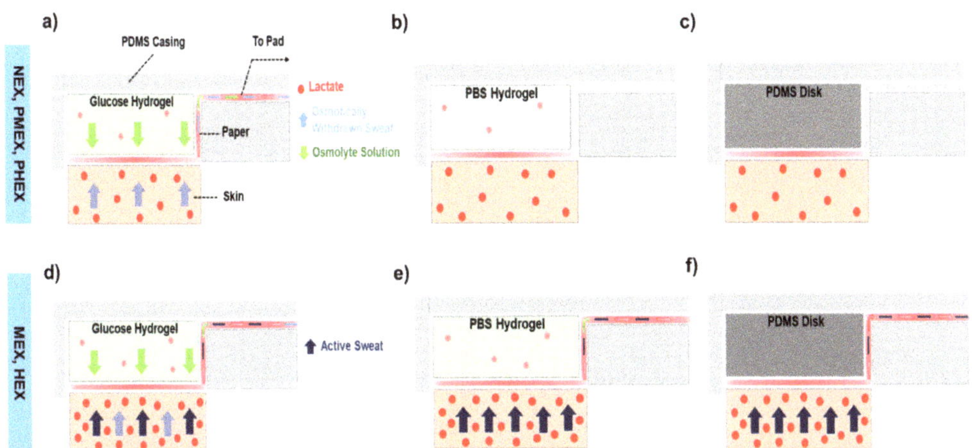



**Figure 4.** Cross-sectional schematics of sweat lactate transport with different pumping materials at the interface with the skin. The pumping materials include (**a**) a 4 M glucose hydrogel patch, (**b**) a PBS hydrogel patch, and (**c**) a PDMS disk patch during HEX, PMEX, and PHEX trials. Schematic showing the sweat lactate transport in (**d**) a 4 M glucose hydrogel patch, (**e**) a PBS hydrogel patch, and (**f**) a PDMS disk patch during MEX and HEX trials. These schematics show that the osmotic sampling of lactate occurs only in a 4 M glucose hydrogel patch. The PBS hydrogel patch and PDMS disk patch sample lactate from active sweat.

We also interpret schematically the on-skin action of the pumping materials during MEX and HEX trials. The 4 M glucose hydrogel patch shows greater fluid and lactate collection on the paper channel than the NEX trials (Figure 2c,e). The additional fluid flow arises due to the inflow of "active" sweat in the patch (Figure 4d). Lactate concentration in the active sweat increases with exercise. Hence, the rise in the amount of lactate content on paper is a direct result of active sweat inflow [1,17,18]. The PBS hydrogel also shows greater fluid and lactate collection on paper than the NEX trials (Figure 2c,e). Since a PBS hydrogel has no fluid withdrawing power, the additional fluid and lactate on paper is derived from the inflow of active sweat (Figure 4e). The same collection mechanism also holds for a PDMS disk patch (Figures 2c,e and 4f).

We also evaluated the amount of lactate sampled by the 4 M glucose patch via osmosis under different physiological conditions. This patch selection was made as 4 M glucose hydrogels possessed the highest (in comparison to PBS hydrogel and PDMS disk) osmotic strength. The efficiency of the 4 M glucose hydrogel toward lactate withdrawal via osmosis is expressed in terms of "osmotic contribution (OC)", which is calculated as:

$$\left( \frac{n_{HG} - n_{PBS}}{n_{PBS}} \right) \times 100 \tag{1}$$

where $n_{HG}$ is the moles of lactate sampled by the 4 M glucose hydrogel and $n_{PBS}$ is the moles of lactate sampled by the PBS hydrogel.

The 4 M glucose hydrogel patch sampled on average ≈40% (NEX), ≈25% (MEX), ≈45% (PMEX), ≈0.5% (HEX), and ≈20% (PHEX) more lactate than a PBS hydrogel patch. The observed variation in the contributions under each condition can be attributed to the difference in the rate of lactate metabolism and to the small differences between the rates of lactate generation by each forearm in each subject (Figure 5).

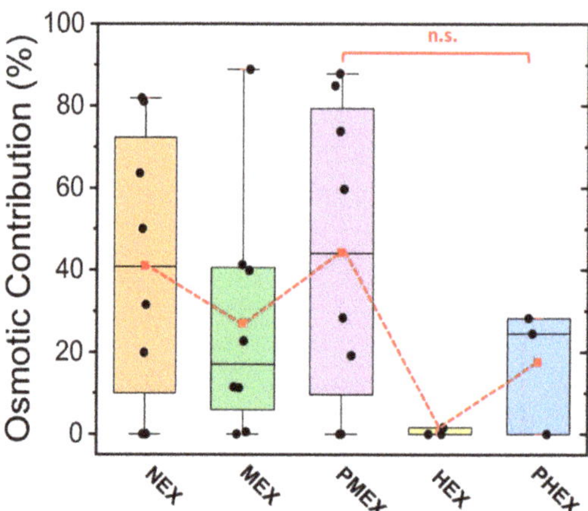



**Figure 5.** Plot showing the effectiveness of the 4 M glucose hydrogel in facilitating lactate withdrawal via osmosis under different physiological conditions. Each dot denotes data from one subject. The dashed line connects the mean values to guide the eye.

Lactate collection via osmosis dominates during NEX, PMEX, and PHEX trials due to the absence of active sweat inflow in the paper channel. MEX and HEX trials lead to the active inflow of sweat in the patch, which increases the lactate content on paper. Moreover, HEX leads to larger sweat and lactate collection than MEX but samples lactate the least via osmosis. Hence, the higher the inflow of active sweat, the less the lactate collection via osmosis. Overall, our wearable patch functions by two concurrent mechanisms for sweat lactate withdrawal: (a) osmosis at low sweat rates and (b) active sweat for medium to high sweat rates.

### 3.4. Relationship between Sweat Lactate and Sweat Volume

We investigated how the sweat lactate content depends on the sampled sweat volume on paper under all five physiological conditions. This helped us to understand the dependency of sweat lactate on the sweat rate. We analyzed this only for the 4 M glucose hydrogel patch for two reasons: (a) 4 M glucose hydrogel has the highest osmotic strength. Hence, its pumping activity was our major focus of interest. (b) The PBS hydrogel patch did not result in sweat collection on paper during NEX, PMEX, and PHEX trials. Hence, calculation of lactate concentration was not always possible for a PBS hydrogel patch. The 4 M glucose hydrogel sampled $\approx$1.8 $\pm$ 0.4 $\mu$L (NEX), $\approx$2.7 $\pm$ 1.3 $\mu$L (MEX), and $\approx$1.6 $\pm$ 0.3 $\mu$L (PMEX), $\approx$11.3 $\pm$ 3.2 $\mu$L (HEX), and $\approx$1.60 $\pm$ 0.5 $\mu$L (PHEX) of fluid on paper. During NEX, MEX, HEX, and PHEX trials, the amount of collected sweat lactate increased with the fluid volume on paper (Figure 6a–c). However, we observed an opposite trend during the PMEX trials (Figure 6c).

The 4 M glucose hydrogel produced an average sweat withdrawal rate (SWR) of 0.03 $\mu$L.cm$^{-2}$.min$^{-1}$, which corresponded to an average sweat volume of $\approx$1.8 $\mu$L after 2 h of testing under NEX. The sampled lactate quantity and sweat volume are in agreement with the results reported earlier by Bhide et al. [58]. A good correlation (Pearson's correlation coefficient, r > 0.7) existed between the sweat lactate amount and fluid volume on paper. This shows that the amount of moles of lactate varies proportionally with sweat rate at rest (Figure 6a). Hence, osmosis leads to joint sweat and lactate withdrawal during NEX trials (Figures 4a and 5).

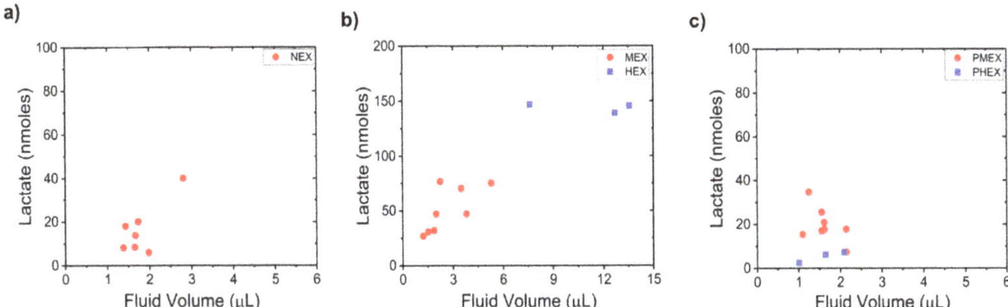

**Figure 6.** Relationship between sampled sweat lactate quantity (under different physiological conditions) and sweat volume on paper for a 4 M glucose hydrogel patch. Plots showing how sweat lactate depends on sweat volume during (**a**) NEX trial, (**b**) MEX and HEX trials, and (**c**) PMEX and PHEX trials. Each dot denotes data from one subject.

The 4 M glucose hydrogel produced an SWR of $\approx$0.10 µL.cm$^{-2}$.min$^{-1}$ and $\approx$0.75 µL.cm$^{-2}$.min$^{-1}$ during MEX and HEX trials, respectively. Such sweat rates are in line with the literature [17,61]. Both sweat and lactate amounts on paper increased during MEX and HEX trials, with HEX collecting the highest amount of lactate (Figure 6b). The observed scattering between the data points is a result of the difference in sweat generation rates of each subject. Around $\approx$1–1.5 µL, and $\approx$7–8 µL of additional sweat flows into the paper during MEX and HEX trials, respectively, in comparison to the NEX trials. This additional fluid is active sweat resulting from exercise (Figures 3 and 4d). A good correlation (r > 0.5) between the sampled lactate amount and sweat volume again proves that sweat lactate amount (moles) is a function of sweat rate [1,17,18]. The 4 M glucose hydrogel produced an SWR of $\approx$0.03 µL.cm$^{-2}$.min$^{-1}$ during both PMEX and PHEX trials. Such an SWR is similar to the NEX trials. The lactate amount possesses a negative correlation (r < 0) with the sampled sweat volume on paper during PMEX trials (Figure 6c). As per the review of Bakker et al., this happens since lactate gets converted to pyruvate as the body cools down after exercise [2]. As a result, the amount of lactate in sweat decreases. A positive correlation existed during PHEX trials due to the extremely low magnitude of lactate amount (with respect to PMEX) and sweat volume.

### 3.5. Estimation of Sweat Lactate Concentration and Its Relationship with Blood Lactate

The amount of sampled lactate and sweat volume on paper were used to calculate the lactate concentration for the 4 M glucose hydrogel patch. The 4 M glucose hydrogel initiates the osmotic withdrawal of sweat upon contact with the skin at a lactate extraction rate (LER) (from skin) of $\approx$0.58 nmole.cm$^{-2}$. min$^{-1}$ (NEX), $\approx$3.21 nmole.cm$^{-2}$. min$^{-1}$ (MEX), $\approx$0.68 nmole.cm$^{-2}$. min$^{-1}$ (PMEX), $\approx$20.30 nmole.cm$^{-2}$. min$^{-1}$ (HEX), and $\approx$0.16 nmole.cm$^{-2}$. min$^{-1}$ (PMEX) (assuming constant sweat generation rate). This corresponded to an average sweat lactate concentration of $\approx$9.38 mM (NEX), 19.30 mM (MEX), 13.08 mM (PMEX), 13.62 mM (HEX), and 3.18 mM (PHEX) (Figure 7a).

The sweat lactate concentration during NEX trial is considered low, as it falls toward the lower end of the normal human sweat lactate concentration range [6,14,36,38]. Lactate concentration during MEX trials was significantly larger (*p* < 0.05) than NEX trials. This happens because (a) exercise induces anaerobic metabolism, which leads to the appearance of lactate in sweat, and (b) exercise causes greater sweat release, which proportionally releases more lactate (Figure 6b). Lactate concentration during PMEX was significantly lower (*p* < 0.05) than that during MEX, since lactate gets converted to pyruvate as the body cools down after exercise. Additionally, this also suggests that base lactate levels were reached within two hours (*p* > 0.05, with respect to NEX). The difference between the lactate concentrations during MEX and HEX trials were insignificant (*p* > 0.05). Although HEX trials lead to greater amount of lactate generation, a higher volume of sweat release slightly decreases the lactate concentration due to dilution (Figures 2e, 3, 6b and 7a) [18,61]. Hence, as the lactate concentration peaks during MEX, we believe that it is the facile physiological zone for estimating the maximum sweat lactate concentration (Figure 7a,b). Sweat lactate concentrations during PHEX trials were significantly lower than the PMEX trials ($p < 0.05$) due to significant lower lactate generation (Figure 2f). This confirms the following two outcomes: (a) the rate of disappearance of lactate from sweat during PHEX trials is faster than PMEX trials, and it happens within 2 h after HEX termination, and (b) two hours is a sufficient time period to attain the base lactate level. Moreover, the lactate concentration readouts of the 4 M glucose hydrogel patches during NEX, MEX, and HEX trials match well with the lactate concentration readouts of existing devices, which were reported under similar exertion levels (rest, medium-intensity exercise, and high-intensity exercise) (Figure 6b) [1,5,12,14,17,23,30,34,36,42–44,47,48,50,53,56,58,61–64]. This validates the efficient functioning of our patch under all physiological conditions.

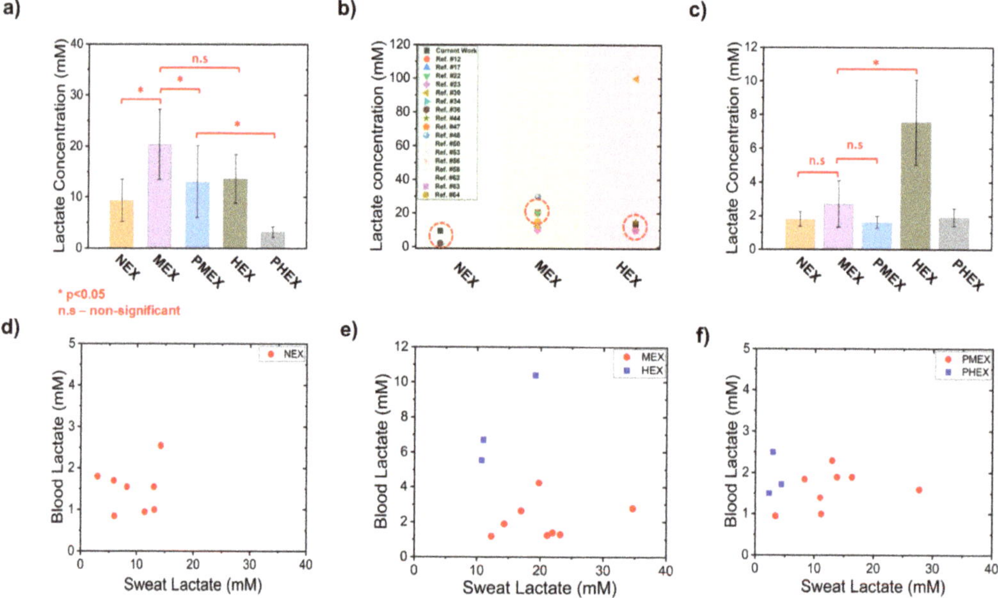

**Figure 7.** Comparative analysis of sweat lactate concentration readout of 4 M glucose hydrogel patch and blood lactate concentration. (**a**) Plot showing the estimated sweat lactate concentration in a 4 M glucose hydrogel patch under different physiological conditions. (**b**) Plot presenting a comparative analysis of the lactate concentration readout from our patch (dotted circular box) to the lactate concentration data from previously published lactate studies. (**c**) Plot showing the measured blood lactate concentration under different physiological conditions. Relationship of sweat and blood lactate concentration during (**d**) NEX trial, (**e**) MEX and HEX trials, and (**f**) PMEX and PHEX trials. * denotes $p < 0.05$. Each dot denotes data from one subject. These plots elucidate the correlation between blood and sweat lactate levels under different physiological conditions. The error bars denote standard deviation from all subjects.

We also measured blood lactate levels and investigated its correlation to sweat lactate under all five physiological conditions (Figure 7c–f). A poor correlation between sweat and blood lactate levels was recorded during NEX trials (Figure 7d). This is expected due to the lack of physical exertion. Hence, as suggested by Anderson et al., the lactate in sweat and blood during NEX is mainly derived from the sweat glands above type I muscle fibers [65]. Type I muscle fibers generate energy mainly by aerobic metabolism. The measured average blood lactate concentration also matches well with what Alam et al. have reported in their review [3]. Blood lactate concentration during MEX trials remained low and similar to NEX trials ($p > 0.05$) (Figure 7c). Hence, MEX does not affect the blood lactate levels. This also results in a poor correlation between blood and sweat lactate levels (Figure 7e). Such trends are expected, since we operate below the lactate threshold (LT) of the human subjects during MEX trials [61]. Goodwin et al. have shown that blood lactate increases exponentially beyond LT [21]. The data also established the source of lactate origination to be mainly the sweat glands above the type IIa muscles and not blood [65]. Type IIa muscle fibers generate energy under both aerobic and anaerobic metabolism. The poor correlation between blood and sweat lactate levels during PMEX exists due to the absence of physical exertion and glycolysis in type I muscle fibers (Figure 7f).

HEX causes a significant rise in the blood lactate levels ($p < 0.05$, with respect to MEX) (Figure 7c). Consequently, the ratio between sweat and blood lactate concentration decreases (with respect to MEX). This results mainly from the following two reasons: (a) HEX makes the subjects exceed their LT (>85% of maximum heart rate) [21]. This causes the lactate to partition into blood from the type IIb muscle fibers. Anderson et al. have shown that the type IIb muscle fibers generate energy purely via anaerobic glycolysis [65]. (b) A drop in the $HCO_3^-$ ion concentration and pH in plasma leads to an inflow of lactate in blood to maintain the acid–base equilibrium [19]. Overall, a prolonged high blood lactate concentration (> 3 mM) induces faster fatigue and is physiologically challenging [65–67]. Hence, individuals can be recognized as being "fit" if they can delay reaching LT or show low blood lactate levels even under HEX [65]. A strong correlation exists between blood and sweat lactate levels during HEX (Figure 7e). The correlation exists due to higher blood lactate levels than those in the MEX trial. Such correlations have been reported by Karpova et al. previously [1]. A poor correlation exists between blood and sweat lactate levels during PHEX trials (Figure 7f) due to the absence of high physical exertion.

## 4. Conclusions

We report here the on-skin functioning of a non-invasive sweat sampling patch that operates under novel osmotic extraction principles. The patch uses a hydrogel disk infused with a solute to raise its chemical potential difference with respect to sweat in the skin. As a result, the hydrogel directly withdraws the sweat from the skin via osmosis and without the necessity of active perspiration. The extracted sweat is transported via a paper channel and evaporated at a pad to maintain the continuous inflow of sweat and its associated biomarkers. Since osmosis is a colligative property inherent to the system, the patch does not require any external power source for long-term operation. The patch will continue to withdraw sweat as long as the chemical potential difference sustains between the hydrogel and sweat. Moreover, this patch design is flexible, non-invasive, and can be comfortably worn on the skin.

We tested the patch with three different pumping media under five different physiological conditions for sweat lactate quantification. Results show that all three pumping media sample lactate on paper during rest, exercise, and post-exercise trials. Lactate was majorly withdrawn by the glucose hydrogel via osmosis when the body remained exertion-free (NEX). The PBS hydrogel and PDMS disk patches sample lactate on paper due to the natural lactate generation rate of the body during NEX trials. The moles of sweat lactate collected increase with exercise intensity and decrease when the body cools down. During exertion (MEX and HEX), lactate is derived primarily from the inflow of active sweat in all three patch systems. Hence, the amount of sweat lactate captured is proportional to the sweat rate. The calculation of sweat lactate concentration was only possible for the glucose hydrogel patch, as there was considerable sweat collection on the paper. The lactate concentration measured by the glucose hydrogel patch correlated well with the lactate concentration readouts of other sensors, which were tested under similar exertion levels. Sweat lactate concentration during HEX remained lower than MEX. This occurs because the rate of sweat production (volume/time) is greater than the rate of lactate generation (moles/time) in sweat during HEX, which eventually leads to dilution and a lower lactate concentration (moles/volume). Hence, sweat lactate concentration should be measured under MEX as it peaks during this physiological state. Blood lactate concentration stayed low and did not show a good correlation with sweat lactate concentration during NEX, MEX, PMEX, and PHEX trials. This occurred due to the absence of physical exertion. In contrast, the blood lactate concentration increased and showed a good correlation with sweat lactate concentration during HEX. This occurred since all subjects surpassed their LT during HEX. The variation in sweat and blood lactate concentration levels under different physiological conditions is presumably an outcome of varying glycolysis rates at different muscle fiber sites.

Overall, our wearable osmotic sweat sampling patch appears to have big potential in facilitating continuous sweat collection for hours and delivering useful health information about human lactate trends under varying physiological conditions. Despite the on-skin studies, our patch requires post-processing measures for estimating the amount of lactate in sweat. After testing, the rectangular strip needs to be cut out and subjected to an assay for lactate determination. Hence, the current version of the patch is not an optimized solution for device operation, since an ideal wearable should deliver continuous monitoring and real-time data generation. Our group is working toward extending the current version of the patch to a continuous, real-time sweat sensing platform for lactate by incorporating enzymatic electrochemical sensors. In addition to lactate, the prototype reported here can be used to sample multiple other sweat based biomarkers as well. We are also working toward transforming our current patch to a lateral flow assay (LFA) platform for cortisol and potassium detection in sweat. We believe that such a prototype, capable of functioning with multiple sensing principles (colorimetric, electrochemical) will provide a better, long-term understanding about the metabolism of different sweat-based biomarkers under both at home and clinical settings.

**Supplementary Materials:** The following are available online at https://www.mdpi.com/article/10.3390/mi12121513/s1. Figure S1: Schematic illustration of the human trial protocol. 4 M glucose hydrogel, PBS hydrogel, and PDMS disk patches were tested on skin during rest, exercise, and post-exercise conditions for sweat lactate estimation. The dry rub test estimates the base lactate level during each physiological stage. Figure S2: Rotational speed of the cycle ergometer and exercise intensity range during human trials. Figure S3: Calibration plot of normalized intensity vs. lactate concentration.

**Author Contributions:** Conceptualization, T.S., M.D.D. and O.D.V.; Data curation, T.S. and J.F.; Formal analysis, T.S., M.D.D. and O.D.V.; Funding acquisition, M.D.D. and O.D.V.; Investigation, T.S., J.F., S.M., C.T.K., M.D.D. and O.D.V.; Methodology, T.S., J.F. and S.M.; Validation, T.S. and S.M.; Writing—original draft, T.S.; Writing—review and editing, T.S., M.D.D. and O.D.V. All authors have read and agreed to the published version of the manuscript.

**Funding:** This research was funded by the NSF-ASSIST Center for Advanced Self Powered Systems of Integrated Sensors and Technologies under the grant number EEC-1160483, and Nano-Bio Materials Consortium (NBMC) in a partnership between the Air Force Research Laboratory (AFRL) and SEMI under agreement number NB18-20-21. The U.S. Government is authorized to reproduce and distribute reprints for Government purposes notwithstanding any copyright notation thereon. The views and conclusions contained herein are those of the authors and should not be interpreted as necessarily representing the official policies or endorsements, either expressed or implied, of Air Force Research Laboratory, the U.S. Government, or SEMI-FlexTech.

**Institutional Review Board Statement:** This study was approved by the Institutional Review Board (IRB) of University of North Carolina (UNC-18–1959 and UNC-19-3065).

**Informed Consent Statement:** Informed consent was obtained from all subjects involved in the study.

**Acknowledgments:** We are thankful to our colleagues—Michael Daniele and Murat Yokus for their insightful advice and assistance. We are also thankful to Sabrina Pietrosemoli Salazar for her assistance with the table of contents image.

**Conflicts of Interest:** The authors declare no conflict of interest.

# References


1. Karpova, E.V.; Laptev, A.I.; Andreev, E.A.; Karyakina, E.E.; Karyakin, A.A. Relationship Between Sweat and Blood Lactate Levels During Exhaustive Physical Exercise. *ChemElectroChem* **2020**, *7*, 191–194. [CrossRef]
2. Bakker, J.; Nijsten, M.W.N.; Jansen, T.C. Clinical Use of Lactate Monitoring in Critically Ill Patients. *Ann. Intensive Care* **2013**, *3*, 12. [CrossRef] [PubMed]
3. Alam, F.; RoyChoudhury, S.; Jalal, A.H.; Umasankar, Y.; Forouzanfar, S.; Akter, N.; Bhansali, S.; Pala, N. Lactate Biosensing: The Emerging Point-of-Care and Personal Health Monitoring. *Biosens. Bioelectron.* **2018**, *117*, 818–829. [CrossRef]
4. Pundir, C.S.; Narwal, V.; Batra, B. Determination of Urea with Special Emphasis on Biosensors: A Review. *Biosens. Bioelectron.* **2016**, *86*, 777–790. [CrossRef] [PubMed]
5. Baker, L.B. Physiology of Sweat Gland Function: The Roles of Sweating and Sweat Composition in Human Health. *Temperature* **2019**, *6*, 211–259. [CrossRef]
6. Sonner, Z.; Wilder, E.; Heikenfeld, J.; Kasting, G.; Beyette, F.; Swaile, D.; Sherman, F.; Joyce, J.; Hagen, J.; Kelley-Loughnane, N.; et al. The Microfluidics of the Eccrine Sweat Gland, Including Biomarker Partitioning, Transport, and Biosensing Implications. *Biomicrofluidics* **2015**, *9*, 031301. [CrossRef] [PubMed]
7. Thomas, N.; Lähdesmäki, I.; Parviz, B.A. A Contact Lens with an Integrated Lactate Sensor. *Sens. Actuators B Chem.* **2012**, *162*, 128–134. [CrossRef]
8. Kim, J.; Valdés-Ramírez, G.; Bandodkar, A.J.; Jia, W.; Martinez, A.G.; Ramírez, J.; Mercier, P.; Wang, J. Non-Invasive Mouthguard Biosensor for Continuous Salivary Monitoring of Metabolites. *Analyst* **2014**, *139*, 1632–1636. [CrossRef]
9. Walenta, S.; Wetterling, M.; Lehrke, M.; Schwickert, G.; Sundfør, K.; Rofstad, E.K.; Mueller-Klieser, W. High Lactate Levels Predict Likelihood of Metastases, Tumor Recurrence, and Restricted Patient Survival in Human Cervical Cancers. *Cancer Res.* **2000**, *60*, 916–921. [PubMed]
10. Cureton, E.L.; Kwan, R.O.; Dozier, K.C.; Sadjadi, J.; Pal, J.D.; Victorino, G.P. A Different View of Lactate in Trauma Patients: Protecting the Injured Brain. *J. Surg. Res.* **2010**, *159*, 468–473. [CrossRef]
11. Derbyshire, P.J.; Barr, H.; Davis, F.; Higson, S.P.J. Lactate in Human Sweat: A Critical Review of Research to the Present Day. *J. Physiol. Sci.* **2012**, *62*, 429–440. [CrossRef]
12. Anastasova, S.; Crewther, B.; Bembnowicz, P.; Curto, V.; Ip, H.M.; Rosa, B.; Yang, G. A Wearable Multisensing Patch for Continuous Sweat Monitoring. *Biosens. Bioelectron.* **2017**, *93*, 139–145. [CrossRef] [PubMed]
13. Mitsubayashi, K.; Suzuki, M.; Tamiya, E.; Karube, I. Analysis of Metabolites in Sweat as a Measure of Physical Condition. *Anal. Chim. Acta* **1994**, *289*, 27–34. [CrossRef]
14. Saha, T.; Fang, J.; Mukherjee, S.; Dickey, M.D.; Velev, O.D. Wearable Osmotic-Capillary Patch for Prolonged Sweat Harvesting and Sensing. *ACS Appl. Mater. Interfaces* **2021**, *13*, 8071–8081. [CrossRef] [PubMed]
15. Meyer, F.; Laitano, O.; Bar-Or, O.; McDougall, D.; Heingenhauser, G.J.F. Effect of Age and Gender on Sweat Lactate and Ammonia Concentrations during Exercise in the Heat. *Braz. J. Med. Biol. Res.* **2007**, *40*, 135–143. [CrossRef] [PubMed]
16. Fellmann, N.; Grizard, G.; Coudert, J. Human Frontal Sweat Rate and Lactate Concentration during Heat Exposure and Exercise. *J. Appl. Physiol.* **1983**, *54*, 355–360. [CrossRef] [PubMed]
17. Buono, M.J.; Lee, N.V.L.; Miller, P.W. The Relationship between Exercise Intensity and the Sweat Lactate Excretion Rate. *J. Physiol. Sci.* **2010**, *60*, 103–107. [CrossRef]
18. Green, J.M.; Bishop, P.A.; Muir, I.H.; Lomax, R.G. Gender Differences in Sweat Lactate. *Eur. J. Appl. Physiol.* **2000**, *82*, 0230. [CrossRef]
19. Patterson, M.J.; Galloway, S.D.R.; Nimmo, M.A. Variations in Regional Sweat Composition in Normal Human Males. *Exp. Physiol.* **2000**, *85*, 869–875. [CrossRef]
20. Sakharov, D.A.; Shkurnikov, M.U.; Vagin, M.Y.; Yashina, E.I.; Karyakin, A.A.; Tonevitsky, A.G. Relationship between Lactate Concentrations in Active Muscle Sweat and Whole Blood. *Bull. Exp. Biol. Med.* **2010**, *150*, 83–85. [CrossRef] [PubMed]
21. Goodwin, M.L.; Harris, J.E.; Hernández, A.; Gladden, L.B. Blood Lactate Measurements and Analysis during Exercise: A Guide for Clinicians. *J. Diabetes Sci. Technol.* **2007**, *1*, 558–569. [CrossRef] [PubMed]
22. Koh, A.; Kang, D.; Xue, Y.; Lee, S.; Pielak, R.M.; Kim, J.; Hwang, T.; Min, S.; Banks, A.; Bastien, P.; et al. A Soft, Wearable Microfluidic Device for the Capture, Storage, and Colorimetric Sensing of Sweat. *Sci. Transl. Med.* **2016**, *8*, 366ra165. [CrossRef]
23. Choi, J.; Kang, D.; Han, S.; Kim, S.B.; Rogers, J.A. Thin, Soft, Skin-Mounted Microfluidic Networks with Capillary Bursting Valves for Chrono-Sampling of Sweat. *Adv. Healthc. Mater.* **2017**, *6*, 1–10. [CrossRef] [PubMed]
24. Choi, J.; Bandodkar, A.J.; Reeder, J.T.; Ray, T.R.; Turnquist, A.; Kim, S.B.; Nyberg, N.; Hourlier-Fargette, A.; Model, J.B.; Aranyosi, A.J.; et al. Soft, Skin-Integrated Multifunctional Microfluidic Systems for Accurate Colorimetric Analysis of Sweat Biomarkers and Temperature. *ACS Sensors* **2019**, *4*, 379–388. [CrossRef] [PubMed]
25. Baysal, G.; Önder, S.; Göcek, İ.; Trabzon, L.; Kızıl, H.; Kök, F.N.; Kayaoğlu, B.K. Microfluidic Device on a Nonwoven Fabric: A Potential Biosensor for Lactate Detection. *Text. Res. J.* **2014**, *84*, 1729–1741. [CrossRef]
26. Ashley, B.K.; Brown, M.S.; Park, Y.; Kuan, S.; Koh, A. Skin-Inspired, Open Mesh Electrochemical Sensors for Lactate and Oxygen Monitoring. *Biosens. Bioelectron.* **2019**, *132*, 343–351. [CrossRef] [PubMed]
27. Mao, Y.; Zhang, W.; Wang, Y.; Guan, R.; Liu, B.; Wang, X.; Sun, Z.; Xing, L.; Chen, S.; Xue, X. Self-Powered Wearable Athletics Monitoring Nanodevice Based on ZnO Nanowire Piezoelectric-Biosensing Unit Arrays. *Sci. Adv. Mater.* **2019**, *11*, 351–359. [CrossRef]

28. Enomoto, K.; Shimizu, R.; Kudo, H. Real-Time Skin Lactic Acid Monitoring System for Assessment of Training Intensity. *Electron. Commun. Japan* **2018**, *101*, 41–46. [CrossRef]
29. Payne, M.E.; Zamarayeva, A.; Pister, V.I.; Yamamoto, N.A.D.; Arias, A.C. Printed, Flexible Lactate Sensors: Design Considerations Before Performing On-Body Measurements. *Sci. Rep.* **2019**, *9*, 13720. [CrossRef]
30. Onor, M.; Gufoni, S.; Lomonaco, T.; Ghimenti, S.; Salvo, P.; Sorrentino, F.; Bramanti, E. Potentiometric Sensor for Non Invasive Lactate Determination in Human Sweat. *Anal. Chim. Acta* **2017**, *989*, 80–87. [CrossRef]
31. Weltin, A.; Kieninger, J.; Enderle, B.; Gellner, A.K.; Fritsch, B.; Urban, G.A. Polymer-Based, Flexible Glutamate and Lactate Microsensors for in Vivo Applications. *Biosens. Bioelectron.* **2014**, *61*, 192–199. [CrossRef] [PubMed]
32. Meyerhoff, C.; Bischof, F.; Mennel, F.J.; Sternberg, F.; Bican, J.; Pfeiffer, E.F. On Line Continuous Monitoring of Blood Lactate in Men by a Wearable Device Based upon an Enzymatic Amperometric Lactate Sensor. *Biosens. Bioelectron.* **1993**, *8*, 409–414. [CrossRef]
33. Tur-García, E.L.; Davis, F.; Collyer, S.D.; Holmes, J.L.; Barr, H.; Higson, S.P.J. Novel Flexible Enzyme Laminate-Based Sensor for Analysis of Lactate in Sweat. *Sens. Actuators B Chem.* **2017**, *242*, 502–510. [CrossRef]
34. Nagamine, K.; Mano, T.; Nomura, A.; Ichimura, Y.; Izawa, R.; Furusawa, H.; Matsui, H.; Kumaki, D.; Tokito, S. Noninvasive Sweat-Lactate Biosensor Emplsoying a Hydrogel-Based Touch Pad. *Sci. Rep.* **2019**, *9*, 10102. [CrossRef] [PubMed]
35. Pribil, M.M.; Laptev, G.U.; Karyakina, E.E.; Karyakin, A.A. Noninvasive Hypoxia Monitor Based on Gene-Free Engineering of Lactate Oxidase for Analysis of Undiluted Sweat. *Anal. Chem.* **2014**, *86*, 5215–5219. [CrossRef] [PubMed]
36. Lin, S.; Wang, B.; Zhao, Y.; Shih, R.; Cheng, X.; Yu, W.; Hojaiji, H.; Lin, H.; Hoffman, C.; Ly, D.; et al. Natural Perspiration Sampling and in Situ Electrochemical Analysis with Hydrogel Micropatches for User-Identifiable and Wireless Chemo/Biosensing. *ACS Sens.* **2020**, *5*, 93–102. [CrossRef]
37. Park, J.; Sempionatto, J.R.; Kim, J.; Jeong, Y.; Gu, J.; Wang, J.; Park, I. Microscale Biosensor Array Based on Flexible Polymeric Platform toward Lab-on-a-Needle: Real-Time Multiparameter Biomedical Assays on Curved Needle Surfaces. *ACS Sens.* **2020**, 1–31. [CrossRef]
38. Bariya, M.; Nyein, H.Y.Y.; Javey, A. Wearable Sweat Sensors. *Nat. Electron.* **2018**, *1*, 160–171. [CrossRef]
39. Nie, Z.; Deiss, F.; Liu, X.; Akbulut, O.; Whitesides, G.M. Integration of Paper-Based Microfluidic Devices with Commercial Electrochemical Readers. *Lab Chip* **2010**, *10*, 3163–3169. [CrossRef] [PubMed]
40. He, W.; Wang, C.; Wang, H.; Jian, M.; Lu, W.; Liang, X.; Zhang, X.; Yang, F.; Zhang, Y. Integrated Textile Sensor Patch for Real-Time and Multiplex Sweat Analysis. *Sci. Adv.* **2019**, *5*, eaax0649. [CrossRef]
41. Sempionatto, J.R.; Nakagawa, T.; Pavinatto, A.; Mensah, S.T.; Imani, S.; Mercier, P.; Wang, J. Eyeglasses Based Wireless Electrolyte and Metabolite Sensor Platform. *Lab Chip* **2017**, *17*, 1834–1842. [CrossRef] [PubMed]
42. Martín, A.; Kim, J.; Kurniawan, J.F.; Sempionatto, J.R.; Moreto, J.R.; Tang, G.; Campbell, A.S.; Shin, A.; Lee, M.Y.; Liu, X.; et al. Epidermal Microfluidic Electrochemical Detection System: Enhanced Sweat Sampling and Metabolite Detection. *ACS Sens.* **2017**, *2*, 1860–1868. [CrossRef] [PubMed]
43. Bhide, A.; Cheeran, S.; Muthukumar, S.; Prasad, S. Enzymatic Low Volume Passive Sweat Based Assays for Multi-Biomarker Detection. *Biosensors* **2019**, *9*, 13. [CrossRef]
44. Jia, W.; Bandodkar, A.J.; Valde, G.; Windmiller, J.R.; Yang, Z.; Ram, J.; Chan, G.; Wang, J. Electrochemical Tattoo Biosensors for Real-Time Noninvasive Lactate Monitoring in Human Perspiration. *Anal. Chem.* **2013**, *85*, 6553–6560. [CrossRef] [PubMed]
45. Lamas-Ardisana, P.J.; Loaiza, O.A.; Añorga, L.; Jubete, E.; Borghei, M.; Ruiz, V.; Ochoteco, E.; Cabañero, G.; Grande, H.J. Disposable Amperometric Biosensor Based on Lactate Oxidase Immobilised on Platinum Nanoparticle-Decorated Carbon Nanofiber and Poly(Diallyldimethylammonium Chloride) Films. *Biosens. Bioelectron.* **2014**, *56*, 345–351. [CrossRef]
46. Abrar, M.A.; Dong, Y.; Lee, P.K.; Kim, W.S. Bendable Electro-Chemical Lactate Sensor Printed with Silver Nano-Particles. *Sci. Rep.* **2016**, *6*, 30565. [CrossRef] [PubMed]
47. Imani, S.; Bandodkar, A.J.; Mohan, A.M.V.; Kumar, R.; Yu, S.; Wang, J.; Mercier, P.P. A Wearable Chemical-Electrophysiological Hybrid Biosensing System for Real-Time Health and Fitness Monitoring. *Nat. Commun.* **2016**, *7*, 11650. [CrossRef] [PubMed]
48. Guan, H.; Zhong, T.; He, H.; Zhao, T.; Xing, L.; Zhang, Y.; Xue, X. A Self-Powered Wearable Sweat-Evaporation-Biosensing Analyzer for Building Sports Big Data. *Nano Energy* **2019**, *59*, 754–761. [CrossRef]
49. Rathee, K.; Dhull, V.; Dhull, R.; Singh, S. Biosensors Based on Electrochemical Lactate Detection: A Comprehensive Review. *Biochem. Biophys. Rep.* **2016**, *5*, 35–54. [CrossRef] [PubMed]
50. Gao, W.; Emaminejad, S.; Nyein, H.Y.Y.; Challa, S.; Chen, K.; Peck, A.; Fahad, H.M.; Ota, H.; Shiraki, H.; Kiriya, D.; et al. Fully Integrated Wearable Sensor Arrays for Multiplexed in Situ Perspiration Analysis. *Nature* **2016**, *529*, 509–514. [CrossRef]
51. Luo, X.; Guo, L.; Liu, Y.; Shi, W.; Gai, W.; Cui, Y. Wearable Tape-Based Smart Biosensing Systems for Lactate and Glucose. *IEEE Sens. J.* **2020**, *20*, 3757–3765. [CrossRef]
52. Roda, A.; Guardigli, M.; Calabria, D.; Maddalena Calabretta, M.; Cevenini, L.; Michelini, E. A 3D-Printed Device for a Smartphone-Based Chemiluminescence Biosensor for Lactate in Oral Fluid and Sweat. *Analyst* **2014**, *139*, 6494–6501. [CrossRef] [PubMed]
53. Chen, M.-M.; Cheng, S.-B.; Ji, K.; Gao, J.; Liu, Y.-L.; Wen, W.; Zhang, X.; Wang, S.; Huang, W.-H. Construction of a Flexible Electrochemiluminescence Platform for Sweat Detection. *Chem. Sci.* **2019**, *10*, 6295–6303. [CrossRef]
54. Yokus, M.A.; Songkakul, T.; Pozdin, V.A.; Bozkurt, A.; Daniele, M.A. Wearable Multiplexed Biosensor System toward Continuous Monitoring of Metabolites. *Biosens. Bioelectron.* **2020**, *153*, 112038. [CrossRef]

55. Biagi, S.; Ghimenti, S.; Onor, M.; Bramanti, E. Simultaneous Determination of Lactate and Pyruvate in Human Sweat Using Reversed-Phase High-Performance Liquid Chromatography: A Noninvasive Approach. *Biomed. Chromatogr.* **2012**, *26*, 1408–1415. [CrossRef]
56. Zhang, Y.; Tao, T.H. A Bioinspired Wireless Epidermal Photoreceptor for Artificial Skin Vision. *Adv. Funct. Mater.* **2020**, *30*, 2000381. [CrossRef]
57. Souza, S.L.; Graça, G.; Oliva, A. Characterization of Sweat Induced with Pilocarpine, Physical Exercise, and Collected Passively by Metabolomic Analysis. *Ski. Res. Technol.* **2018**, *24*, 187–195. [CrossRef] [PubMed]
58. Bhide, A.; Lin, K.; Muthukumar, S.; Prasad, S. On-Demand Lactate Monitoring towards Assessing Physiological Responses in Sedentary Populations. *Analyst* **2021**, *146*, 3482–3492. [CrossRef] [PubMed]
59. Shay, T.; Saha, T.; Dickey, M.D.; Velev, O.D. Principles of Long-Term Fluids Handling in Paper-Based Wearables with Capillary–Evaporative Transport. *Biomicrofluidics* **2020**, *14*, 034112. [CrossRef] [PubMed]
60. Yokus, M.A.; Saha, T.; Fang, J.; Dickey, M.D.; Velev, O.D.; Daniele, M.A. Towards Wearable Electrochemical Lactate Sensing Using Osmotic-Capillary Microfluidic Pumping. In Proceedings of the 2019 IEEE Sensors, Montreal, QC, Canada, 27–30 October 2019; pp. 1–4.
61. Klous, L.; de Ruiter, C.J.; Scherrer, S.; Gerrett, N.; Daanen, H.A.M. The (in)Dependency of Blood and Sweat Sodium, Chloride, Potassium, Ammonia, Lactate and Glucose Concentrations during Submaximal Exercise. *Eur. J. Appl. Physiol.* **2021**, *121*, 803–816. [CrossRef]
62. Zhao, C.; Li, X.; Wu, Q.; Liu, X. A Thread-Based Wearable Sweat Nanobiosensor. *Biosens. Bioelectron.* **2021**, *188*, 113270. [CrossRef] [PubMed]
63. Xuan, X.; Pérez-Ràfols, C.; Chen, C.; Cuartero, M.; Crespo, G.A. Lactate Biosensing for Reliable On-Body Sweat Analysis. *ACS Sens.* **2021**, *6*, 2763–2771. [CrossRef] [PubMed]
64. Vinoth, R.; Nakagawa, T.; Mathiyarasu, J.; Mohan, A.M.V. Fully Printed Wearable Microfluidic Devices for High-Throughput Sweat Sampling and Multiplexed Electrochemical Analysis. *ACS Sens.* **2021**, *6*, 1174–1186. [CrossRef] [PubMed]
65. Anderson, G.S.; Rhodes, E.C. A Review of Blood Lactate and Ventilatory Methods of Detecting Transition Thresholds. *Sports Med.* **1989**, *8*, 43–55. [CrossRef] [PubMed]
66. Green, J.M.; Pritchett, R.C.; Crews, T.R.; McLester, J.R.; Tucker, D.C. Sweat Lactate Response between Males with High and Low Aerobic Fitness. *Eur. J. Appl. Physiol.* **2004**, *91*, 1–6. [CrossRef] [PubMed]
67. Broskey, N.T.; Pories, W.J.; Jones, T.E.; Tanner, C.J.; Zheng, D.; Cortright, R.N.; Yang, Z.W.; Khang, N.; Yang, J.; Houmard, J.A.; et al. The Association between Lactate and Muscle Aerobic Substrate Oxidation: Is Lactate an Early Marker for Metabolic Disease in Healthy Subjects? *Physiol. Rep.* **2021**, *9*, e14729. [CrossRef]

*micromachines*

*Article*

# Optimization and Fabrication of Multi-Level Microchannels for Long-Term Imaging of Bacterial Growth and Expansion

Hsieh-Fu Tsai [1,2,*], Daniel W. Carlson [1], Anzhelika Koldaeva [3], Simone Pigolotti [3] and Amy Q. Shen [1,*]

1   Micro/Bio/Nanofluidics Unit, Okinawa Institute of Science and Technology Graduate University, Onna-son, Okinawa 904-0495, Japan; daniel.carlson@oist.jp
2   Department of Biomedical Engineering, Chang Gung University, Taoyuan 333, Taiwan
3   Biological Complexity Unit, Okinawa Institute of Science and Technology Graduate University, Onna-son, Okinawa 904-0495, Japan; anzhelika.koldaeva1@oist.jp (A.K.); simone.pigolotti@oist.jp (S.P.)
*   Correspondence: hftsai@cgu.edu.tw (H.-F.T.); amy.shen@oist.jp (A.Q.S.); Tel.: +886-3-2118800 (ext. 3079) (H.-F.T.)

**Abstract:** Bacteria are unicellular organisms whose length is usually around a few micrometers. Advances in microfabrication techniques have enabled the design and implementation of microdevices to confine and observe bacterial colony growth. Microstructures hosting the bacteria and microchannels for nutrient perfusion usually require separate microfabrication procedures due to different feature size requirements. This fact increases the complexity of device integration and assembly process. Furthermore, long-term imaging of bacterial dynamics over tens of hours requires stability in the microscope focusing mechanism to ensure less than one-micron drift in the focal axis. In this work, we design and fabricate an integrated multi-level, hydrodynamically-optimized microfluidic chip to study long-term *Escherichia coli* population dynamics in confined microchannels. Reliable long-term microscopy imaging and analysis has been limited by focus drifting and ghost effect, probably caused by the shear viscosity changes of aging microscopy immersion oil. By selecting a microscopy immersion oil with the most stable viscosity, we demonstrate successful captures of focally stable time-lapse bacterial images for $\geq$72 h. Our fabrication and imaging methodology should be applicable to other single-cell studies requiring long-term imaging.

**Keywords:** multi-level microfluidic device; live cell imaging; long-term microscopy imaging; focus drifting; immersion oil viscosity; bacterial population dynamics; single-cell studies; *E. coli*; mother machine; computational fluid dynamics

**Citation:** Tsai, H.-F.; Carlson, D.W.; Koldaeva, A.; Pigolotti, S.; Shen, A.Q. Optimization and Fabrication of Multi-Level Microchannels for Long-Term Imaging of Bacterial Growth and Expansion. *Micromachines* **2022**, *13*, 576. https://doi.org/10.3390/mi13040576

Academic Editors: Violeta Carvalho, Senhorinha Teixeira and João Eduardo P. Castro Ribeiro

Received: 21 March 2022
Accepted: 5 April 2022
Published: 7 April 2022

**Publisher's Note:** MDPI stays neutral with regard to jurisdictional claims in published maps and institutional affiliations.

## 1. Introduction

Even simple bacteria such as *Escherichia coli* (*E. coli*) give rise to non-trivial phenomena as their population grows. For example, it has been observed that individuals are characterized by measurable phenotypic differences that persist through generations [1]. These differences, in particular fluctuations of cell division times, affect the population as a whole [2–4]. Further, when bacteria grow in space, they can align with their neighbor and generate complex spatial patterns [5,6]. These dynamics can affect the genetic diversity of the population [7]. Some of these effects can be quantitatively studied by means of microfluidic devices. Indeed, advances in microfabrication have enabled creation of microdevices with confined geometry and controlled microenvironments to study microbial dynamics at single-cell resolution, see, e.g., [8–15]. These devices can also be used to mimic microstructures, such as those found in soil, using controlled flow and microenvironmental conditions [16]. More complex microenvironments can be mimicked in multi-level microdevices for studying interaction between bacteria and human physiology, e.g., bacterial interaction with intestinal villi epithelial cells, in a quantitative manner [17–19].

However, building microdevices for bacterial studies presents its own challenges. For one, entrapping micron-sized bacteria that are sensitive to external flows requires optimized hydrodynamically confined design with feasible microfabrication methods [15,20–23]. A careful hydrodynamical design is essential for nutrient perfusion, waste removal, and prevention of bacterial biofilm formation to avoid microchannel blockage [24]. In addition, a robust stable imaging system is necessary to achieve long-term single-cell studies [25].

Microchannels hosting cells serve as a model system to mimic and study cell growth and competition in broad topics, such as biofilm formation in the soil and stem cell proliferation in the intestinal duct [19,26–28]. In the original work involving microchannels to entrap and study the outgrowth of single bacterium with one open end, also known as the mother machine, single cell genetics was investigated but competitions among bacterial clones could not be examined due to the limitation of diffusion-based nutrient delivery with one open and one closed-end microchannel design [8]. Hashimoto et al. developed microchannels with two open ends where bacterial growth rate was found to increase in certain clonal populations compared to those of the ensemble colony [2]. However, the microchannel dimensions (1.5–3 µm in width) in these studies were limited by the microfabrication processes (resolution-limiting lithography and isotropic wet etching) and the flow cell was not optimized to prevent microbubble formation, leading to unstable flow field in which high shear gradient could cause filamentation of bacteria [29].

In this study, we devise a straightfoward microfabrication process to create multi-level microstructures by employing a maskless aligner, a spin coater, and a hotplate. Our hydrodynamically-optimized microchannel design facilitates bacteria entrapment and culturing for more than 80 h. With this device, we have successfully imaged bacteria without focus drift by using a stable low viscosity immersion oil for over 60 h.

## 2. Materials and Methods

### 2.1. Design and Fabrication of Multi-Level Microchannel in Silicon Mold

The multi-level microchannel was designed in AutoCAD (AutoDesk, San Rafael, CA, USA) and exported to GDS format for maskless lithography using KLayout. The design consists of three layers: the first alignment mask layer, the flow channel layer, and the growth channel layer (Figure 1). The flow channel layer (for nutrient delivery and bacterial removal) was 15 µm in depth and the growth channel layer (for bacterial entrapment, outgrowth, and long-term imaging) was 1 µm in depth.

To create the microstructure mold, we developed a lithography approach using a single negative photoresist (mr-DWL-5, Micro Resist Technology GmbH, Berlin, Germany) and a maskless writer (DL-1000, Nanosystem Solutions, Okinawa, Japan). The hard baking temperature and exposure dosage protocol was based on the manufacture instructions. Briefly, after lithographic exposure, the wafer was hard baked and developed in propylene glycol methyl ether acetate (PGMEA, Sigma-Aldrich, Burlington, MA, USA), washed thoroughly with isopropanol, and dried with nitrogen air. The true depth of the microstructures was confirmed with a stylus profilometer (DektakXT, Bruker, Billerica, MA, USA).

### 2.2. Fabrication and Assembly of the PDMS/PMMA Microdevice

The silicone mold containing the multi-level microstructures was passivated with Trichloro(1H, 1H, 2H, 2H-perfluorooctyl)silane (Sigma-Aldrich, Burlington, MA, USA) in a vacuum dessicator for 2 h to render the surface hydrophobic and ease removal of mold. The passivated mold was placed in a custom polytetrafluoroethylene holder made in-house to cast 4 mm-high poly(dimethylsiloxane) precursor (PDMS, Sylgard 184, Dow Corning, Midland, MI, USA), a biocompatible transparent silicone rubber, containing the monomer:curing agent = 10:1 ratio mixed with an orbital mixer (ARE-310, Thinky, Tokyo, Japan). Then, 50 g of mixed PDMS precursor was poured on the multi-level mold, degassed in vacuum, covered with a transparent acrylic sheet, and cured at 80 °C for at least 4 h to crosslink PDMS into gel form. Individual PDMS microdevice was diced, inlet and outlet holes were punched on the device with a puncher (21 gauge, Accu-punch MP, Syneo, West Palm Beach, FL, USA).

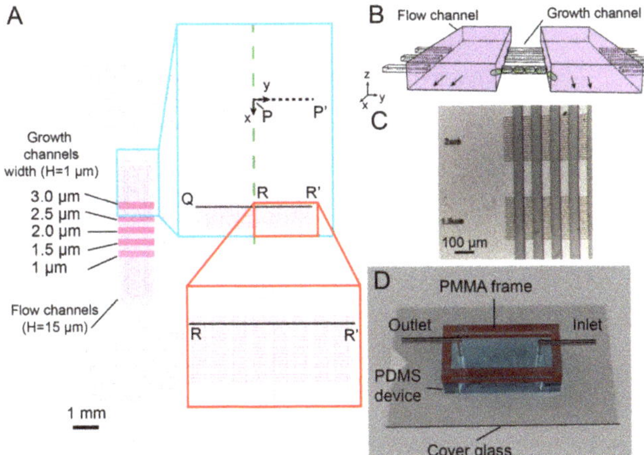



**Figure 1.** The microdevice containing multi-level microchannnels for long-term bacterial expansion imaging. (**A**) Schematic of the microchannel design in the PDMS device. Two layers are shown in purple (flow channels for nutrient delivery, 15 µm in height) and pink (growth channels for bacterial growth, 1 µm in height). The flow channels are joined by a tree-like flow stabilizing structure and a flow resistor to the inlet and outlet of the device. The green dashed line shows the symmetry axis used in OpenFOAM numerical simulation. The black dashed line (P-P′) indicates the region and the coordinate system used for numerical simulations in the flow channels. The zoomed-in part of the red box displays the growth channels (3 µm-wide) housing the bacteria. The black line segment (Q-R-R′) represents the growth channel analyzed in the numerical simulation (**B**). The perspective diagram of two flow channels (purple) and three growth channels illustrates their 3D relationship. The bacteria are cultured in the growth channels and old bacteria are expelled and sheared off in flow channels. The diagram is not to scale for sake of clarity. (**C**) Episcopic stereo-microscopy image of the silicon mold showing the high accuracy in aligning the two layers of the PDMS microchannels. (**D**) Snapshot of the assembled PMMA/PDMS microfluidic device for the bacterial experiments. The top reservoir encircled by the PMMA frame are filled with liquid buffer. Small air-bubbles are prevented from entering the chip and the hydrostatic pressure differences are mitigated to prevent secondary flow disturbing the bacteria.

Due to the small working distance (0.13 mm) of 100× microscopy objectives required for bacterial imaging, high-precision coverglass (no.1.5H, 60 × 24 × 0.175 mm, Paul Marienfeld GmbH, Lauda-Königshofen, Germany) was used. The coverglasses were mounted on a 3D printed washing stand made in-house and ultrasonically cleansed in a detergent bath (TFD4, Franklab, Montigny-le-Bretonneux, France) and ultrapure water (MilliQ, Millipore, Billerica, MA, USA) before drying at 80 °C.

The PDMS microdevice treated by an ambient air plasma (PDC-001-HP, Harrick Plasma, Ithaca, NY, USA) was irreversibly bonded to a coverglass. The microdevice was tilted 5–10° to the glass edge to mitigate the interference patterns caused by microstructure induced diffractions, ensuring the focus locking mechanism on the microscope (Perfect Focus System, Nikon, Tokyo, Japan). A 2 mm-thick polymethylmethacrylate (PMMA) sheet was cut with a $CO_2$ laser cutter (VLS2.30, Universal Laser Systems, Scottsdale, AZ, USA) and bonded with the PDMS microdevice with a dual energy double sided tape (5302A, Nitto Denko Corporation, Osaka, Japan) [30], completing the assembly of the PDMS/PMMA microdevice. To avoid biofilm formation and clogging of the 1 micron-high growth channels, a pristine microdevice was used for each experiment.

### 2.3. Bacterial Cultivation and Maintenance

The *E. coli* strain MG1655 derived from K-12 wild-type strain was used in this study. MG1655 was cultivated and maintained in Luria–Bertani agar or LB broth (LB Lennox, Becton Dickinson, Franklin Lakes, NJ, USA). To assist tracking bacteria of different lineages with fluorescence microscopy, MG1655 strain were transformed with a low copy plasmid constitutively expressing a green fluorescence protein (GFP) or mCherry red fluorescence protein (pUA66 PrpsL-GFP Kan$^R$ or pUA66 PrpsL-mCherry Kan$^R$). The pUA66 plasmid harbors a cassette containing a kanamycin resistance gene and a promoter sequence for 30S ribosomal protein upstream of the fluorescent protein open reading frame (ORF). The pUA66 PrpsL-GFP was a gift from Dr. Pamela Silver from Harvard University (Addgene plasmid #165606; accessed date: 2022/04/05; http://n2t.net/addgene:105606; RRID: Addgene_105606) [31]. Details on pUA66 PrpsL-mCherry production can be found in our recent work [7].

The pUA66 plasmid was transformed in MG1655 using standard chemical transformation with 0.1 M calcium chloride and 42 °C heat shock. The transformed bacterial colonies bearing the plasmid were selected with 50 µg mL$^{-1}$ kanamycin (Nacalai Tesque, Kyoto, Japan) in the LB broth or agar. The bacteria strain was preserved under −80 °C following overnight outgrowth in LB broth and addition of sterile glycerol to a final concentration of 15 wt%.

### 2.4. Numerical Simulation

Both finite element method (FEM) and finite volume method (FVM) are commonly used numerical methods in computational fluid dynamics (CFD), differing in their spatial discretization schemes. Both methods perform similarly for a given accuracy, but FVM may require less memory than FEM under certain implementations [32–34]. Mature commercial software, such as the FEM solver COMSOL Multiphysics® are very user friendly and suitable for rapid design iteration of complex systems. Therefore, we leveraged COMSOL for initial design iteration of the overall device, and used the FVM solver OpenFOAM considering a reduced geometry with a refined mesh for a final high-fidelity analysis of the hydrodynamic performance [35].

We assumed that the aqueous LB broth and dilute *E. coli* suspension in LB broth used in our experiments were Newtonian fluids, corresponding to the continuity equation (Equation (1)) and Navier-Stokes equation (Equation (2)),

$$\nabla \cdot \mathbf{u} = 0, \tag{1}$$

$$\rho \left( \frac{\partial \mathbf{u}}{\partial t} + \mathbf{u} \cdot \nabla \mathbf{u} \right) = -\nabla p + \mu \nabla^2 \mathbf{u} + f_b, \tag{2}$$

where $\mathbf{u}$, $p$, and $f_b$ represent the velocity vector, pressure, and body forces (per unit volume) that act on the Newtonian fluid, respectively. Due to the small length scales associated with the microfluidic device and the maximum mass flow rate $Q$ (10 µL min$^{-1}$) on the inlet in our study, Equation (2) can be simplified and solved as steady-state laminar flow without effects from the body force and temporal terms. The Reynolds number $Re = \rho U W / \mu \approx 11$ with $U = Q/A$, $A$ is the cross section area of the microchannel. We further imposed zero pressure boundary condition at the outlet, and a no-slip boundary ($\mathbf{u} = 0$) at the channel walls. For OpenFOAM we also introduced a symmetry condition at the midplane (see Figure 1A).

We first investigated the entire 3D model with COMSOL for rapid design iteration of the flow channels. We kept the full detail of the geometry but partitioned it with nested subdomains about the growth channels to handle meshing the length-scale disparity, culminating in a total element count of $8 \times 10^7$ under the COMSOL "fine" mesh sizing. The finest element edge length to gap ratio was 1/40 for both the flow channels and the growth channels. Steady-state laminar flow through the device was solved with a COMSOL batch routine running on 128 processors. To probe the flow profile in the growth channels, we extracted an $x - y$ slice of the volume at $z = 0.5$ μm (mid-plane of the growth channels).

For detailed validation of flow profiles in the growth channels, we employed FVM-based OpenFOAM software for a reduced geometry mesh, achieving a cell edge length to gap ratio of $1/530$ in the growth channels and $1/180$ in the flow channels. The finer resolution in the growth channel provided a better description of the flow profile.

### 2.5. Experimental Setup and Live Cell Microscopy

To expel microbubbles, the microchannels in the PDMS/PMMA microdevice were passivated first with 99.5% ethanol (Nacalai Tesque, Kyoto, Japan) and then with ultra-pure water. The microchannel was then passivated with a solution composed of 2 mg mL$^{-1}$ bovine serum albumin (BSA, Nacalai Tesque, Kyoto, Japan) and 0.5 mg mL$^{-1}$ salmon sperm DNA (Thermo Fisher Scientific, Waltham, MA, USA) for 30 min before washed away by phosphate buffered saline (PBS). The top reservoir enclosed by the PMMA frame was filled with PBS as well to avoid entrapment of air-bubbles at the inlet or the outlet.

Prior to seeding bacteria in the microchannels, MG1655 *E. coli* bacteria outgrown overnight in LB broth were inoculated to a new tube of 4 mL LB broth with 0.1 mM kanamycin. The *E. coli* suspension was grown at 37 °C and 200 rpm in an orbital shaker until the bacteria reached the log phase with OD$_{600}$ = 0.2. Then, 1 mL of bacterial suspension was centrifuged at 3000× $g$ for 10 min and resuspended in a 50 μL LB broth. Then, 10 μL of bacterial suspension was injected into the PDMS/PMMA device with a micropipet. The top reservoir was filled with PBS to prevent hydrostatic pressure difference in the system [30], allowing bacteria to stay in the growth channels without disturbance.

A PDMS/PMMA microdevice was affixed on a holder in a stage-top incubator (WKSM, Tokai Hit, Shizuoka, Japan) on an inverted motorized epi-fluorescence microscope equipped with focus locking mechanism (Ti-E, Nikon, Tokyo, Japan). A dummy PDMS/PMMA microdevice was placed on the holder as well and affixed with a K-type thermocouple (ANBE MST Co., Kanagawa, Japan). The stage-top incubator increased the temperature of the microdevice to 37 °C within 10 min and maintained this temperature onward.

A set of tubing with stainless steel tubes (21 gauge, New England Small Tube, Litchfield, NH, USA) were attached to a 25 mL glass syringe (Trajan Scientific and Medical, Victoria, Austria) filled with M9 media supplemented with 1 μM rifapentine, 0.1 mM kanamycin, 2 mM glucose, and 1× MEM vitamins (Sigma-Aldrich, Burlington, MA, USA) and an empty 20 mL plastic syringe (Terumo, Tokyo, Japan) [7]. Rifapentine were added to inhibit biofilm formation without affecting the bacterial activity [36]. The stainless steel tube connected to the media filled syringe was inserted into the inlet and the tube connected to an empty syringe was inserted into the outlet. The M9 infusion at 1.6 μL min$^{-1}$ using a multichannel syringe pump (neMESYS, Cetoni GmbH, Korbussen, Germany) started after the bacteria inhabited stably in the growth channel. The perfusion rate doubled every 2 h until it reached 10 μL min$^{-1}$. The gradual ramping of flow rate was essential to prevent sudden pressure change that could expel the bacteria from the growth channels.

For long-term imaging of bacteria, an 100× oil immersion objective (Plan apo $\lambda$, Nikon, Tokyo, Japan) with 1.5× intermediate magnification was used to take time-lapse images after appropriate immersion oil was placed between the objective and the bottom of the PDMS/PMMA microdevice. The depth of field of 100× objective was only 0.39 μm so a focus locking system was used to mitigate thermal drift over long-term imaging (PFS, Nikon, Tokyo, Japan). The bacterial expansion in the growth channels were imaged with mercury lamp excitation (Intensilight, Nikon, Tokyo, Japan) together with GFP or mCherry filter cubes (Semrock, Rochester, NY, USA). The images were acquired using a scientific complementary oxide semiconductor camera (sCMOS, Prime95B, Teledyne Photometrics, Tucson, AZ, USA) at an interval of 3 min.

*2.6. Viscosity and Refractive Index Measurements of Immersion Oils*

We investigated the dynamic viscosity and refractive index of commonly used immersion oils, including Nikon NF2, Nikon F, Nikon F2, Olympus Immoil-F, and Millipore 104699 oil. The dynamic viscosity of the immersion oils were measured using a strain-controlled rotational rheometer (ARES-G2, TA Instruments, New Castle, DE, USA) with a stainless steel 50 mm cone-plate geometry (1° angle). Immersion oils of 650 μL were dispensed using a positive displacement micropipette (Microman, Gilson, Middleton, WI, USA). The rheometer fixture was covered by a solvent trap and all the measurements were performed at $37 \pm 0.1$ °C controlled via the Advanced Peltier System (TA Instruments, New Castle, DE, USA). The shear viscosity $\eta$ was measured at steady state with the shear rate $\dot{\gamma}$ being varied from 0.01 to 100 s$^{-1}$ with logarithmically increasing steps. To create aged oil samples, the pristine immersion oils were first dispensed into 1.5 mL microcentrifuge tubes with loose lids and later incubated for 60 hours on a 37 °C dry bath. The aged oil samples were then loaded and measured on the ARES-G2 rheometer following the same protocol as the pristine ones.

The refractive index of immersion oils were measured by an automatic refractometer (Abbemat MW, Anton Paar GmbH, Graz, Austria) at 23 °C and 589.3 nm (nD). The refractometer was calibrated by two-point adjustment with industrial reference standards (Kyoto Electronics Co., Ltd, Kyoto, Japan). Half a milliliter of pristine or aged immersion oil samples were dispensed into the sample well and the temperature was allowed to equilibrate before the results were taken. Two-way ANOVA with Tukey's post-hoc test was conducted and null hypothesis was tested.

## 3. Results

*3.1. Simplified Multi-Level Microstructure Fabrication by Maskless Lithography*

The multi-level microchannel design consisted of a 15 μm-depth flow channel layer (purple, Figure 1A) for delivering nutrient and removing bacteria, overlaid on top of 1 μm-depth growth channels (pink, Figure 1A) with varied widths (1 μm to 3 μm) in which bacteria were retained and outgrown in a stable fashion (Figure 1B). The top flow channel layer contained 16 flow channels (Length L × Width W = 4500 × 50 μm) which branched uniformly into tree-like microstructures from inlet and outlet with flow resistors. The 30 μm-long growth channels with different widths (1 μm to 3 μm, 0.5 μm incremental depth, 10 channels per set with 10 μm inter-spacing) were designed so that bacteria could outgrow and expel from both ends, in contrast to the single-ended design of the mother machine [8,37].

The large aspect-ratio differences of the two channel layers (Figure 1B) often require microfabrications with two processes: one for the flow channel layer and the other for the growth channel layer [2,10,38–40]. We developed a simple lithography approach using a single negative photoresist and a maskless writer (DL-1000, Nanosystem Solutions, Okinawa, Japan). Although there were two layer structures, the small 1 μm thickness of the first layer created minimum contrast, making it difficult to be detected by the on-board camera in the maskless writer. To resolve this issue, we used an additional layer with cross marks for alignment. First, a 5 μm-thick photoresist was spin-coated on a 100 mm silicon wafer. Several 50 × 200 μm cross marks were placed at the center of the wafer and exposed at 405 nm by the maskless writer. To fabricate the 1 μm-high growth channel layer, the mr-DWL-5 photoresist was diluted with gamma butyrolactone (Sigma-Aldrich, Burlington, MA, USA) at 7.5:1 (*w/w*) ratio and spin-coated at 7500 rpm 30 s. The flow channel was aligned to the cross marks and exposed with 2× subpixel exposure followed by a post-exposure bake at 95 °C for 5 min.

After the wafer cooled down to room temperature, without development of the flow channel layer, the mr-DWL-5 photoresist (15 μm) was spin-coated on the wafer at 750 rpm for 30 s and soft baked at 95 °C for 5 min. The flow channel pattern was also aligned to the cross pattern, exposed with the maskless writer, and baked at 95 °C for 5 min. The entire wafer was then developed in PGMEA and washed thoroughly with isopropanol and dried with nitrogen air. No visible defect was found in the microstructures made with the diluted photoresist. By aligning to the cross marks, good alignment could be achieved despite repeated photoresist coating (Figure 1C). The alignment process and exposure process were carried out directly in the maskless writer, therefore simplifying the workflow and the mold was created within one working day. Finally, PDMS microdevices were fabricated by standard molding process from the aforementioned mold and bonded to a coverglass with acrylic frame attached to the PDMS (details in the Section 2.2 in Materials and Methods).

### 3.2. Optimal Loading and Stable Culture of Bacteria by Balancing Hydrostatic Pressure and Optimizing Flow Resistance in the Microdevice

Due to the small height of the growth channels, it is difficult to introduce a single layer of bacteria inside the growth channels during loading. By balancing the hydrostatic pressure through filling liquid buffer in the top reservoir encircled by the PMMA frame (Figure 1D) [30], hydrostatic pressure-driven secondary flow could be prevented so that bacteria would swim into the growth channels in a stable manner.

In addition, pressure differences across two flow channels could cause pressure-driven flows to remove bacteria from the growth channels [2,10]. To address this challenge, the flow resistors and tree-like microstructures were introduced in our design as they were essential for uniform and stable flow distribution to keep the bacteria from escaping in the open-ended growth channels (Figure 1A,B).

We further optimized and validated the microchannel design by performing computational fluid dynamics analysis. First, we investigated the full domain of the microfluidic design and optimized the flow uniformity across all 16 flow channels by the flow resistors and tree-like microstructures using FEM-based COMSOL simulation (Figure 1A). Flow velocity differences across different flow channels were less than 0.8% (Figure 2A). However, due to the large length-scale differences between the flow channel and the growth channel, mesh joining, and quality assurance became challenging, resulting in a coarse spatial resolution around the growth channels.

To better resolve flow in the growth channels, we next numerically investigated a reduced geometry using FVM-based OpenFOAM by assuming symmetry across the mid-plane of $x$–$y$ plane (green dashed line, Figure 1A). We removed the outlet contraction such that only the inlet expansion and eight flow channels of length $x = 810$ µm remained in the simulation. We considered only four rows of the 3 µm-wide growth channels starting at $x = 452$ µm. The reduced geometry was then taken as the input to the snappyHexMesh utility, whereby an initial uniform background mesh was "snapped" to the extents of the geometry and incrementally refined to form the desired hexahedral mesh. A no-slip boundary condition was imposed on all channel walls, and uniform velocity and zero pressure conditions were applied to the inlet and outlets, respectively. Three iteratively finer meshes were generated this way, up to a final cell count of $1.33 \times 10^8$ (mesh M3). Compared to the global COMSOL solution, the velocity profile noise floor about the growth channels was 3.5 times lower for M3. The velocity profile in the flow channels resolved by OpenFOAM (red symbols) coincided nicely with those by COMSOL (black curve), see Figure 2A.

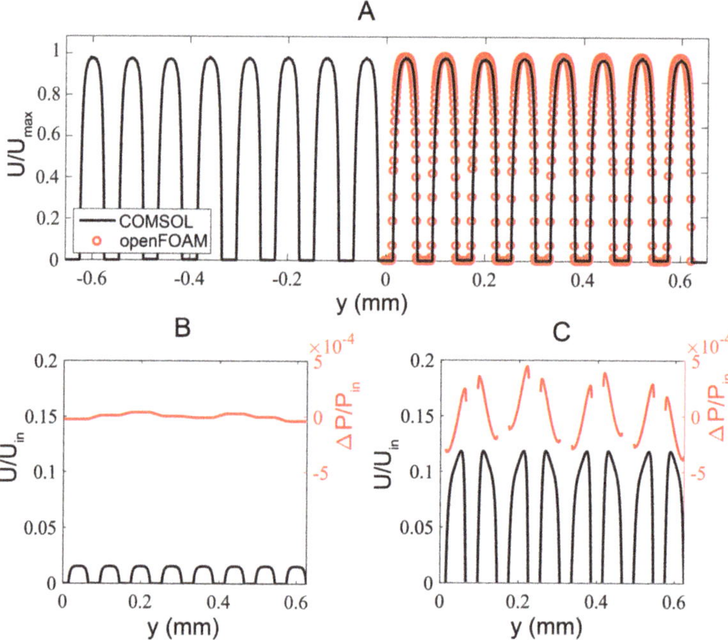

**Figure 2.** Numerical simulation results. (**A**) Flow profile comparison obtained from the full domain COMSOL solution (corresponding to Q-R′ segment in Figure 1A) and the reduced geometry OpenFOAM case, corresponding to R-R′ at $x = 452$ μm, $z = 7.5$ μm. (**B**) Flow and pressure profiles extracted from mid-plane of a growth channel (along R-R′ in Figure 1A; $x = 452$ μm, $z = 0.5$ μm). Secondary flow effects are negligible resulting in a nearly flat pressure distribution along the growth channels. The low pressure difference ensures that the bacteria are retained in the growth channels while strong shear takes place at the junction of growth channels and flow channels. (**C**) Flow and pressure profiles extracted at the $z$ mid-plane immediately after the expansion section (along P-P′ in Figure 1A; $x = 0$ μm, $z = 7.5$ μm).

Investigating the flow at the mid plane of a growth channel (R–R′ at $z = 0.5$ μm in Figure 1A), the flow profile in individual flow channel was uniform and laminar (black curve in Figure 2B). Furthermore, secondary flow effects across each flow channel segments from two flow channels were negligible as evidenced by the nearly flat pressure distribution along the R–R′ profile (red curve in Figure 2B). The low pressure difference ensured that bacteria were retained in the growth channels while the high shear rate at the junction of the flow channel and growth channel enabled easy removal of the bacteria expelled from the two open-ends of the growth channels.

Figure 2C displayed the flow and pressure profiles in eight symmetrical flow channels at the mid-plane immediately after the inlet tree-like expansion ($x = 0$ μm, $z = 7.5$ μm). The uniform flow velocity and comparable pressure between individual channels suggested that the tree-like microstructures helped to distribute the flow uniformly among the flow channels.

By hydrodynamic stabilization through the flow resistors and the tree-like microstructures in the microfluidic channels (Figure 1A) and the bubble-preventing world-to-chip interface design (Figure 1D), we have successfully developed a robust microfluidic device for bacteria culture and growth studies. The bacteria could be retained in the growth microchannels of 3 μm in width (top panel in Figure 3) or 1 μm in width (bottom panel in Figure 3) from the initial seeding and expanded to form clonal populations after perfusion for 60 h.

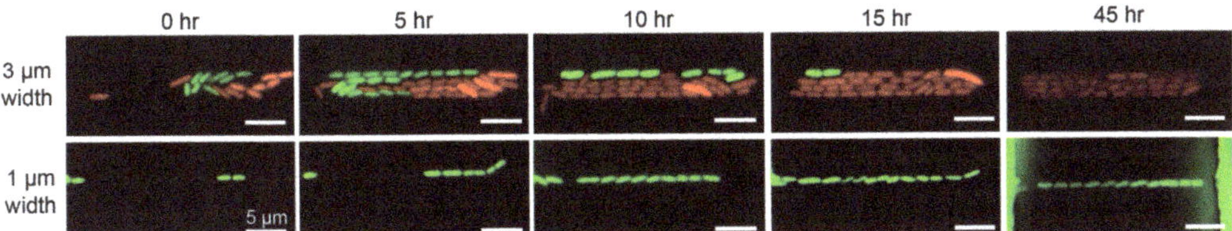



**Figure 3.** Stable bacterial expansion in 3 μm and 1 μm wide growth microchannels. Images are captured from 0 to 45 h in this specific study. (**Top panel**) Two bacterial populations individually carrying green or red fluorescence proteins compete in 3 μm-wide channels. After 15 h, the descendants of red fluorescence clone occupy the entire microchannel. (**Lower panel**) Outgrowth of *E. coli* carrying green fluorescence in a 1 μm-wide channels throughout 45 h. See also Ref. [7].

### 3.3. Choice of Immersion Oil for Long-Term Imaging of Bacterial Expansion in the Microdevice

Although the optimized microfluidic device we developed had enabled the bacteria to be cultured and grow consistently for more than three days (see Supplementary Video S1–S3), the microscopy focus often drifted after 10 hours during the long-term imaging session when the recommended immersion oil (Nikon F type) was used for the Nikon Ti-E microscope system.

The focus locking mechanism is essential for long-term microscopy imaging of biological specimen. Commercial microscope system equipped with focus locking mechanism utilizes an infrared light projection and detect its reflection at a glass–air or glass–liquid interface [41]. Any drift caused by thermal variation in the optical system or mechanical flexing of the sample can be corrected by refocusing along z-direction relatively to the interface with the motorized focusing stage. The 100× oil immersion objective (numerical aperture NA = 1.45) used in this study only has a depth of field of 0.39 μm and the *E. coli* bacteria are around 1 μm in size, therefore, if the focus locking mechanism is suboptimal or the glass–liquid interface contrast is not strong enough for detection, e.g., interfered by the refractive index difference between glass (1.516) and PDMS (1.43) [42], the image can easily drift out of focus or oscillate around the focal plane (ghost effect). The poor focus in images is the major cause of image segmentation failure in single cell tracking [43].

To understand the role of the immersion oil in the focus locking function of the microscopy imaging system, we investigated the dynamic viscosity and refractive index of commonly used immersion oils, including Nikon NF2, Nikon F, Nikon F2, Olympus Immoil-F, and Millipore 104699 oil. The dynamic viscosity of the immersion oils were measured using a strain-controlled rotational rheometer (ARES-G2, TA Instruments, New Castle, DE, USA) with a stainless steel 50 mm cone-plate geometry (1° angle). Figure 4A only displayed the flow curves of three representative immersion oils in their pristine and aged states. The dashed line in Figure 4A represents the limit of measurable viscosity defined by the instrumental torque limit [44,45],

$$\eta > \frac{F_\tau T_{min}}{\dot{\gamma}}, \tag{3}$$

where for cone-plate geometry, $F_\tau = 3/(2\pi R^3)$, $R = 25$ mm, and the minimal measurable torque on ARES-G2 rheometer $T_{min} = 0.1$ μN m. All the measured viscosity data were higher than the detectable viscosity at respective shear rates, suggesting that measurements were not affected by the inertia of the rheometer.

We found that the viscosity of the Nikon F type immersion oil increased by 262.5% after 60 h of aging under 37 °C (similar imaging condition in long-term bacteria expansion studies), see Figure 4B. Other immersion oils, such as Olympus Immoil-F and Nikon NF2, exhibited 18.2% and 34.5% of viscosity increases after aging. However, the Nikon F2 type immersion oil and Millipore 104699 immersion oil did not show significant viscosity changes after aging (Figure 4B).

Further characterization of the optical properties (refractive index) of all the immersion oils before and after 60 h of aging showed little difference, suggesting that the focus drift was mainly caused by the viscosity increase in the immersion oil over the long-term imaging period (Figure 4B).

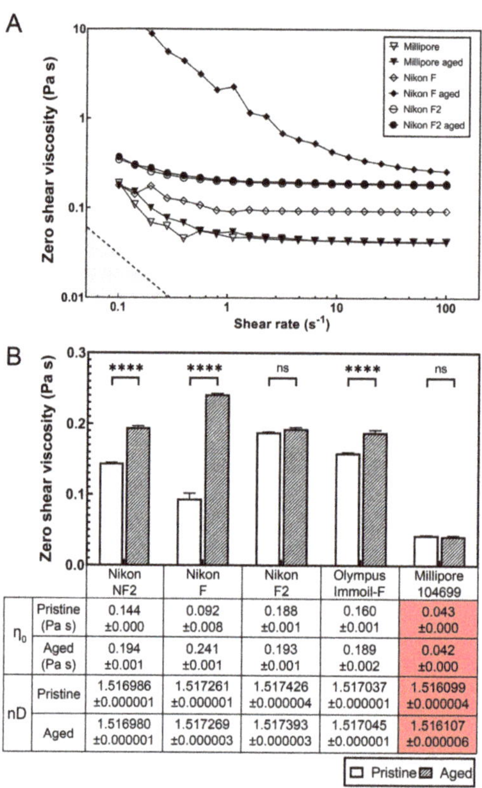

**Figure 4.** Dynamic viscosity and refractive index of microscopy immersion oils. (**A**) Shear viscosity-shear rate plot of pristine microscopy immersion oil (open symbols) and aged (60 h, closed symbols) at 37 °C measured by a rotational shear rheometer. Nikon F oil (diamond symbols) shows a significant viscosity increase after aging while Nikon F2 and Millipore oil display minimum viscosity difference between pristine and aged samples. (**B**) Zero shear viscosity approximated by best-fit flow calculated from three measurements. The mean zero shear viscosities are listed in the table. The error bars indicate standard deviation. **** denotes $p < 0.0001$; ns (non significant) denotes $p > 0.05$. The refractive indexes nD (measured at 589.3 nm, 23 °C) of pristine or 60 h-aged microscopy immersion oils are listed in the table.

We hypothesize that during the long-term imaging experiment, certain immersion oils are prone to aging, consequently their viscosities increase over time. Although the microscopy imaging system attempted to correct the focal length, the immersion oil with increased viscosity dragged the sample during Z stage movement, leading to error. As a result, the microscope focus gradually drifted over time even with focus locking mechanisms, see Figure 5A,B with Nikon type F immersion oil. Poorly focused images eventually contributed to inaccuracy or failure of data analysis. When using an immersion oil with low viscosity and no aging-associated changes, such as Millipore 104699 immersion oil, the focus drift was eliminated, see Figure 5C,D with Millipore immersion oil. Clearly focused images of bacterial expansion were acquired over more than three days (Supplementary

Video S3). Note that individual bacteria with clear boundary could be seen clearly, even though the intensity was a bit lower than those in Figure 5B. Use of microscopy immersion oils whose viscosity does not change much with aging, such as Millipore immersion oil, has provided good focus stability as required for long-term imaging.

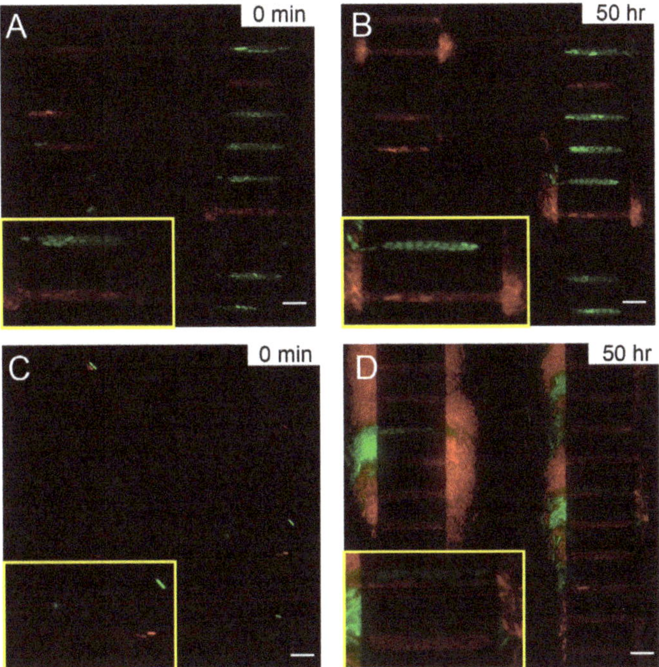


**Figure 5.** Imaging single bacteria expansion with Nikon type F immersion oil at (**A**) 0 min and (**B**) 50 h, and with Millipore immersion oil (**C**) at 0 h and (**D**) 50 h. The scale bars correspond to 10 μm. The yellow insets are enlarged views of two growth channels to highlight the focus quality.

## 4. Discussion

We have designed and validated numerically that the open-ended microchannels with various widths are suitable for cultivating and tracking bacterial expansion in long-term studies in a high throughput manner. In particular, previous studies were limited to single-width microchannels due to microfabrication limitations. These microdevices were only suitable to study growth expansion of one bacterium [8,40] or clonal expansion of three lanes of bacteria [2], limiting the experimental throughput.

Note that the microdevice in this study is designed for bacterial expansion and not for trapping of single bacterium using nanostructures [46,47] or microdroplets [48]. The open-ended microchannels in our design confine the degree of freedom of bacterial expansion while allowing bacteria to interact with each other. The bacteria grow and align horizontally into organized stripe patterns along the wall and eventually get expelled into the flow channel (Figure 3). A study of the clonal expansion and its mathematical model can be found in Ref. [7].

Although *E. coli* bacteria are randomly loaded into the growth channels in our microdevice by their active swimming and passive diffusion, single or multiple bacteria can form the colony in each growth channel. How the growth dynamics and cell–cell interactions in confined geometry may be affected by multiple bacterial strains with different phenotypes will be an interesting topic worth future investigations.

Drying of the immersion oil has been an issue during microscopy imaging, as dried oil must be prevented from depositing on the microscope lens [49,50]. In addition to the drying effect, our studies also reveal that certain immersion oils become more viscous over time, leading to focus drift and ghost effect during long-term imaging. Our results suggest that the choice of low-viscosity immersion oil without aging-associated viscosity change is vital for long-term bacterial imaging which requires exceptional focus locking performance. Future microbial or single cell imaging studies that require long-term observations with oil immersion lens can benefit from these findings.

Combining hydrodynamically-optimized microchannels and the use of optimal immersion oil, we have developed a high throughput platform to study bacterial growth and expansion in confined geometry. Bacteria without genetic manipulation other than fluorescence expression can be stably cultured in the microchannels. For future work, addition of active control components in the microdevice, such as microvalves, can provide active manipulation or trapping of bacteria [40]. Due to the use of wild type bacteria, long-term culture in microsystems inevitably leads to biofilm formation (Figure 5) despite channel passivation and use of low dosage antibiotic rifapentine. In the future, inhibition of biofilm formation by new antibiotics and new anti-fouling treatment might further increase the imaging duration of our platform.

Furthermore, our design strategies and device fabrication techniques can be applied to general studies of single-cell growth and cell–cell interactions in confined geometry [51]. For example, confined channels can mimic the porous microstructures in the soil to study how bacteria grow, adapt, and evolve in settings otherwise difficult to observe in nature [6,16,28,52,53]. In addition, the mechanical interaction due to single cell growth is not limited to prokaryotes, such as bacteria. In eukaryotic tissues, such as the epiderm or the intestinal duct, stem cells reside in specific stem cell layers, outgrow, and differentiate as the tissue develops [26,54]. Designing and optimizing microfluidic platforms to model these systems can shed new insights into their cellular dynamics.

## 5. Conclusions

In this study, we developed a process to align and fabricate multi-level and multi-width microchannels for long-term bacterial expansion studies. We optimized the microchannel design by balancing hydrostatic pressure and flow resistance in the microdevice, supported and validated by CFD analysis. As a result, we have achieved a stable culture of bacteria in open-ended microchannels. The hydrodynamical optimization of our microdevice can be adapted for designing new devices to study bacterial dynamics in microenvironments and a variety of other single-cell studies. Our findings on material property changes of immersion oils due to aging and evaporation are beneficial for tracking long-term live cell dynamics using microscopy imaging.

**Supplementary Materials:** The following supporting information can be downloaded at: https://www.mdpi.com/article/10.3390/mi13040576/s1, Video S1: The time-lapse movie of the expansion of bacteria bearing green or red fluorescence taken with Nikon F immersion oil. The scale bar corresponds to 10 μm. The focus slowly deviates and eventually causes single bacteria segmentation difficulty after around 15 h. Video S2: The time-lapse movie of the expansion of bacteria bearing green or red fluorescence taken with Millipore immersion oil. The scale bar indicates 10 μm. The focus locking is more stable throughout the experiment and allows long-term tracking and analysis. Video S3: The time-lapse movie of the expansion of *E. coli* expressing GFP for 80 h with stable focus taken with Millipore immersion oil. The scale bar indicates 10 μm.

**Author Contributions:** Conceptualization, H.-F.T., A.K., S.P. and A.Q.S.; Methodology, H.-F.T., A.K.; Numerical simulation, H.-F.T. and D.W.C.; Writing—original draft preparation, H.-F.T. and D.W.C.; Writing—review and editing, A.K., A.Q.S. and S.P.; Supervision, A.Q.S.; Funding acquisition, A.Q.S. All authors have read and agreed to the published version of the manuscript.

**Funding:** The authors thank Okinawa Institute of Science and Technology Graduate University (OIST) for its financial support with subsidy funding from the Cabinet Office, Government of Japan. H-F.T. had been financially supported by the Proof-of-Concept Program from Technology Development &

Innovation Center at OIST. The authors are also grateful for the support provided by the Scientific Computing and Data Analysis section of the Research Support Division at OIST.

**Institutional Review Board Statement:** Not applicable.

**Informed Consent Statement:** Not applicable.

**Data Availability Statement:** The data presented in this study are available on request from the corresponding author.

**Acknowledgments:** The authors thank Paola Laurino for gifting the MG1655 *E. coli* bacterial strain. We thank Emily Maria, Chua-Zu Huang, and Kang-Yu Chu for fruitful discussion. We also thank the computing resources provided by the Scientific Computing and Data Analysis Section at OIST. We appreciate Nikon Japan for gifting various immersion oils for characterization and testing.

**Conflicts of Interest:** The authors declare no conflicts of interest. The funders had no role in the design of the study; in the collection, analyses, or interpretation of data; in the writing of the manuscript, or in the decision to publish the results.

## Abbreviations

The following abbreviations are used in this manuscript:

| | |
|---|---|
| ANOVA | Analysis of variance |
| CFD | Computational fluid dynamics |
| FEM | Finite element method |
| FVM | Finite volume method |
| GFP | Green fluorescence protein |
| PDMS | Poly(dimethyl)siloxane |
| PMMA | Polymethylmethacrylate |

# References


1. Susman, L.; Kohram, M.; Vashistha, H.; Nechleba, J.T.; Salman, H.; Brenner, N. Individuality and slow dynamics in bacterial growth homeostasis. *Proc. Natl. Acad. Sci. USA* **2018**, *115*, E5679–E5687. [CrossRef] [PubMed]
2. Hashimoto, M.; Nozoe, T.; Nakaoka, H.; Okura, R.; Akiyoshi, S.; Kaneko, K.; Kussell, E.; Wakamoto, Y. Noise-driven growth rate gain in clonal cellular populations. *Proc. Natl. Acad. Sci. USA* **2016**, *113*, 3251–3256. [CrossRef] [PubMed]
3. Lin, J.; Amir, A. The effects of stochasticity at the single-cell level and cell size control on the population growth. *Cell Syst.* **2017**, *5*, 358–367. [CrossRef] [PubMed]
4. Pigolotti, S. Generalized Euler-Lotka equation for correlated cell divisions. *Phys. Rev. E* **2021**, *103*, L060402. [CrossRef]
5. Cho, H.; Jönsson, H.; Campbell, K.; Melke, P.; Williams, J.W.; Jedynak, B.; Stevens, A.M.; Groisman, A.; Levchenko, A. Self-Organization in High-Density Bacterial Colonies: Efficient Crowd Control. *PLoS Biol.* **2007**, *5*, e302. [CrossRef]
6. Sheats, J.; Sclavi, B.; Cosentino Lagomarsino, M.; Cicuta, P.; Dorfman, K.D. Role of growth rate on the orientational alignment of Escherichia coli in a slit. *R. Soc. Open sci.* **2017**, *4*, 170463. [CrossRef]
7. Koldaeva, A.; Tsai, H.F.; Shen, A.Q.; Pigolotti, S. Population genetics in microchannels. *Proc. Natl. Acad. Sci. USA* **2022**, *119*, e2120821119. [CrossRef]
8. Wang, P.; Robert, L.; Pelletier, J.; Dang, W.L.; Taddei, F.; Wright, A.; Jun, S. Robust Growth of Escherichia coli. *Curr. Biol.* **2010**, *20*, 1099–1103. [CrossRef]
9. Wessel, A.K.; Hmelo, L.; Parsek, M.R.; Whiteley, M. Going local: technologies for exploring bacterial microenvironments. *Nat. Rev. Microbiol.* **2013**, *11*, 337–348. [CrossRef]
10. Meyer, M.T.; Subramanian, S.; Kim, Y.W.; Ben-Yoav, H.; Gnerlich, M.; Gerasopoulos, K.; Bentley, W.E.; Ghodssi, R. Multi-depth valved microfluidics for biofilm segmentation. *J. Micromech. Microeng.* **2015**, *25*, 095003. [CrossRef]
11. Zhang, X.Y.; Sun, K.; Abulimiti, A.; Xu, P.P.; Li, Z.Y. Microfluidic System for Observation of Bacterial Culture and Effects on Biofilm Formation at Microscale. *Micromachines* **2019**, *10*, 606. [CrossRef] [PubMed]
12. Zhou, W.; Le, J.; Chen, Y.; Cai, Y.; Hong, Z.; Chai, Y. Recent advances in microfluidic devices for bacteria and fungus research. *Trends Analyt. Chem.* **2019**, *112*, 175–195. [CrossRef]
13. Busche, J.F.; Möller, S.; Stehr, M.; Dietzel, A. Cross-Flow Filtration of Escherichia coli at a Nanofluidic Gap for Fast Immobilization and Antibiotic Susceptibility Testing. *Micromachines* **2019**, *10*, 691. [CrossRef] [PubMed]
14. Morales Navarrete, P.; Yuan, J. A Single-Layer PDMS Chamber for On-Chip Bacteria Culture. *Micromachines* **2020**, *11*, 395. [CrossRef]
15. Hardo, G.; Bakshi, S. Challenges of analysing stochastic gene expression in bacteria using single-cell time-lapse experiments. *Essays Biochem.* **2021**, *65*, 67–79. [CrossRef]

16. Cai, P.; Sun, X.; Wu, Y.; Gao, C.; Mortimer, M.; Holden, P.A.; Redmile-Gordon, M.; Huang, Q. Soil biofilms: microbial interactions, challenges, and advanced techniques for ex-situ characterization. *Soil Ecol. Lett.* **2019**, *1*, 85–93. [CrossRef]

17. Kim, H.J.; Li, H.; Collins, J.J.; Ingber, D.E. Contributions of microbiome and mechanical deformation to intestinal bacterial overgrowth and inflammation in a human gut-on-a-chip. *Proc. Natl. Acad. Sci. USA* **2016**, *113*, E7–E15. [CrossRef]

18. Polini, A.; del Mercato, L.L.; Barra, A.; Zhang, Y.S.; Calabi, F.; Gigli, G. Towards the development of human immune-system-on-a-chip platforms. *Drug Discov. Today* **2019**, *24*, 517–525. [CrossRef]

19. Burmeister, A.; Hilgers, F.; Langner, A.; Westerwalbesloh, C.; Kerkhoff, Y.; Tenhaef, N.; Drepper, T.; Kohlheyer, D.; von Lieres, E.; Noack, S.; et al. A microfluidic co-cultivation platform to investigate microbial interactions at defined microenvironments. *Lab Chip* **2019**, *19*, 98–110. [CrossRef]

20. Christ, K.V.; Turner, K.T. Design of hydrodynamically confined microfluidics: controlling flow envelope and pressure. *Lab Chip* **2011**, *11*, 1491–1501. [CrossRef]

21. Hol, F.J.H.; Dekker, C. Zooming in to see the bigger picture: Microfluidic and nanofabrication tools to study bacteria. *Science* **2014**, *346*, 1251821. [CrossRef] [PubMed]

22. Qiu, X.; Huang, J.H.; Westerhof, T.M.; Lombardo, J.A.; Henrikson, K.M.; Pennell, M.; Pourfard, P.P.; Nelson, E.L.; Nath, P.; Haun, J.B. Microfluidic channel optimization to improve hydrodynamic dissociation of cell aggregates and tissue. *Sci. Rep.* **2018**, *8*, 2774. [CrossRef] [PubMed]

23. Kao, Y.T.; Kaminski, T.S.; Postek, W.; Guzowski, J.; Makuch, K.; Ruszczak, A.; von Stetten, F.; Zengerle, R.; Garstecki, P. Gravity-driven microfluidic assay for digital enumeration of bacteria and for antibiotic susceptibility testing. *Lab Chip* **2020**, *20*, 54–63. [CrossRef] [PubMed]

24. van Teeffelen, S.; Shaevitz, J.W.; Gitai, Z. Image analysis in fluorescence microscopy: Bacterial dynamics as a case study. *BioEssays* **2012**, *34*, 427–436. [CrossRef] [PubMed]

25. Skylaki, S.; Hilsenbeck, O.; Schroeder, T. Challenges in long-term imaging and quantification of single-cell dynamics. *Nat. Biotechnol.* **2016**, *34*, 1137–1144. [CrossRef] [PubMed]

26. Bach, S.P.; Renehan, A.G.; Potten, C.S. Stem cells: the intestinal stem cell as a paradigm. *Carcinogenesis* **2000**, *21*, 469–476. [CrossRef]

27. O'Donnell, A.G.; Young, I.M.; Rushton, S.P.; Shirley, M.D.; Crawford, J.W. Visualization, modelling and prediction in soil microbiology. *Nat. Rev. Microbiol.* **2007**, *5*, 689–699. [CrossRef]

28. Coyte, K.; Tabuteau, H.; Gaffney, E.; Foster, K.; Durham, W. Microbial competition in porous environments can select against rapid biofilm growth. *Proc. Natl. Acad. Sci. USA* **2016**, *114*, 161–170. [CrossRef]

29. Jahnke, J.P.; Smith, A.M.; Zander, N.E.; Wiedorn, V.; Strawhecker, K.E.; Terrell, J.L.; Stratis-Cullum, D.N.; Cheng, X. "Living" dynamics of filamentous bacteria on an adherent surface under hydrodynamic exposure. *Biointerphases* **2017**, *12*, 02C410. [CrossRef]

30. Tsai, H.F.; IJspeert, C.; Shen, A.Q. Voltage-gated ion channels mediate the electrotaxis of glioblastoma cells in a hybrid PMMA/PDMS microdevice. *APL Bioeng.* **2020**, *4*, 036102. [CrossRef]

31. Stirling, F.; Bitzan, L.; O'Keefe, S.; Redfield, E.; Oliver, J.W.K.; Way, J.; Silver, P.A. Rational Design of Evolutionarily Stable Microbial Kill Switches. *Mol. Cell* **2017**, *68*, 686–697.e3. [CrossRef] [PubMed]

32. Molina-Aiz, F.; Fatnassi, H.; Boulard, T.; Roy, J.C.; Valera, D. Comparison of finite element and finite volume methods for simulation of natural ventilation in greenhouses. *Comput. Electron. Agric.* **2010**, *72*, 69–86. [CrossRef]

33. Lopes, D.; Agujetas, R.; Puga, H.; Teixeira, J.; Lima, R.; Alejo, J.; Ferrera, C. Analysis of finite element and finite volume methods for fluid-structure interaction simulation of blood flow in a real stenosed artery. *Int. J. Mech. Sci.* **2021**, *207*, 106650. [CrossRef]

34. Jeong, W.; Seong, J. Comparison of effects on technical variances of computational fluid dynamics (CFD) software based on finite element and finite volume methods. *Int. J. Mech. Sci.* **2014**, *78*, 19–26. [CrossRef]

35. Weller, H.G.; Tabor, G.; Jasak, H.; Fureby, C. A tensorial approach to computational continuum mechanics using object-oriented techniques. *Comput. Phys.* **1998**, *12*, 620–631. [CrossRef]

36. Funari, R.; Bhalla, N.; Chu, K.Y.; Söderström, B.; Shen, A.Q. Nanoplasmonics for Real-Time and Label-Free Monitoring of Microbial Biofilm Formation. *ACS Sens.* **2018**, *3*, 1499–1509. [CrossRef] [PubMed]

37. Kaiser, M.; Jug, F.; Julou, T.; Deshpande, S.; Pfohl, T.; Silander, O.; Nimwegen, E. Monitoring single-cell gene regulation under dynamically controllable conditions with integrated microfluidics and software. *Nat. Commun.* **2018**, *9*, 212. [CrossRef]

38. Bellouard, Y.; Said, A.; Dugan, M.; Bado, P. Fabrication of high-aspect ratio, micro-fluidic channels and tunnels using femtosecond laser pulses and chemical etching. *Opt. Express* **2004**, *12*, 2120–2129. [CrossRef]

39. Tao, S.; Desai, T. Microfabrication of Multilayer, Asymmetric, Polymeric Devices for Drug Delivery. *Adv. Mater.* **2005**, *17*, 1625–1630. [CrossRef]

40. Li, H.; Torab, P.; Mach, K.E.; Surrette, C.; England, M.R.; Craft, D.W.; Thomas, N.J.; Liao, J.C.; Puleo, C.; Wong, P.K. Adaptable microfluidic system for single-cell pathogen classification and antimicrobial susceptibility testing. *Proc. Natl. Acad. Sci. USA* **2019**, *116*, 10270–10279. [CrossRef]

41. Bian, Z.; Guo, C.; Jiang, S.; Zhu, J.; Wang, R.; Song, P.; Zhang, Z.; Hoshino, K.; Zheng, G. Autofocusing technologies for whole slide imaging and automated microscopy. *J. Biophotonics* **2020**, *13*, e202000227. [CrossRef] [PubMed]

42. Schneider, F.; Draheim, J.; Kamberger, R.; Wallrabe, U. Process and material properties of polydimethylsiloxane (PDMS) for Optical MEMS. *Sens. Actuator A Phys.* **2009**, *151*, 95–99. [CrossRef]

43. Koho, S.; Fazeli, E.; Eriksson, J.; Hänninen, P. Image Quality Ranking Method for Microscopy. *Sci. Rep.* **2016**, *6*, 28962. [CrossRef] [PubMed]
44. Johnston, M.T.; Ewoldt, R.H. Precision rheometry: Surface tension effects on low-torque measurements in rotational rheometers. *J. Rheol.* **2013**, *57*, 1515–1532. [CrossRef]
45. Ewoldt, R.H.; Johnston, M.T.; Caretta, L.M., Experimental Challenges of Shear Rheology: How to Avoid Bad Data. In *Complex Fluids in Biological Systems: Experiment, Theory, and Computation*; Springer: New York, NY, USA, 2015; pp. 207–241. [CrossRef]
46. Guo, P.; Hall, E.W.; Schirhagl, R.; Mukaibo, H.; Martin, C.R.; Zare, R.N. Microfluidic capture and release of bacteria in a conical nanopore array. *Lab Chip* **2012**, *12*, 558–561. [CrossRef]
47. Di Giacomo, R.; Krödel, S.; Maresca, B.; Benzoni, P.; Rusconi, R.; Stocker, R.; Daraio, C. Deployable micro-traps to sequester motile bacteria. *Sci. Rep.* **2017**, *7*, 45897. [CrossRef]
48. Dichosa, A.E.K.; Daughton, A.R.; Reitenga, K.G.; Fitzsimons, M.S.; Han, C.S. Capturing and cultivating single bacterial cells in gel microdroplets to obtain near-complete genomes. *Nat. Prot.* **2014**, *9*, 608—621. [CrossRef]
49. Sacher, R. Microscope Immersion Oil. *Micros Today* **2000**, *8*, 33–35. [CrossRef]
50. Cooke, T.; Ltd., S. Microscope immersion oil with thixotropic properties. *J. Sci. Instrum.* **1958**, *35*, 476–476. [CrossRef]
51. Si, F.; Le Treut, G.; Sauls, J.T.; Vadia, S.; Levin, P.A.; Jun, S. Mechanistic Origin of Cell-Size Control and Homeostasis in Bacteria. *Curr. Biol.* **2019**, *29*, 1760–1770.e7. [CrossRef]
52. Kilbertus, G. Microhabitats in soil aggregates. Their relationship with bacterial biomass and the size of the procaryotes present. *Rev. d'Ecol. Biol. Sol* **1980**, *17*, 543–557.
53. Gage, D.J. Analysis of infection thread development using Gfp-and DsRed-expressing Sinorhizobium meliloti. *J. Bacteriol.* **2002**, *184*, 7042–7046. [CrossRef] [PubMed]
54. Blanpain, C.; Fuchs, E. Epidermal Stem Cells of the Skin. *Annu. Rev. Cell Dev. Biol.* **2006**, *22*, 339–373. [CrossRef] [PubMed]

MDPI
St. Alban-Anlage 66
4052 Basel
Switzerland
Tel. +41 61 683 77 34
Fax +41 61 302 89 18
www.mdpi.com

*Micromachines* Editorial Office
E-mail: micromachines@mdpi.com
www.mdpi.com/journal/micromachines